Development and Formulation of Veterinary Dosage Forms

DRUGS AND THE PHARMACEUTICAL SCIENCES

DRUGS AND THE PHARMACEUTICAL SCIENCES

A Series of Textbooks and Monographs

1. Pharmacokinetics, *Milo Gibaldi and Donald Perrier*
2. Good Manufacturing Practices for Pharmaceuticals: A Plan for Total Quality Control, *Sidney H. Willig, Murray M. Tuckerman, and William S. Hitchings IV*
3. Microencapsulation, *edited by J. R. Nixon*
4. Drug Metabolism: Chemical and Biochemical Aspects, *Bernard Testa and Peter Jenner*
5. New Drugs: Discovery and Development, *edited by Alan A. Rubin*
6. Sustained and Controlled Release Drug Delivery Systems, *edited by Joseph R. Robinson*
7. Modern Pharmaceutics, *edited by Gilbert S. Banker and Christopher T. Rhodes*
8. Prescription Drugs in Short Supply: Case Histories, *Michael A. Schwartz*
9. Activated Charcoal: Antidotal and Other Medical Uses, *David O. Cooney*
10. Concepts in Drug Metabolism (in two parts), *edited by Peter Jenner and Bernard Testa*
11. Pharmaceutical Analysis: Modern Methods (in two parts), *edited by James W. Munson*
12. Techniques of Solubilization of Drugs, *edited by Samuel H. Yalkowsky*
13. Orphan Drugs, *edited by Fred E. Karch*
14. Novel Drug Delivery Systems: Fundamentals, Developmental Concepts, Biomedical Assessments, *Yie W. Chien*
15. Pharmacokinetics: Second Edition, Revised and Expanded, *Milo Gibaldi and Donald Perrier*
16. Good Manufacturing Practices for Pharmaceuticals: A Plan for Total Quality Control, Second Edition, Revised and Expanded, *Sidney H. Willig, Murray M. Tuckerman, and William S. Hitchings IV*
17. Formulation of Veterinary Dosage Forms, *edited by Jack Blodinger*
18. Dermatological Formulations: Percutaneous Absorption, *Brian W. Barry*
19. The Clinical Research Process in the Pharmaceutical Industry, *edited by Gary M. Matoren*
20. Microencapsulation and Related Drug Processes, *Patrick B. Deasy*
21. Drugs and Nutrients: The Interactive Effects, *edited by Daphne A. Roe and T. Colin Campbell*

Development and Formulation of Veterinary Dosage Forms

Second Edition, Revised and Expanded

edited by

Gregory E. Hardee
ISIS Pharmaceuticals
Carlsbad, California

J. Desmond Baggot
Univeristy of Zimbabwe
Harare, Zimbabwe

CRC Press
Taylor & Francis Group
Boca Raton   London   New York

CRC Press is an imprint of the
Taylor & Francis Group, an **informa** business

CRC Press
Taylor & Francis Group
6000 Broken Sound Parkway NW, Suite 300
Boca Raton, FL 33487-2742

First issued in paperback 2019

© 1998 by Taylor & Francis Group, LLC
CRC Press is an imprint of Taylor & Francis Group, an Informa business

No claim to original U.S. Government works

ISBN-13: 978-0-8247-9878-9 (hbk)
ISBN-13: 978-0-367-40059-0 (pbk)

A CIP record for this book is available from the British Library.

Library of Congress Cataloging-in-Publication Data available on application

Visit the Taylor & Francis Web site at
http://www.taylorandfrancis.com

and the CRC Press Web site at
http://www.crcpress.com

Preface

Over a decade has elapsed since the first edition of *Formulation of Veterinary Dosage Forms* was published. During that time, many advances in pharmaceutical technology have been made, and drug registration requirements have changed considerably. In the second edition, new information on the development of veterinary dosage forms is included, and emphasis has been added to the analytical science necessary to design, evaluate, and control this new dosage form and medicinal technology. Consideration is given to whether these improved veterinary dosage forms facilitate their administration to domestic animals and decrease local irritation and residual effects.

In this edition, the stages of drug formulation development and decision-making processes are sequentially presented. Veterinary dosage forms differ from human dosage forms in that their administration to a range of animal species is a primary consideration. Whether the target species are companion or food-producing animals largely influences the nature of the studies that have to be carried out to support drug product registration. The target species determines whether a veterinary dosage form will be designed for individual animal therapy or for disease prevention in a herd or flock of animals. The pharmacological class of drug, in addition to target species, determines the design of veterinary dosage forms. Priority is generally given to classes of drugs that are used in disease prevention and farm animal production.

Selection of the dosage form appropriate for a particular application is covered in the first chapter, in which the preclinical data necessary to select and support registration of the dosage form are extensively detailed. The emergence of biotechnology has in many ways been felt first in the animal health field. As such, we have seen a change in the type of new chemical entities (NCEs) being proposed for veterinary dosage forms. Two new chapters, "Protein/Peptide Veterinary Formulations" and "Bioavailability Bioequivalence Assessment," discuss the peculiarities associated with these classes of NCEs.

The advancement of technology and regulatory requirements has also impacted upon the type and sophistication of analytical data necessary for the selection and development of a new veterinary dosage form. A chapter dealing with the appropriateness and validity of the testing methods utilized to characterize the active ingredient and dosage form has been added. This chapter includes important conceptual information regarding the principles upon which the final dosage form specifications are determined, justified, and fixed.

We believe that this edition will be a significant contribution for those concerned with the provision and availability of new animal medicines—regulators, researchers, pharmaceutical developers, manufacturers, and veterinary practitioners. It will become the standard reference in the areas of veterinary dosage form development and registration. We wish to thank the authors for their enthusiasm in contributing their expertise.

Gregory E. Hardee
J. Desmond Baggot

Contents

Contributors

J. Desmond Baggot, M.V.M., Ph.D., D.Sc., F.R.C.V.S. Professor of Preclinical Veterinary Studies, University of Zimbabwe, Harare, Zimbabwe

Melanie R. Berson, D.V.M. Veterminary Medical Officer, Office of New Animal Drug Evaluation, Center for Veterinary Medicine, United States Food and Drug Administration, Rockville, Maryland

Russell Bey, Ph.D. Professor, Veterinary Microbiology, Department of Veterinary PathoBiology, College of Veterinary Medicine, University of Minnesota, St. Paul, Minnesota

Scott A. Brown, D.V.M., Ph.D. Associate Director, Animal Health Drug Metabolism, Pharmacia & Upjohn, Inc., Kalamazoo, Michigan

David W. Cook Director, Product Development, Syrvet Inc., Waukee, Iowa

Gary R. Dukes, Ph.D. Specifications Development Manager, Pharmaceutical Development, Pharmacia & Upjohn, Inc., Kalamazoo, Michigan

Thomas H. Ferguson, Ph.D.   Research Scientist, Animal Science Product Development, Elanco Animal Health, A Division of Eli Lilly and Company, Greenfield, Indiana

Todd P. Foster, Ph.D.   Research Scientist, Pharmaceutical Development, Pharmacia & Upjohn, Inc., Kalamazoo, Michigan

Nathalie Garcon, Ph.D.   Manager, Vaccine Development, SKB Biologicals, Rixensart, Belgium

Gerald B. Guest, D.V.M.   Center for Veterinary Medicine, United States Food and Drug Administration, Rockville, Maryland

David A. Hahn, Ph.D.   Senior Scientist, Pharmaceutical Development, Pharmacia & Upjohn, Inc., Kalamazoo, Michigan

Paul R. Klink, Ph.D.   Research Scientist, Animal Science Product Development, Elanco Animal Health, A Division of Eli Lilly and Company, Greenfield, Indiana

Gary Olaf Korsrud, B.S.A., M.Sc., Ph.D.   Research Scientist, Centre for Veterinary Drug Residues, Canadian Food Inspection Agency, Saskatoon, Saskatchewan, Canada

Gérard Lambert, Ph.D.   Bureau of Veterinary Drugs, Food Directorate, Health Protection Branch, Health Canada, Ottawa, Ontario, Canada

James D. MacNeil, B.Sc., Ph.D.   Head, Centre for Veterinary Drug Residues, Canadian Food Inspection Agency, Saskatoon, Saskatchewan, Canada

Judy A. Magruder, M.B.A.   Research Scientist, Implant Development, Alza Corporation, Palo Alto, California

Marilyn N. Martinez, Ph.D.   Pharmacologist, Office of New Animal Drug Evaluation, Center for Veterinary Medicine, United States Food and Drug Administration, Rockville, Maryland

Randy Simonson, Ph.D.   Director of Research and Development, Bayer Animal Health, Worthington, Minnesota

Man Sen Yong, Ph.D., D.A.B.T.   Chief, Human Safety Division, Bureau of Veterinary Drugs, Food Directorate, Health Protection Branch, Health Canada, Ottawa, Ontario, Canada

Introduction: Veterinary Drug Availability

GERALD B. GUEST
Center for Veterinary Medicine, United States Food and Drug Administration, Rockville, Maryland

Although the United States (U.S.) and the more developed nations of the remainder of the world are blessed with a variety of pharmaceuticals, feed additives, and biological products to treat, prevent, and control animal diseases, there is a healthy desire among persons involved in animal health issues to increase our animal medicine chest. The interest stems from the desire to efficiently produce food that is safe and plentiful and from the desire to have more and better government-approved products available for the prevention and treatment of diseases of dogs, cats, and horses and for an increasing variety of minor animal species. For the animal health industry, increased drug availability means broader markets, increased revenues, and an opportunity to better serve their customers. For the veterinarian, more animal health products means that he or she is better able to treat the usual and the unusual conditions, and to prevent animal disease and suffering. No doubt, we are all winners when new technology and industrial and regulatory initiatives hasten the availability of safe and effective animal health products.

A number of factors influence the availability of drugs used in animals: the fiscal health of the animal drug industry in general; the pressures and demands of national and local laws and regulatory agencies; and the long-term willingness of the animal drug industry to invest the time and money necessary to explore new technologies and to bring new products to the marketplace.

1

The animal health products industry is small. In 1994 the total sales of all animal health products in the U.S. were estimated to be $3.1 billion. This figure represents a 10% increase from the 1993 figure of $2.8 billion. The total sales for dosage form pharmaceuticals and food additives for animals in 1994 were $2.6 billion (1).

The Animal Health Institute (AHI), the national trade association representing manufacturers of animal health products, reported in December 1993 for industries regulated by the Food and Drug Administration that animal prescription, over-the-counter (OTC), and feed-use drugs, animal revenues were $2.3 billion. Whereas, revenues from human prescription and OTC drugs totaled $61.1 billion. Sales of human medical devices were reported at $39.4 billion in the same survey (2).

In addition to the overall small size of the animal drug industry by comparison to the human health market, it has been reported that 87% of animal health products have annual sales of under $1 million. Only 5% of all animal health products have revenues totaling more than $5 million annually (2). At the same time as the animal health industry was reporting total sales of products in 1994 to be $3.1 billion, an Animal Health Institute survey indicated that the total research and development expenditures for animal health products for 1994 were $421 million (3).

At the time of this writing, the Animal Health Institute and Dr. Stephen F. Sundlof, Director of the Center for Veterinary Medicine (CVM) of the U.S. Food and Drug Administration (FDA) have made improved product availability the No. 1 goal in both the industry and the U.S. government, respectively. In their attempt to bring about greater animal drug availability, AHI, as a member of the Coalition of Animal Health, which is made up of major food animal producer and veterinary groups, as well as companies that manufacture animal health products, have said that the problem is that animal health product approvals have declined while research and development costs have risen. The Coalition for Animal Health is supporting legislation to increase animal health product availability by changes to the government's approval procedures for animal pharmaceuticals. Legislation entitled the Animal Drug Availability Act of 1996 has been signed into law. The new law modifies the Federal Food, Drug, and Cosmetic Act to:

1. Lessen the requirements for demonstrating drug effectiveness for most new products
2. Simplify requirements for approval of drug combinations
3. Encourage the use of independent experts to help resolve scientific disputes between the animal drug sponsor and the FDA
4. Expand export opportunities for animal drugs
5. Increase the accountability of the FDA to make timely decisions on animal drug applications.

This legislation should be viewed as an important tool for change both in the government and in the animal health industry. Government regulators will have the opportunity to reassess longstanding approval requirements with a view toward more efficient review and evaluation of new animal drug products. Drug sponsors should use the opportunity to develop new strategic plans that will take advantage of changes to the new animal drug approval process.

Not only is animal drug availability influenced by federal, state, and local regulatory requirements for approval and licensing and by the animal health industry's pursuit of new technologies, what drugs are available also depends on the laws that affect the practice of veterinary medicine. On Oct. 22, 1994, President Bill Clinton signed into law a change to the Food, Drug, and Cosmetic Act (FD&C) which would give greater flexibility of drug choices to licensed veterinarians. Since the establishment of the animal drug amendments to the FD&C Act in 1968, as mandated by federal law, veterinarians were prohibited from using a drug in ways other than the use stated on the FDA-approved labeling. In contrast, a physician has no such prohibition. The Animal Medicinal Drug Use Clarification Act (AMDUCA) the President signed will allow extralabel use of animal and human health products in animals by veterinarians. Certainly, this change makes additional drug therapies available for use in a variety of animals (4).

In the United States animal drug products are monitored closely even after their approval for marketing. New information on an older drug product can result in changes to labeling or, in extreme cases, can result in a product being removed from the market. Several important animal drugs have been removed from the market by the government or by the drug's sponsor in recent years because of food safety concerns.

In the early 1980s, chloramphenicol, an animal drug approved for use only in companion animals, was found to be widely used illegally in food-producing animals. The FDA was concerned about the potential adverse effect in man should persons be exposed to residues of the drug in meat, milk, or eggs. Chloramphenicol is known to cause often fatal aplastic anemia in susceptible individuals. The dosage form most widely used in an extralabel manner for treating food animals at that time was chloramphenicol oral solution. This form of the drug was banned from the market on Jan. 23, 1986 (5). Other dosage forms of the drug remain available today for the treatment of diseases in non-food-producing animals.

The drug dimetridazole is a similar case. Dimetridazole was approved by the FDA for use in turkeys for the prevention and treatment of blackhead (histomoniasis, infectious enterohepatitis) and for improving rate of weight gain and feed efficiency. In the 1980s, new information before the FDA indicated serious safety concerns because of persistence of drug residues in edible animal tissues. These safety concerns along with the widespread, unapproved use in swine for treating intestinal diseases resulted in dimetridazole being removed from the market in the U.S. The approval was revoked on July 6, 1987 (5).

Ipronidazole, a very closely related drug used for the treatment of the same diseases, was voluntarily removed from the market by the drug's sponsor following the removal of dimetridazole. Dimetridazole and ipronidazole had been marketed by competing drug companies. With neither product now available, the control and treatment of blackhead outbreaks in turkeys is particularly difficult.

Another example of removal of drug products from the market because of food safety concerns occurred with furazolidone and nitrofurazone. These nitrofuran drugs had been approved for a wide variety of uses in poultry and swine. The FDA took action to remove nitrofurans from the market because of concern about the carcinogenicity of these products, and the effect that drug residues in meat, milk, or eggs might have on human health. The announcement of the decision to ban the uses of furazolidone and nitrofurazone in food-producing animals was made on Aug. 23, 1991. A number of these drugs are still approved for use in non-food-producing animals.(6)

This list of removed animal drugs would not be complete without a mention of the production uses of diethylstilbestrol (DES) in food-producing animals. The first use of DES was allowed by the FDA in 1947. DES pellets were implanted under the skin of the neck of poultry. However, it has been estimated that even when the practice was at its peak only about 1% of the chickens produced in the U.S. were treated with DES. Beginning in the 1950s, residues of DES were detected in poultry skin, liver, and kidneys. No one knew how much or how little DES in the form of residues in food might be harmful. In 1961, the use of DES in poultry was suspended.

DES had been approved for use as a growth promotant in cattle and sheep in 1954. Again, owing to concerns about food safety, the use of DES in cattle and sheep was withdrawn in 1979. Newer safe and effective products have taken the place left by the removal of DES. A number of growth-promoting ear implants are currently approved for use in cattle and sheep. In addition to this technology, poultry, as well as ruminants, benefit from feed-use growth promoting animal drugs (5).

It should be noted that governments differ in their approaches to regulation of animal drugs. A notable influence on the availability of animal drugs in the U.S. is the so-called Delaney Amendment of the Food, Drug, and Cosmetic Act. The Delaney amendment speaks to the regulation of human food additives and drug residues that are known to be carcinogens in man or test animals. The FDA's removal of DES, dimetridazole, and the nitrofurans in the U.S. is directly related to the carcinogenicity issue. In other parts of the world, a number of developed nations that do not have similar laws concerning carcinogens, continue to allow the use of dimetridazole and the nitrofurans in food-producing animals. However, DES is not used legally in any country in the world.

On Aug. 18, 1995, the FDA approved a new fluoroquinolone antibacterial drug, sarafloxacin hydrochloride, for use in chickens and turkeys. This prescription veterinary drug is to be used in the drinking water to control mortality in broiler chickens and growing turkeys infected with *Escherichia coli*. Because fluoroquinolones are a new class of drugs for treating infections in humans and animals, there were extensive discussions between the Center for Veterinary Medicine and the center's counterparts in the human-medicines side of the FDA, with the Centers for Disease Control (CDC), and with the U.S. Department of Agriculture (USDA) prior to the drug's being approved. The concern is with the possible development of microbial resistance to these drugs and the possible result that disease in man and animals may be more difficult to treat if this resistance occurs on a widespread basis. Based on these concerns, the policy of the FDA's Center for Veterinary Medicines is that this drug should not be used in an extralabel manner in major food-producing species such as cattle, swine, chickens, and turkeys (7). The implication arising from this action is that even as new products become available to the veterinarian and animal producers, there may be restrictions that limit the use of these products. Public health–related issues will surely continue to influence drug availability (7).

The reality of drug availability is that the number of products leaving the marketplace for various reasons reduces the medicinal arsenal of the veterinarian and the animal producer. At the same time, the number of new entities that are being approved by regulatory agencies in recent years has slowed.

The AHI reports that it takes an average of 11 years to bring an animal pharmaceutical to market and that only one in 7500 chemicals ever make it through the process of feasibility, development, and approval (2). The AHI stated that their research shows that the FDA has approved only six truly novel animal drugs for food-producing animals in the years 1990 to 1995 (8). A novel drug in the AHI's definition is one that contains active ingredients that have never been approved.

Clearly, the interface between science and law which influences the availability of animal pharmaceuticals is complex. While innovation, new technologies, reduced regulatory requirements, and enabling laws are important to drug availability, one cannot lose sight of the fact that new products must first be safe and effective. The obligation to deliver such products to the marketplace falls on industry and government alike. The role of neither group should be allowed to dominate.

REFERENCES

1. "Animal Health Product Sales Top the $3 Billion Mark in 1994," News Release, Animal Health Institute, Alexandria, Va. May 10, 1995.
2. AHI Quarterly. Animal Health Institute, Alexandria, Va., Vol. 15, No. 1.

3. "R&D Expenditures for Animal Health Products Exceed $421 Billion Mark," News Release, Animal Health Institute, Alexandria, Va., May 10, 1995.
4. Brody MD. Congress Entrusts Veterinarians with Discretionary Extra-Label Use. J Am Vet Med Assoc 1994; 205(10):1366–1370
5. Why veterinary drugs are withdrawn. FDA Vet 1993; VIII(V):1–3
6. Nitrofuran approval withdrawn. FDA Vet 1991; VI(VI):1–2.
7. Update on fluoroquinolones. FDA Vet 1995; X(VI):1–2.
8. Coalition for Animal Health, Animal Drug Availability Act of 1995. (Brochure.) Animal Health Institute Research, February 1995.

1

Basis for Selection of the Dosage Form

J. DESMOND BAGGOT
University of Zimbabwe, Harare, Zimbabwe

SCOTT A. BROWN
Pharmacia & Upjohn, Inc., Kalamazoo, Michigan

I. INTRODUCTION

Each species of domestic animal has certain distinguishing features, some of which contribute to variations in its handling of a drug. Dietary habit appears to provide the most satisfactory basis for grouping species in a general way. Herbivorous species consist of the horse and ruminant animals (cattle, sheep, and goats), omnivorous species (pig), and carnivorous species (dog and cat). In terms of physiological function, the digestive system is the principal distinguishing feature between herbivorous and carnivorous species. Other distinguishing features, which could be considered as allied to dietary habit, are the activity of the hepatic microsomal enzymes and the urinary pH reaction. In these respects, the pig resembles more closely the carnivorous species. Within each group the individual species are distinct, so extrapolation of pharmacological data from one species to another may not be valid. However, with an understanding of comparative pharmacology, information derived from studies in one species can be applied for predictive purposes to another species. The confidence of such predictions is largely determined by a knowledge of the physicochemical properties of the drug substance which, in turn, determine its pharmacokinetic behavior and fate in the body.

The translocation process for drugs is common to all mammalian species. Since passive diffusion is the mechanism by which drugs penetrate biological

7

membranes, lipid solubility and degree of ionization are the main properties of a drug substance that goven its translocation—i.e., absorption, distribution, and mechanism of elimination. The blood plasma is the body fluid into which drugs are absorbed and by which they are conveyed throughout the body for distribution to other tissues. Drugs distribute nonselectively to tissues: only a small fraction of the dose administered reaches the site of action. The pattern of distribution is largely determined by the degree of perfusion of tissue, molecular structure, and, in a general way, lipid solubility of the drug substance. The liver and kidneys, which are highly perfused and represent the principal organs of elimination for the majority of therapeutic agents, continually receive a major fraction of the amount of drug in the plasma. Because of the central role of the plasma in translocation processes, the plasma concentration of a drug is usually directly related to the concentration in the immediate vicinity of the site action—i.e., the biophasic concentration. Consequently, the plasma concentration versus time profile for a drug reflects the temporal course of its action. Factors influencing the concentration of a drug in the plasma include the size of the dose, formulation of the drug preparation, route of administration, extent of distribution and plasma protein binding, and rate of elimination.

To ensure selection of the most efficient dosage form and that reasonable predictions can be made with regard to the performance of formulations and drugs the physicochemical, pharmacological, and physiological influences on drug response are discussed in this chapter.

A. Drug Classification

Drugs can be broadly classified according to the system of the body on which they exert their primary action. This is generally qualified by the principal effect produced. Further classification of a drug can be based on the type(s) of receptor with which the drug interacts (activates or inhibits) or on chemical structure. Because generalization is inherent in drug classification, several exceptions are inevitable.

Knowledge of the precise classification of a drug allows prediction to be made of the pharmacological effects that are likely to be produced and provides a basis for the selection of drugs for concurrent use. When combination therapy is considered desirable, drugs that have different though complementary mechanisms of action on the same body system should be selected. Although a drug acts primarily on one system of the body, the resultant effects may affect several systems.

Antimicrobial drugs act selectively on microorganisms, but their action is not confined to pathogenic microorganisms. They are classified on the basis of chemical structure and proposed mechanism of antimicrobial action. The usual dosing rate of an antimicrobial drug is based on the quantitative susceptibility, which is detremined in vitro, of pathogenic microorganisms and on the pharmacokinetic properties of the drug. Antimicrobial drugs do not normally produce pharmacological effects in that they do not interact with drug receptors. Some antimicrobial drugs may, however, alter the rate of elimination or increase the toxicity of pharmacological agents administered concurrently (drug interaction).

Anthelmintic drugs have a relatively selective action on helminth parasites in the host animal. With the notable exception of the organophosphorus compounds, which inactive cholinesterase enzymes, anthelmintic drugs do not normally produce pharmacologic effects. The recommended doses of anthelmintic drugs take cognizance of their margin of safety in the target animal species. Classification of anthelmintic drugs is based on their chemical structure.

II. SPECIES COMPARISONS OF ANATOMY AND PHYSIOLOGY

A. Digestive System

The anatomical arrangement of the gastrointestinal tract and dietary habit are features that can serve to distinguish between the domestic animal species. Since the urinary pH reaction is determined mainly by the composition of the diet, the usual pH range differs between herbivorous (horse, cattle, sheep, and goat—alkaline) and carnivorous (dog and cat—acid) species, while urinary pH can vary over a wide range in omnivorous (pig) species.

The pig, dog, and cat are monogastric (single-stomached) species. The physiology of digestion and drug absorption processes are, in general, similar in these species, and are not unlike those in the human. The stomachs of human beings and dogs are lined with three main types of mucosal tissue: cardiac, gastric (oxyntic), and pyloric. The pig stomach is lined with the same mucosal types but differs in that cardiac mucosa, the glands of which secrete mucus and bicarbonate ion (Holler, 1970), constitutes a much larger relative area of the stomach lining. The gastric mucosa proper contains the compound tubular glands which secrete hydrochloric acid (parietal or oxyntic cells) and pepsinogen (neck chief cells). The strongly acidic reaction of the gastric contents (usual pH range is 3 to 4) can inactivate certain drugs, such as penicillin G and erythromycin. This type of inactivation can usually be overcome by modifying the dosage form.

Gastric emptying is perhaps the most important physiological factor controlling the rate of drug absorption, since, in monogastric species, the small intestine is the principal site of absorption. A drug in solution can be expected to be well absorbed if it is stable (i.e., neither chemically nor enzymatically inactivated) in the stomach, lipid-soluble, and not completely ionized in the small intestine. An effective pH of 5.3 in the microenvironment of the mucosal surface of the small intestine, rather than the reaction of the intestinal contents (pH 6.6), is consistent with observations on the absorption of drugs that are organic electrolytes. In the normal intestine, weak acids with pK_a values above 3 and bases with pK_a less than 7.8 have been shown to be very well absorbed (Hogben et al., 1959). Changes in the intestinal blood flow can alter the rate of absorption of lipid-soluble drugs (Ther and Winne, 1971; Rowland et al., 1973).

The horse is also a monogastric species but is a herbivore and, under natural conditions of management, feeds continuously. Unlike other monogastric species, a major portion of the stomach of the horse is lined with stratified squamous epithelium. Although the mean pH of gastric contents (pH 5.5) is higher than that in the pig and dog, the pH reaction can vary widely (1.13 to 6.8) in horses (Schwarz et al., 1926). Furthermore, gastric contents may by their nature hinder accessibility of drug molecules to the mucosal lining for absorption. A major fraction of an oral dose of drug may be adsorbed onto the contents and conveyed to the large intestine for absorption. The primary site of protein digestion to amino acids is the small intestine (Kern et al., 1974). The pH reaction of ingesta in the ileum of the horse is 7.4. The metabolic, digestive, and secretory functions of the gastrointestinal tract of the horse were reviewed by Alexander (1972). The relative capacity of the components of the gastrointestinal tract in various animal species is presented (Table 1). The capacity of the stomach is small in the horse compared with the pig and dog, but the converse situation applies to the large intestine. Two interesting though unrelated features of the equine species are that horses do not possess a gallbladder and are unable to vomit. Microbial digestion of carbohydrates to volatile fatty acids (acetoacetic mainly, propionic, and butyric acid—the proportions formed vary with the diet) and of remaining dietary protein to peptides and amino acids takes place in the cecum and colon. The mean pH reaction of the large intestinal contents

Table 1 Relative Capacity of Components of Digestive Tract of Domestic Animal Species

Component	Relative capacity (%)			Relative capacity (%)	
	Horse	Pig	Dog	Component	(sheep and goat)
Stomach	8.5	29.2	62.3	Rumen	52.9
Small intestine	30.2	33.5	23.3	Reticulum	4.5
Cecum	15.9	5.6	1.3	Omasum	2.0
Large colon	38.4			Abomasum	7.5
		31.7	13.1		
Small colon and rectum	7.0			Small intestine	20.4
				Cecum	2.3
				Colon and rectum	10.4

is 6.6. It has been estimated that up to 50% of the soluble and virtually 100% of the insoluble carbohydrate in the diet is presented to the large intestine for digestion and subsequent absorption. The fibrous component of the diet is digested primarily in the large intestine, although horses digest fiber less efficiently than ruminant species.

The anatomical arrangement of the gastrointestinal tract clearly distinguishes ruminant (cattle, sheep, and goats) from monogastric species (horses, pigs, dogs, and cats). The forestomach, which refers to the rumen and reticulum (collectively called the reticulorumen) and omasum, is a voluminous compartment lined with stratified squamous epithelium in which microbial fermentation takes place continuously. The approximate capacities of the adult reticulorumen are 100 to 225 L in cattle and 10 to 25 L in sheep and goats. The reticulorumen occupies approximately 60% of the total capacity of the gastrointestinal tract; the omasum, the third compartment of the forestomach, occupies about 2%. Digestive juice is not secreted in the forestomach. Microbial digestion converts carbohydrates into volatile fatty acids (acetate, propionate, and butyrate); the gases carbon dioxide and methane are produced. Bacterial hydrolysis of dietary protein through peptides to amino acids takes place in the reticulorumen; ammonia is the principal soluble nitrogenous constituent of ruminal fluid. The forestomach contents vary from liquid to semisolid consistency, and the pH reaction is normally maintained within a relatively narrow range (pH 5.5 to 6.5), in spite of the high concentration of volatile fatty acid produced. This is accomplished by buffers secreted in the alkaline saliva (pH 8.0 to 8.4) and, it appears, directly by the forestomach epithelium. The volume of saliva produced in cattle is 90 to 190 L/day. Even though salivary secretion is continuous, the flow increases during feeding and rumination. In addition to its buffering activity saliva has an antifoaming property, which serves to prevent dietary bloat. Despite the stratified squamous nature of its epithelial lining, the rumen has been shown to have considerable absorptive capacity (Phillipson and McAnally, 1942; Masson and Phillipson, 1951). After comminution by both microbial digestion and rechewing, the liquid portion of reticuloruminal contents, in which small particles of feed are suspended, is pumped by the omasum into the abomasum, or true stomach. Based on average values of salivary flow and volume of the rumen liquid pool, the turnover rate for reticuloruminal fluid was estimated to be 2.0/day for cattle and 1.1 to 2.2/day for sheep (Hungate, 1966). The omasum aids in the transfer of ingesta from the reticulum to the abomasum, removes water and electrolytes from ingesta, and reduces the size of particulate matter in the contents. Omasal transfer of ingesta is regulated by the volumes of fluid in the reticulorumen and abomasum. The abomasum, which occupies approximately 4% to 5% of the capacity of the gastrointestinal tract in adult cattle and 7.5% to 10% in sheep and goats, is the only part of the ruminant stomach that se-

cretes digestive juices. Secretions from the fundic area of the abomasum contain hydrochloric acid, pepsin, and, in young suckling preruminant animals,
rennin (a milk-coagulating enzyme). The reaction of abomasal contents does not
vary much and is usually close to pH 3.0 (Masson and Phillipson, 1952).

Due to the large volume of ruminal fluid, a drug can attain only a low concentration in the reticulorumen whether administered in solution or as a solid
dosage form. This diluting effect may decrease the rate, but not necessarily the
extent, of absorption. The nonionized, lipid-soluble form of weak organic electrolytes, particularly acids because of the acidic reaction of ruminal fluid (pH
5.5 to 6.5), should normally be well absorbed from the reticulorumen. When
aspirin (in a solid dosage form) was administered orally to cows, the systemic
availability of salicylate was 50% to 70% and the drug was slowly absorbed,
mean absorption half-life was 2.9 h, from the reticulorumen (Gingerich et al.,
1975). The 12-hour dosage interval for aspirin in adult cattle is related to the
rate of absorption of salicylate rather than the rate of elimination, which is rapid.
The indigenous microflora may metabolize certain drugs by hydrolytic and reductive reactions. This would substantially decrease the systemic availability of
these drugs. Lipid-soluble organic bases, administered parenterally, diffuse from
the systemic circulation into ruminal fluid as a part of their normal pattern of
distribution. In this acidic fluid, they may become "trapped" by ionization,
depending on their pK_a values. Slow absorption of these drugs could result in
the persistence of low concentrations for an extended period. However, metabolism by the liver (first-pass effect) may offset their persistence in the body. The
theoretical equilibrium distribution, expressed as concentration ratio, of weak
organic acids and bases of differing pK_a values between saliva (pH 8.2) or ruminal fluid (pH 5.5 to 6.5) and plasma (pH 7.4) is presented graphically (Fig.
1).

Reticular groove closure, which can be induced in adult ruminant animals
by prior oral administration of sodium bicarbonate or copper sulfate solution
or lysine-vasopressin (0.3 IU/kg, IV) in sheep, permits the direct passage of a
drug in solution from the cardiac orifice of the rumen to the abomasum. Using
this technique, an orally administered drug in solution would bypass the rumen
and be immediately available for absorption from the abomasum. Organic acids, such as the nonsteroidal anti-inflammatory drugs (NSAID), should be rapidly absorbed. Spontaneous closure of the reticular groove may occur reflex-
ively in some animals given an anthelmintic solution orally and could decrease
the clinical efficacy of the drug on gastrointestinal helminth parasites. Ruminal
fluid delays the absorption of a drug both by the diluting effect of the large
volume and by delaying the onward passage of the drug to the abomasum and
small intestine.

Because of marked differences in the anatomical arrangement of the gastrointestinal tract and associated digestive physiology between herbivorous (horse
and ruminant animals) and nonherbivorous (pig, dog, and cat) species of ani-

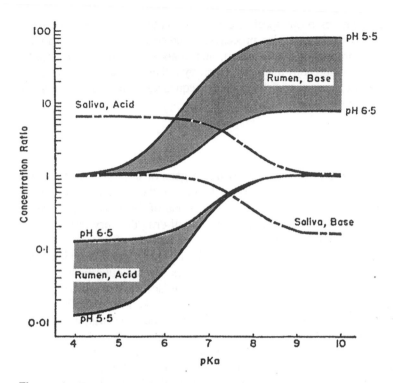

Figure 1 Expected equilibrium distribution between saliva or rumen contents and plasma of acids and bases of differing pKa. Concentration ratio is the ratio of the salivary or ruminal concentration to concentration free in the plasma, calculated separately for acids and bases, for saliva of pH 8.2 and rumen contents over a range of ph 5.5 to 6.5, assuming plasma is pH 7.4. (From Dobson, 1967.)

mals, variations can be expected to occur in both the rate and extent of absorption of orally administered drugs. The "first-pass" effect applies to all species and, due to the generally higher capacity of the liver of herbivorous species to metabolize lipid-soluble drugs by microsomal oxidative reactions, is likely to decrease the systemic availability of these drugs to a greater extent than in the nonherbivorous species. In a comparative study of salicylate absorption following oral administration of sodium salicylate in gelatin capsules (Davis and Westfall, 1972), the peak concentrations of the drug in plasma were considerably higher in pigs and dogs than in ponies and goats. Although based on limited evidence, gastrointestinal absorption may be described as fast and relatively complete in the pig, dog, and cat; may be variable in rate, and may occur in two phases separated by the time taken for intestinal contents to pass from the stomach to the cecum in the horse; slow in the ruminant species. The systemic availability of drug over a 24-h period in herbivorous species may approach that over a

12-h period in nonherbivorous species, depending on the influence of the first-pass effect. Microbial fermentation processes in the cecum and colon of horses and in the reticulorumen of cattle, sheep, and goats are qualitatively similar. Disturbance of the indigenous microbial flora, by either a digestive disorder or an antimicrobial drug, can have deleterious consequences on the nutritional well-being of herbivorous species.

B. Skin

The skin is one of the most easily accessible organs of the mammalian body. Because of this, its use as a route for drug delivery to the animal has been suggested for centuries. However, the skin is also one of the most impenetrable barriers of the body. Thus, the transdermal route of drug delivery has not been used to any extent in the past except for a limited number of drugs.

Skin is a complex tissue composed of different layers of cells which have very different physical and chemical characteristics. As the outermost layer of the epidermis, the stratum corneum presents the major barrier to most drugs (Fig. 2). This layer of the skin is composed of dead, keratinized cells which are remarkably desiccated. Because of these characteristics, the stratum corneum rather than the dermal or viable epidermal tissues of the skin provides the rate-limiting barrier to dermal absorption of drugs. In fact, the dermis and viable epidermis are 1000 and 10,000 times more permeable to drugs than the stratum corneum (Chien, 1987). Because the cornified epithelium is by far the most rate-limiting barrier for absorption of drugs, in vitro drug diffusion through dead or isolated epidermis has been proven to be equivalent to percutaneous absorption through living skin (Chien, 1987). It should be remembered that the skin is a metabolically active tissue, and systemic availability of parent drug may underestimate the amount of drug absorbed into the skin due to metabolism in a first-pass situation (Carver and Riviere, 1989).

In formulating a drug to produce local or systemic effects, between-species differences might be expected because of the nature of the skin of different animals (Huber and Reddy, 1978). Humans and horses have highly developed and effective sweat glands. The cow, pig, sheep, dog, and cat do not have the ability to sweat profusely. Sheep, goats, and cattle exude large quantities of lipoid material from sebaceous glands to protect their skin. Pigs have an extensive layer of keratin, which must be considered when designing dermatologic preparations. For example, Wohrl (1977) has found that levamisole "spot-on" formulation was of only limited value in pigs, whereas it has been most effective in other species (see Chap. 1, Sec. III.E.; and Chap. 2, Sec. III).

The epidermis of the dog is thin and uncomplicated compared with the epidermis of man (Wester and Maiback, 1987). In contrast to human skin, canine skin is thickest in regions where there is the most hair, and thinnest on the abdomen and footpads. There are more folds in dog skin; these folds are almost

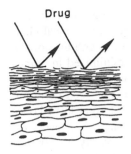

Drug

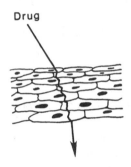

Drug

Drug

Figure 2 As the outermost layer of the epidermis, the stratum corneus presents the major barrier to most drugs.

scalelike, with the hair follicles arising from the depressions in the folds. Canine skin has from 100 to 600 hair groups per square centimeter of skin, with two to 15 hairs per group. Cats have a more dense hair coat, with 800 to 1600 hair groups comprised of approximately 20 hairs each per square centimeter of skin.

Adnexa of the skin act as easy pathways for diffusion through the stratum corneum. Adnexal structures are particularly important for very large molecules, ions, and polar compounds with many functional groups (e.g., hydroxyl or amino groups), such as cortisol. In these instances, the physicochemical characteristics of the drug substantially decrease its penetration through the corneal layers of the skin (Chien, 1982). The effect of adnexa is minimized in human beings by the low proportion of surface area that is comprised of adnexa (0.1% of body surface area in human beings; Chien, 1982). The result is a rapid initial pathway that does not contribute much to the amount of drug absorbed once the drug concentrations in the circulation achieve steady state. However, the concentration of drug within the adnexal structures may be vastly higher than that attained in the rest of the stratum corneum (Chien, 1982). The adnexal pathway may be much more important in animals because the density of hair follicles and sebaceous glands is much higher. The rapid initial influx of drug from the adnexal pathway may contribute a substantially greater fraction to the total amount of drug absorbed through canine or feline skin.

C. Female Reproductive Cycle

In domestic animals the female reproductive cycle is referred to as the estrous cycle. It varies with the species in several respects which include duration of cycle, length of estrus (sexual receptivity), and time of ovulation (Table 2). The cat is unique among the domestic animal species in that ovulation is induced by coitus. In seasonal breeding species (the mare, ewe, doe [goat], and queen), the time of year during which estrous cycles occur are strongly influenced by the photoperiod. Both the mare and queen become anestrous in late autumn due to decreasing daylight, and cycles are reestablished with increasing daylight. The converse situation applies to ewes and does. The plane of nutrition can affect the onset of estrous cycles in seasonal breeding species. The various stages of the estrous cycle and maintenance of pregnancy are under the control of steroid (sex) hormones, mainly progesterone and estradiol-17β (nonpregnant) or estrone (pregnant). Ovarian activity is regulated by hormones produced by the anterior pituitary gland. These hormones, follicle-stimulating hormone (FSH) and luteinizing hormone (LH), act synergistically. The main function of FSH is to promote the growth of follicles. The ovulatory process and luteinization of the granulosa, which is essential to the formation of the corpus luteum, are the primary functions of LH. The secretion of both FSH and LH by the anterior pituitary gland is under hypothalamic control, specifically gonadotropin-releasing hormone (GnRH) although the release of LH is much more sensitive than that of FSH (Stabenfeldt and Edqvist, 1984).

Pharmacological intervention at any stage of the reproductive cycle, whether to induce ovulation in mares, ewes, or cows or to suppress estrus or prevent ovum implantation in bitches or queens, is based on changing the plasma concentrations of the hormones that affect the particular process. To be successful

Table 2 Average Length of Various Stages of Reproductive Cycles of Domestic Animals

Species	Duration of estrous cycle	Length of estrus	Time of ovulation	Time fertilized ova enter uterus (after conception)	Time of implantation (after conception)	Type of placenta	Length of pregnancy
Mare[a]	21 days	5–6 days	Last day of estrus	3–4 days	30–35 days	Epitheliochorial	345 days
Cow	21 days	18 h	12 h after end of estrus	3–4 days	30–35 days	Epitheliochorial	280 days
Ewe[a]	17 days	36 h	30 h after beginning of estrus	3–5 days	15–18 days	Syndesmochorial	147 days
Doe[a] (goat)	20 days	40 h	30–36 h after beginning of estrus	4 days	20–25 days	Syndesmochorial	147 days
Sow	21 days	45 h	36–40 h after beginning of estrus	3–4 days	14–20 days	Epitheliochorial	113 days
Bitch	In estrus at 7–8 month intervals depending on breed	Proestrus, 9 days; estrus, 7–9 days	First or second day of estrus	5–6 days	15 days	Endotheliochorial	64 days
Queen*	16 days (nonbred) (pseudo-pregnancy lasts 36 days)	5–6 days	Induced 24–32 h after coitus	4 days	13 days	Endotheliochorial	65 days

Source: Modified from Hansel and McEntee (1977).
[a]Seasonally polyestrous.

a good understanding of the temporal pattern of the various hormone concentrations in plasma is essential. In addition to observing the effects produced by a drug substance administered, the relevant hormone concentrations in plasma should be monitored. This provides data to support the dosage recommendation and a means for comparing the biological efficacy of different dosage forms of a drug substance (e.g., GnRH).

D. Mammary Gland

The mammary gland presents a unique opportunity for administration of veterinary dosage formulations to milk-producing animals. The bovine udder is richly supplied with blood mainly through the external pudendal arteries and supplemented by a subsidiary supply, cranially through the subcutaneous abdominal artery, and caudally via the perineal artery. The ratio of the volume of blood circulating through the mammary gland to volume of milk produced has been estimated to be 670:1, at a moderate level of milk production. This provides ample opportunity for the unbound fraction of lipid-soluble drugs to passively diffuse from the systemic circulation into milk. Numerous small veins leaving the parenchyma anastomose and converge around the base of the udder into a circular vessel that is drained by three veins: the large subcutaneous abdominal, the external pudendal, and the perineal vein.

The largest differences in mammary gland physiology are in the relative volume of milk that is produced in various species and in the composition of the milk (proportion of butterfat to whey). Dairy cattle produce the most milk on a total-weight basis, although dairy goats produce more milk on a percent of body weight basis. Goat's milk has a larger proportion of butterfat to whey than cow's milk, although wide differences exist among breeds of dairy cattle (Table 3).

Milk is produced by alveolar cells that form milk by a secretion process. In essence, nutrients are brought into the alveolar cells from the blood supply at the basal and lateral membranes, and milk is formed and flows from the apical region of the cells. After milking, the lactiferous ducts are in a collapsed state. This leads to retention of newly secreted milk in a lobule-alveolar system until pressure from milk secretion becomes great enough to force the ducts open and permit milk flow into the cisterns of the gland and teat. As long as the intramammary pressure remains below 25 to 40 mm Hg, milk secretion is a continuous process. Thereafter, milk production slows. At the time of milking oxytocin is released, causing the myoepithelial cells surrounding the alveoli and ducts to contract, forcing milk into the large ducts and the gland cistern. The intramammary pressure doubles nearly instantly, and then falls as the milking process occurs (Fulper, 1991). Thus, bulk flow is always into the milk and out

Table 3   Composition (g/dl) of Milk of Various Species

Species	Fat	Protein	Lactose	Ash
Cow:				
Ayrshire	4.1	3.6	4.7	0.7
Brown Swiss	4.0	3.6	5.0	0.7
Holstein	3.5	3.1	4.9	0.7
Jersey	5.5	3.9	4.9	0.7
Shorthorn	3.6	3.3	4.5	0.8
Zebu	4.9	3.9	5.1	0.8
Nanny-goat	3.5	3.1	4.6	0.8
Ewe	10.4	6.8	3.7	0.9
Sow	7.9	5.9	4.9	0.9
Mare	1.6	2.6	6.1	0.5
Bitch	9.5	9.3	3.1	1.2
Queen	7.1	10.1	4.2	0.5

of the animal when milk letdown and the process of milking occurs. This limits the ability of any formulation to diffuse from the site of administration (i.e., teat cistern) into the parenchyma of the gland without some physical mixing, often recommended by manufacturers of intramammary products to increase distribution to the entire gland.

Although no anatomic barrier exists between systemic blood and the milk-forming cells, a somewhat restrictive function barrier exists. This is because there are few capillary fenestrae to allow large molecular weight compounds to pass and the process of nutrient uptake by the mammary alveolar cells is relatively specific. In addition, there are active processes that can transport some drugs (e.g., penicillin G) across the blood-milk barrier, resulting in plasma:milk concentration ratios that are poorly predicted by standard passive diffusion paradigms such as the Henderson-Hasselbalch equation (Schadewinkel-Scherkl et al., 1993). However, unless a drug is secreted by an active transport mechanism into the milk, penetration of systemically administered compounds into the mammary gland requires that they be relatively lipid-soluble to diffuse through the "blood-mammary gland" barrier.

Milk is generally slightly acidic (pH 6.0 to 6.5) compared with physiologic pH, although mastitic milk tends to have a somewhat higher pH (approximately 7.0). Nevertheless, because the pH is less than that of the systemic circulation, weak bases with pK_a values greater than milk pH will be selectively trapped in their ionized form in milk rather than plasma. Conversely, weak acids with pK_a values less than normal milk will be selectively trapped in mastitic milk.

E. Liver and Kidneys

The liver and kidneys are the principal organs of elimination for drugs and drug metabolites. The rate at which these organs eliminate drugs is determined by blood flow, availability of drug in the systemic circulation for elimination (influenced by plasma protein binding), activity of hepatic drug-metabolizing enzymes, and efficiency of renal excretion mechanisms.

In domestic animal species, the liver constitutes 1.25% to 2.5% of live body weight, and hepatic arterial body flow represents 26% to 29% of cardiac output. The kidneys constitute 0.25 to 0.5% of live body weight and receive 22% to 24% of cardiac output. It follows that in proportion to organ weight (mass), the blood flow to the kidneys is three to four times that to the liver. This multiple is larger (ca. 7) in humans, since the liver:kidney weight ratio is higher than in domestic animals. The greater blood supply to the kidneys enables this organ to efficiently perform its primary function, which is to regulate the composition and volume of the body fluids.

Although the liver and kidneys mainly eliminate drugs, either organ may delay the elimination of some drugs. The liver accomplishes this through enterohepatic circulation, whereby a drug excreted in bile may be reabsorbed from the small intestine, while the kidneys contribute through pH-dependent passive tubular reabsorption. Since only a fraction of the drug that reaches either organ is conserved (reabsorbed) by these processes, their effect on the overall rate of elimination decreases as elimination proceeds.

Through the first-pass effect, the liver decreases the systemic availability of orally administered drugs that undergo extensive hepatic metabolism. This may be particularly significant in herbivorous species, since drug absorption from the gastrointestinal tract may be slower and hepatic microsomal oxidative reactions generally proceed more rapidly than in nonherbivorous species. There are wider species variations in the bioavailability, which refers to the rate and extent of absorption (or, more precisely, systemic availability), of drugs when administered orally than parenterally. This underlies the requirement for differences in oral dosing rate, defined as dose/dosage interval, among domestic animal species since the therapeutic range of plasma concentrations of a drug is generally the same in the various species.

The liver metabolizes drugs and certain endogenous substances by a variety of microsomal enzyme-mediated oxidative reactions and glucuronide conjugation. The drug-metabolizing enzymes are located in the lipophilic membranes of the smooth-surface (devoid of ribosomes) endoplasmic reticulum. The liver also contributes to the metabolism of drugs by other metabolic pathways—both phase I (oxidative, reductive, and hydrolytic reactions) and phase II (sulfate and glycine conjugation, acetylation, and methylation) reactions (Williams, 1967). Based on a knowledge of the functional group in a drug molecule, the probable metabolic pathway(s) can generally be predicted (Table 4). The metabolic pathways for drugs are qualitatively similar in domestic animal species but differ widely among species in the rate at which they proceed. The rate at which a drug is metabolized, particularly by phase I reactions, is characteristic of both the drug and the animal species. This applies even among the ruminant species (cattle, sheep, and goats). In general, however, metabolism of lipid-soluble drugs by phase I reactions proceeds more rapidly in herbivorous species than in nonherbivorous species.

Some predictions can be made regarding certain phase II (conjugation or synthetic) reactions. Requirements of these reactions are the presence in a drug molecule of a functional group that is suitable for conjugation, an endogenous reactant, and a transferring enzyme. Some examples of endogenous reactants and transferring enzymes are uridine diphosphate (UDP) glucuronic acid/UDP-glucuronyltransferase (microsomes), phosphoadenosyl phosphosulfate/sulfotransferase (cytosol), acetyl-CoA/N-acetyltransferase (cytosol), and S-adenosylmethionine/transmethylases (cytosol).

In contrast to phase I metabolic reactions, which appear to take place in all domestic animal species, some phase II (synthetic) reactions are either defective or absent in certain species (Table 5). The cat synthesizes glucuronide conjugates slowly due to a deficiency in the microsomal transferring enzyme UDP-

Table 4 Probable Biotransformation Pathways for Drugs

Functional group	Biotransformation pathways
Aromatic ring	Hydroxylation
Hydroxyl	
aliphatic	Chain oxidation; glucuronic acid conjugation; sulfate conjugation (to a lesser extent)
aromatic	Ring hydroxylation; glucuronic acid conjugation; sulfate conjugation; methylation
Carboxyl	
aliphatic	Glucuronic acid conjugation
aromatic	Ring hydroxylation; glucuronic acid conjugation; glycine conjunction
Primary amines	
aliphatic	Deamination
aromatic	Ring hydroxylation; acetylation; glucuronic acid conjugation; methylation; sulfate conjugation
Sulfhydryl	Glucuronic acid conjugation; methylation; oxidation
Ester linkage Amide bond }	Hydrolysis

Source: Baggot (1977).

Table 5 Domestic Animal Species with Defects in Certain Conjugation Reactions

Species	Conjugation reaction	Major target groups	State of synthetic reaction
Cat	Glucuronide synthesis	-OH, -COOH-NH$_2$, =NH, -SH	Present, slow rate
Dog	Acetylation	Ar-NH$_2$	Absent
Pig	Sulfate conjugation	Ar-OH, Ar-NH$_2$	Present, low extent

Source: Baggot (1977).

glucuronyl transferase. The decreased rate at which this metabolic reaction proceeds in cats considerably enhances the potential toxicity of several drugs (including the NSAIDs) and xenobiotics in the feline species. Although the dog and the fox appear to be unable to acetylate the aromatic amino group of drugs, this does not delay the elimination of these drugs since alternative metabolic pathways compensate.

Two or more metabolic reactions can occur simultaneously or may proceed sequentially, when the initial reaction converts the drug to a metabolite that has a functional group that is suitable for the subsequent (generally conjugation) reaction. Conjugates of drugs and phase I metabolites are highly polar, generally inactive pharmacologically, and, with a few exceptions (e.g., the N^4 acetyl derivative of some sulfonamides), more water-soluble than the parent drug. Glucuronide conjugates are particularly suitable for carrier-mediated (active) transport into urine and/or bile.

Prodrugs are compounds that are inactive per se, but which undergo metabolism to an active form tn vivo. Prodrugs may be administered to enhance stability of the dosage form to increase stability in the gastrointestinal tract or to alter the rate of presentation of the active moiety to the patient. An example of a prodrug is enalapril, which is metabolized to the active drug enalaprilat. Since prodrugs depend on in vivo metabolism to the active form, the onset of action may be delayed. In addition, conversion to the active compound may be significantly influenced by individual animals, disease processes, or species differences. Changes in the route of administration may cause profound differences in active drug formation, particularly if one of the routes includes first-pass metabolism.

Compounds that are active but short-lived within the animal and are converted to one or more metabolites that confer the observed biological activity of the product, are often loosely termed "prodrugs." An example of this kind of compound is ceftiofur, which, when administered intravenously, has a half-life of less than 5 min but is metabolized to the equipotent metabolite, desfuroylceftiofur, which has a half-life in various species of 5 to 12 h (Brown et al., 1991, 1995; Craigmill et al., in press; Jaglan et al., 1994). Similar to classic prodrugs, the biological activity of these types of compounds is due almost entirely to their active metabolite(s). In other words, metabolic conversion precedes activity.

Changes in the gastrointestinal flora and/or hepatic metabolizing enzymes can theoretically alter the rate and/or extent of conversion of a prodrug to its active moiety. However, since most of these prodrug biotransformations are mediated by nearly ubiquitous and nonspecific esterases, there is little evidence that these changes alter the clinical efficacy or toxicity of these compounds.

Some active drugs are rapidly converted to metabolites with similar or different degrees of activity. Although not technically prodrugs (because the administered compounds have activity themselves), these types of formulations may be considered similar to prodrugs if the majority of in vivo activity is produced by the metabolite rather than the administered compound. Ceftiofur sodium is an example of such a drug. The half-life of parent ceftiofur is less than 5 min after IV administration, but the half-life of the primary metabolite, desfuroylceftiofur, is 8 to 10 h in cattle (Brown et al., 1991). Desfuroylceftiofur is equipotent to ceftiofur against most major pathogens of veterinary importance (Salmon et al., 1995). It is the activity of desfuroylceftiofur that provides the efficacy of ceftiofur, and it is the long half-life of desfuroylceftiofur that allows ceftiofur to be effective when administered once a day to animals.

Many other drugs produce active metabolites which contribute to the observed in vivo activity of the administered drug. For instance, procainamide (PCA) is metabolized to N-acetylprocainamide (NAPA), which is approximately equally active as an antiarrhythmic as PCA. Both PCA and NAPA contribute to the efficacy observed after administration of PCA. It is for this reason that some experts have suggested monitoring both PCA and NAPA when adjusting dosage regimens in humans, and it is the contribution of NAPA to efficacy that causes the therapeutic dose of PCA to be significantly smaller in human beings than in dogs. In other words, the therapeutic range of PCA in humans is 4 to 12 μg PCA/ml or 20 to 30 μg PCA + NAPA/ml. Dogs do not metabolize PCA to NAPA as efficiently, so the ratio of PCA to NAPA is different, and the therapeutic range (for humans) of 4 to 12 μg PCA/ml does not provide PCA + NAPA in the range of 20 to 30 μg/ml.

For several years, drug metabolism was assumed to be synonymous with detoxification. It is becoming increasingly evident that some drugs and xenobiotics are metabolically transformed to reactive intermediates that are toxic to various organs. Such toxicity may only become apparent when high doses of the chemical substances are administered and the usual metabolic pathways (particularly phase II reactions) become saturated or are compromised by disease states. This phenomenon is well known for acetaminophen (paracetamol) in cats and humans, and may explain the hepatotoxicity of aflatoxin, a mycotoxin produced by *Aspergillus* spp. in contaminated grain.

Polar drugs and drug metabolites (particularly conjugates, i.e., products of phase II metabolic reactions) are eliminated by excretion in the urine generally and/or in bile. Since these compounds have limited capacity to passively diffuse through lipoidal membranes, their distribution is restricted to extracellular fluid.

Renal excretion comprises the following mechanisms:

1. Glomerular filtration (a passive, nonsaturable process) of molecules that are free (unbound) in the circulating blood
2. Carrier-mediated excretion (an active, saturable process) of certain polar organic compounds (including drug conjugates) by the proximal tubule cells
3. pH-dependent passive reabsorption by the distal nephron of the non-ionized, lipid-soluble moiety of weak organic electrolytes

The renal excretion mechanisms that are involved in the elimination of drugs and drug metabolites are determined largely by the physicochemical properties of the compound. Extensive (>80%) binding to plasma proteins decreases the availability of drugs for glomerular filtration but does not hinder carrier-mediated active tubular secretion. Carrier-mediated excretion is, however, subject to competitive inhibition by substances (organic anions/cations) of generally similar character. While a drug may enter tubular fluid by both glomerular filtration and proximal tubular secretion, its renal clearance may nonetheless be low owing to substantial reabsorption of the drug in the distal nephron. Since reabsorption takes place by passive diffusion, it is influenced by the concentration of the drug and its degree of ionization in distal tubular fluid. The latter is determined by the pK_a of the drug and the urinary pH reaction. Urinary pH is determined mainly by the composition of the diet and varies among species. The usual urinary reaction in carnivorous animals is acidic (pH 5.5 to 7.0), whereas in herbivorous species it is alkaline (pH 7.2 to 8.4). When a significant fraction (>20%) of the dose of a weak organic electrolyte is eliminated by renal excretion, urinary pH will affect the rate of excretion (and half-life) of the drug. The excretion of weak organic acids will be enhanced under alkaline and decreased under acidic urinary conditions. The converse applies to weak organic bases.

The glomerular filtration rate (GFR) has a reasonably consistent and predictable influence on the rate of excretion of drugs that are eliminated entirely by renal excretion. Based on inulin clearance, mean values of GFR (ml/min/kg) are 1.63 in horses; 1.8 to 2.4 in cattle, sheep, and goats; and 3.5 to 4.0 in dogs and cats. The systemic clearances of drugs that are eliminated solely by glomerular filtration (e.g., aminoglycoside antibiotics) have relative values that are higher in dogs and cats and lower in horses than in domestic ruminant species.

Some drugs (e.g., cardiac glycosides, tetracyclines) and glucuronide conjugates of a variety of lipophilic drugs and endogenous substances (bilirubin, steroid hormones) are excreted by the liver into bile. Compounds excreted in bile have molecular weights exceeding 300 and are relatively polar. Organic anions (includes glucuronide conjugates) and cations are actively secreted by hepatic cells into bile by carrier-mediated transport processes that appear to be similar to those in the proximal renal tubule. Drugs and drug metabolites excreted in bile pass into the duodenum, from which some (depending on their lipid solubility) may be reabsorbed by passive diffusion. Glucuronide conjugates may be hydrolyzed by β-glucuronidase, an enzyme present in intestinal microorganisms, liberating the parent compound (or phase I metabolite), which could then be reabsorbed (enterohepatic circulation). When a significant fraction of the dose undergoes enterohepatic circulation, elimination of the drug is delayed. It is usual for such drugs (and their metabolites) to be gradually eliminated by renal excretion.

Gallbladder bile is a mildly acidic fluid (pH 6.5 to 7.2) that is released intermittently owing to cholecystokinin (intestinal hormone) stimulation, which causes contraction of the gallbladder. Since the horse does not possess a gallbladder, bile flow is continuous in the equine species. The rate of bile secretion in domestic animal species is 12 to 24 ml/kg/day; the lower end of the range applies to the dog and cat, and the upper end to the horse.

F. Hepatic Portal System

All mammals and most avian species have a well-developed hepatic portal system. In this system, blood flow from the alimentary tract (except the oral cavity and the rectum) travels through the liver prior to entering the systemic circulation. As a result, all of the solute that is absorbed from the gastrointestinal tract is exposed to the liver's metabolic and excretory processes prior to reaching the systemic circulation, where the xenobiotic can be diluted by systemic blood volume and redirected to other tissues of the body. This "first-pass" through the liver exposes the liver to potentially high concentrations of drug that is absorbed from the gastrointestinal tract and presents the possibility that the compound be extensively metabolized and/or excreted before reaching the systemic circulation. As a result, drugs that are administered orally and are highly metabolized by the liver may have very low bioavailability (i.e., systemic absorption) of the parent compound. As an example, propranolol administered orally has a bioavailability of less than 25%, indicative of the extensive first-pass metabolism of the parent drug to inactive metabolites prior to its reaching the systemic circulation.

G. Renal Portal System (Avian and Reptilian)

Birds and reptiles have a unique renal portal system that drains the rear portion of the animal. Thus, drugs administered parenterally in the lower extremities of these animals will pass through the renal tissue before entering the systemic circulation. Parenteral injection of some drugs into the rear limbs of birds results in low systemic availability of the compounds compared with administration into the forelimbs or pectoral muscles of the birds. The absorption pattern of compounds such as β-lactam antibiotics will vary with the site of injection.

H. Homeotherms/Poikilotherms

Traditional thought on the processes of absorption, distribution, metabolism, and excretion centers around homeotherm physiology in that the metabolic processes are relatively constant regardless of the environmental conditions in which the animal resides. In contrast, the metabolic processes in poikilotherms (i.e., cold-blooded animals) are exquisitely susceptible to changes in environmental conditions. For example, the elimination of gentamicin, as reflected by the elimination half-life, is much more rapid in homeotherms than in poikilotherms (Brown and Riviere, 1991). As a result of such temperature-sensitive elimination processes in poikilotherms, metabolism and pharmacokinetic studies are often carried out at more than one environmental temperature to evaluate the susceptibility of the drug's elimination to expected differences in temperature (Mader et al., 1985). In addition, in salmon, which, as adults, live in salt water but spawn and grow as fingerlings in fresh water, metabolism and excretion studies must be conducted under both environmental conditions to understand a drug's fate and effects in changed body situations.

I. Poultry/Fish/Reptiles

Although the anatomy and physiology of the respiratory system are generally similar in domestic animals, the avian system differs in several aspects. Avian lungs are small and attached to the ribs. They are mainly passive in action and expand with the thoracic cage. The avian lung is attached to air sacs. The respiratory rate of birds may vary from 46 to 380 respirations per minute (Huber and Reddy, 1978), compared with approximately 12 respirations per minute for the horse, and up to 30/min for the cow. The respiration rate of dogs varies considerably, depending on whether they need to cool their body by panting.

III. ROUTES OF ADMINISTRATION AND DOSAGE FORMS

A. Oral Dosage Forms

Oral dosage forms account for a large proportion of drug preparations. Drug absorption from oral dosage forms generally decreases in the order: solutions > suspensions and pastes/gels > capsules > tablets of various types > powders. The time of feeding relative to oral dosing can markedly influence the availability of a drug for absorption from an oral dosage form. It follows that both the dosage form of a drug and the temporal relationship between feeding and oral dosing affect the bioavailability of the drug. The influence of these factors varies with the species of animal, particularly between herbivores and nonherbivores. The amount (or type) of feed can affect the activity, by influencing the duration of exposure, of an orally administered anthelmintic drug on gastrointestinal parasites.

Solutions, emulsions, and suspensions are the liquid dosage forms used for oral administration of drugs to animals. The selection of a dosage form is based on the physicochemical properties of the drug to be formulated, the relative duration (immediate or sustained) of the effect desired, and the species of animal for which the dosage form is intended. Liquid dosage forms are generally administered by nasogastric (stomach) tube to horses; by mouth (as a drench) to cattle, sheep, and goats; and by mouth (sometimes with the aid of a syringe) to dogs and cats.

1. Solutions

Oral solutions present a drug in a form that is most readily available for absorption. The bioavailability (i.e., the rate and extent of absorption) of the drug would be expected to be maximal from an oral solution. Factors that could decrease bioavailability include complexation or micellization with ingredients of the oral formulation, degradation of the drug in the acidic environment in the stomach of monogastric species or metabolism by microorganisms in the reticulorumen, interaction with contents of the gastrointestinal tract, and metabolism by the gastrointestinal mucosa or the liver before reaching the systemic circulation (first-pass effect). Most of these factors would also apply to other types of oral dosage form.

A particular advantage that oral solutions have over solid dosage forms is that the drug is already in solution and evenly distributed throughout the preparation. This facilitates administration of the drug at a dose appropriate for the individual animal—i.e., on a mg/kg body weight basis. It also increases the efficiency of dosing a large number of animals—e.g., a flock of sheep from a multidose reservoir with a drench gun. The diluting effect of the liquid vehicle decreases irritation to the gastric mucosa which could occur if the drug was administered as a solid dosage form. Oral solutions have certain disadvantages, however. A drug in solution is the dosage form that is usually most susceptible to degradation. Because of this, accelerated and room temperature stability studies must be performed to ensure that the drug product has a satisfactory shelf-life.

Since solutions are, in many instances, fair to good microbial growth media, particular care must be taken with ensuring that bacterial count at manufacture is low and that pathogenic microorganisms are totally absent. This may be effected by imposing microbiological contamination specifications on all materials used in the preparation of the product. Pope et al. (1978) described how microbial contamination in a manufacturing facility and of a drug product was minimized by subjecting contaminated raw material to irradiation pasteurization prior to its receipt at the facility.

When selecting a preservative for inclusion in a drug solution the following points should be considered:

1. Antimicrobial effectiveness throughout the shelf-life of the drug product
2. Lack of irritation if the drug solution is applied topically, especially important in the case of ophthalmic preparations
3. Systemic toxicity is not produced

Since phenolic compounds are often used as preservatives and their major pathway of metabolism is conjugation with glucuronic acid, such solutions, particularly when administered repeatedly, can produce toxicity in cats. This is because cats slowly form glucuronide conjugates due to a deficiency in microsomal UDP-glucuronyl transferase (Robinson and Williams, 1958).

2. Emulsions

An emulsion is a thermodynamically unstable heterogeneous system consisting of at least one immiscible liquid intimately dispersed in another in the form of droplets. Oral emulsions (oil/water) are generally manufactured to make the active oil phase easier to manage and more palatable, and the bioavailability of the drug substance is sometimes increased. When dispersed in corn oil and then emulsified (o/w), the rate and extent of absorption of griseofulvin, a poorly water-soluble drug, are increased (Bates and Sequeira, 1975). The increased bioavailability of griseofulvin was attributed to decreased gastrointestinal motility including gastric emptying coupled with contraction of the gallbladder caused by linoleic and oleic acids which are liberated during digestion of the corn oil. The combination of effects on gastrointestinal motility and bile flow produced by the emulsified corn oil increased the bioavailability of griseoful-vin. It is likely that the bioavailability of other poorly water-soluble compounds would be increased when administered as oil/water emulsions.

Emulsions do, however, present problems with stability, not only of the emulsion per se but also of the active ingredient (Pope, 1980a, b). Changes in the solubility of emulsifiers and their hydrophilic-lipophilic balance, the viscosity of the emulsion, and the partitioning of ingredients should always be considered. Since emulsions are highly susceptible to microbial attack, a preservative should be included in the formulation to ensure a satisfactory shelf-life. The preservation of emulsions presents special problems which include the potential loss of the active preservative to either the oil phase, through dissociation, or complexation with emulsifiers or other ingredients of the formulation. It is only the free, undissociated preservative in the aqueous phase of an emulsion that has preservative activity.

3. Suspensions

A suspension is a two-phase system composed of a solid material dispersed in a liquid. There are a number of reasons why a suspension is considered the most suitable oral-dosage form for a particular drug. Certain drugs are chemically unstable in solution but stable as a salt or derivative in suspension. Suspensions, like solutions, are easier to administer than solid dosage forms and allow precision in dosage on a unit weight basis. The latter, however, depends on the consistency of the suspension, which is a heterogeneous system in which components may settle while measuring the dose. To delay settling in the system, a suspending agent that will increase viscosity is usually added. Care must be taken in selecting the viscosity-increasing suspending agent since adverse effects could be produced in some species of animals. For example, carboxymethyl-cellulose has been implicated in causing allergic reactions in humans, horses, and cattle (DeWeck and Schneider, 1972).

Variables, in addition to the rate of settling, associated with suspension formulations include flocculation and changes in particle size of the suspended drug with time. Assessment of these variables and how they relate to shelf-life of suspensions have been discussed by Pope (1980a,b).

4. Pastes and Gels

Pastes and gels are semisolid oral dosage forms that are particularly suitable for administration by the owners to horses or cats. Application of a paste to the distal forelimbs of cats is an alternative to oral administration. The formulation must be such that the paste is syringeable over a wide range of ambient temperatures; moderately tenacious so that it will adhere to the tongue or, in the case of cats, to the site of application; and tasteless or suitably flavored for horses. A paste or gel that is too fluid in consistency or unpalatable leads to imprecise dosage. Companion animals appear to be particularly sensitive to taste and will resist swallowing unpleasant-tasting substances.

Provided an oral paste or gel is well formulated, a relatively precise dose of the drug can be administered since the delivery device is a pre-loaded calibrated syringe. The rate of drug absorption from a paste would be expected to be slower than from a solution, but faster than from a solid dosage form. Classes of drugs that could be formulated as oral pastes or gels include anthelmintics, some antimicrobial agents, and NSAIDs. Pastes and gels afford ease and safety of administration to animals.

5. Capsules

Unlike other dosage forms, the capsule is strictly a unit dose container or, more precisely, a tasteless, easily administered and digested container for different materials such as powders, granules, pellets, suspensions, emulsions, or oils. Unit dose in this context refers to the amount of drug contained within the capsule rather than the total dose to be administered to the animal being treated. Commercially available capsules, which are intended mainly for use in humans, often contain an inappropriate dose for most animal species.

There are two basic types of capsules. They differ in the method of manufacture, type of gelatin shell, and the materials they may contain. Release rate and onset of action of the drug will differ between the two types. The hard gelatin capsule is usually used for solid-fill formulations and the soft gelatin capsule for liquid- or semisolid-fill formulations. In selecting the type of capsule to be used, consideration must be given to the following:

1. The hard gelatin capsule can be purchased as a body and cap and easily filled in the manufacturing facility (Hostetler and Bellard, 1970). Soft gelatin capsules are different, however, in that the capsule shell is formed in situ around the fill material. Specialized equipment is required for this process (Stanley, 1970).

2. In some instances, a poorly water-soluble drug may be formulated as a solution in a suitable vehicle and encapsulated in a soft gelatin shell. Vehicles that may be used fall into two categories:
   a. Water-immiscible, volatile and nonvolatile, such as vegetable oils and mineral oils, aromatic and aliphatic hydrocarbons, chlorinated hydrocarbons, ethers, esters, alcohols, ketones, fatty acids, etc.
   b. Water-miscible, nonvolatile, such as nonionic surfactants, polyethylene glycol, glycerol, and glycol esters.

The vehicle, on being released in the gastrointestinal tract, may liberate the active ingredient in a form that is more readily available for absorption than if a solid-fill formulation were used. The effect of dosage form on the serum concentrations of indoxole, an NSAID, was studied by Wagner et al. (1966). They observed that the serum concentration response decreased in the order: emulsion ≅ soft gelatin capsule > aqueous suspension > powder-filled capsule. Hom and Miskel (1970) have also shown significantly more rapid dissolution of various soft gelatin capsules than tablet formulations of the same active ingredient. Soft gelatin capsules should probably be used in preference to hard gelatin capsules only when the fill is liquid or bioavailability of the drug from the hard gelatin capsule formulation does not meet the requirements.

When deciding whether to use a gelatin capsule or another type of drug delivery formulation, the following criteria should be considered:

1. Gelatin capsules are, in general, more expensive than tablets because of the need to purchase, or in situ manufacture, the shell. Since filling the capsules is far slower than manufacturing tablets, the labor costs of production are higher.
2. Capsules offer an advantage in that the particle size and distribution of the original starting compound is rarely altered by the final filling process. In formulating tablets, however, the powder granule is subjected to physical stresses that may alter the primary particle size and, in turn, its bioavailability characteristics. Similarly, the compression into tablets of coated pellets or granules designed for controlled release may rupture the prepared material, with the result that the primary release characteristics of the individual dose units might be altered. When these pellets/granules are packed in a capsule, the release pattern will be randomly distributed about a previously determined mean value. Performance of the capsule dosage form will, consequently, be more predictable. Individual enteric-coated dose units packed in a capsule may show a more prolonged release pattern because of the broader random movement of pellets from the low pH of gastric fluid to the higher pH of intestinal contents. This could only be easily achieved by encapsulation.
3. Capsules are an effective means for deterring the taste and odor of an unpalatable drug substance. Other oral dosage forms of the drug might not be tolerated by an animal.
4. The gelatin capsule will protect the contents from light, but not from oxygen or moisture. Hence, information on the stability and moisture sensitivity of the formulation should be obtained before selecting the gelatin shell (capsule) dosage form.

6. Tablets and Boluses

Conventional tablets and modified-release boluses are solid-dosage forms that are widely used for drug administration to small animals (dogs and cats) and ruminant animals (cattle, sheep, and goats), respectively. For use in horses, conventional but not modified-release tablets are generally crushed and administered as an aqueous solution or suspension by nasogastric tube. Modified-release tablets, designed mainly for use in humans, must be given intact and can be administered to horses as well as small animals. However, the dose required for horses often precludes their use in this species.

Tablets have certain advantages over liquid dosage forms. A tablet contains an equivalent dose of active drug in a more compact form and usually presents the fewest problems with regard to stability. However, bioavailability of a drug can vary widely among tablet formulations. Because of the wide range of body weights and total dose requirements of different animal species, the strength of a tablet (amount of drug contained therein) largely determines its suitability for use in a particular species. There is an element of uncertainty with regard to the retention of a tablet or bolus by an animal. Dogs are notorious for ejecting tablets, sometimes discreetly, within minutes of their administration while ruminant species quite often eructate boluses. However, boluses are generally satisfactory for administration to a small group of animals, but their administration to a large herd of cattle or flock of sheep is time-consuming. In this circumstance, other dosage forms of the drug, such as oral solutions or suspensions, parenteral solutions or topical preparations, are less cumbersome to administer. Powders or granules, applied as top dressing to the feed, are the most convenient but generally least reliable method of drug administration.

7. *Protective Coating*

To effect optimal drug delivery to the principal site of absorption or site of action in the intestine, the active ingredient in a solid dosage form should be protected from the highly acidic environment in the stomach or the microbial metabolizing activity in the rumen from which onward passage of a drug product may be delayed for up to 12 h. Protection of a drug may be achieved either by encapsulation or by tablet coating.

Drugs that are degraded by the acidic environment in the stomach or that cause irritation of the mucosal lining or nausea and vomiting should be enteric-coated. Such drugs will produce similar undesirable effects in monogastric animal species as in humans. Since horses are unable to vomit, the effects produced may be more pronounced than in other species. The effectiveness of orally administered anthelmintics that act in the intestine but are subject to degradation in the stomach would be decreased unless protected from degradation.

Drugs administered orally to ruminant species pass directly into ruminal contents, which have a large volume (several liters), liquid to semisolid consistency, and moderately acidic reaction (pH 5.5 to 6.5). Microbial fermentation processes and metabolism of xenobiotics (including drugs) by reductive and hydrolytic reactions take place in the rumen. Moreover, the turnover rate of reticuloruminal fluid is slow so that onward passage of drugs to the abomasum (which is physiologically equivalent to the stomach in monogastric species) and intestine is considerably delayed. This combination of circumstances increases the potential for degradation of drugs, particularly those that act in the intestine. The protective coating of solid dosage forms, by decreasing dissolution, might delay degradation while stimulation of reticular groove closure, which can be chemically induced, would permit orally administered solutions of drugs to bypass the rumen. It would appear that protective coating of drug dosage forms has less application for ruminant than for monogastric species.

Complete protection from degradation by the forestomach and abomasal contents can be achieved by encapsulation of an active ingredient. Miller and Gordon (1972), using an encapsulation technique, showed that the effectiveness of a feed additive larvicide, 2-chloro-1-(2,4,5-trichlorophenyl) vinyl dimethyl phosphate, fed to cattle was increased. This was attributed to the higher fecal concentration of the compound which is active against larvae of the housefly (Musca domestica).

8. Powders and Granules

Powders and granules provide, in general, a better environment for maintaining stability of the active drug than liquid dosage forms. However, when administered in the feed, the dose ingested and bioavailability of the drug can vary widely. The animal may eat only a portion of the medicated feed, and the feed ingested may influence the absorption pattern of the drug and affect (generally decrease) its bioavailability.

When a powder or granule formulation is applied as a top dressing to or mixed in the feed, it must be palatable to the animal. Moreover, the animal must be feeding, which implies that a powder or granule formulation administered in the feed is generally limited to prophylactic medication. Inappetance or indifference to feeding is a usual feature of illness in animals.

Uncertainty as to the dose (amount of drug) ingested and the variation in bioavailability can be decreased by administering a powder or granule formulation dissolved or suspended in a liquid vehicle as a drench to ruminant animals or by nasogastric tube to horses. However, stability of the active ingredient in the liquid vehicle must be considered, and dosing of individual animals is required. The packaging of powders or granules as unit dose sachets for different animal species provides convenience for owners when small numbers of animals are to be dosed. The opposite is the case when the dosage form is intended for administration in the feed to a large number of animals.

9. Modified-Release Products

Modified-release, long-acting oral products have been described by a variety of terms which are often, incorrectly, used interchangeably. A sustained-release dosage form provides an initial amount of drug sufficient to produce a desired therapeutic blood concentration, as well as additional drug that is released at a zero-order rate to maintain the therapeutic concentration for a defined period.

A prolonged-release dosage form does not maintain a steady-state blood concentration. However, it maintains the blood concentration above the minimum therapeutic concentration for a longer duration than does a conventional single-dose formulation. A repeat-action dosage form provides long action by providing a single usual dosage and a second single dose at some later time. For example, the core of the tablet may be enteric-coated. The outer portion of the tablet provides the initial release while release from the interior core is delayed until it reaches the intestinal contents. Many of the physical and chemical methods for achieving modified-release are discussed in subsequent chapters and by Pope (1978), Ballard (1978), Lee and Robinson (1978), Sincula (1978), and Chandrasekaran and co-workers (1978). Whether a drug should be considered a candidate for modified-release depends on the following factors.

For nonruminant animals, drugs with a half-life in the range of 4 to 6 h are usually ideal. If the half-life is greater than 8 h, modified release is ruled out because gastric transit time would be the limiting factor. For drugs with a half-life ≤ 1 h, an extremely high dose size would be necessary to sustain for 12 h. If the potency of the drug is very high, safety, rather than the dose size for 12 h of sustained release, could be the limiting factor.

In nonruminant animals, the drug has to be absorbed evenly throughout the gastrointestinal tract. If absorption is site-specific in for instance, a portion of the upper intestine (riboflavin in the human), prolonging release from the dosage form would be useless unless the dosage form could be maintained in the stomach throughout its effective release life. Methods by which this may be achieved are reviewed in Chapter 5 and include hollow bits, nonpyloric passage, and buoyant devices.

10. Rumen Retention Devices

Systemic drug delivery is possible by oral administration of a drug delivery device that remains in the reticulorumen, due to its density or geometry, for a prolonged period of time. The physiology of the gastrointestinal tract of the ruminant does not dictate a minimum retention time for the oral dosage form. Ruminant slow-release products are formulated for zero buoyancy and retainment of sieve action (Pope, 1975, 1978). Regurgitation is prevented by designing the formulation so that zero buoyancy lodges the dose in the reticulum and, if dense enough, will remain there. Passage through the forestomach is prevented by ensuring that the formulation does not disintegrate to small particles, since there appears to be a limiting size of particle that can pass the neck of the omasum (Pope, 1975, 1978). Hence, controlled release in ruminants may be effected over periods varying from weeks to years. This, of course, offers definite advantages in overcoming mineral intake deficiencies (Dewey et al., 1958; Marston, 1962; Skerman et al., 1959; Andrews et al., 1958; Moor and Smyth, 1958; Allen et al., 1979; Givens et al., 1979), in controlling bloat (Laby, 1974) and parasites (Anderson and Laby, 1979; Christie et al., 1978). The types of formulation, their size and density requirements, and the release rate parameters to be considered have been discussed previously (Pope, 1978). Problems that may arise include prolonged drug residues (tissue depletion is prolonged because the rate-limiting step is drug delivery), uncertainty of when drug delivery ends, and abbatoir problems with the stainless steel often used to weight the boluses.

Slow-release mechanisms include matrix disintegration and diffusion through a semipermeable membrane from stationary matrix (Gyurick, 1988). In matrix disintegration, the dissolution of the matrix aids release of drug. If the controlled-release bolus is of an erosion type, distinct differences in bolus erosion patterns exist in age classes of animals. The smaller rumen apparently forces the bolus into a more abrasive position and more bolus erosion occurs. Also, the administration of additional boluses changes the erosion pattern in the rumen. Slow-release sulfamethazine boluses release sulfamethazine over 5 days; albendazole capsules with expandable plastic wings release albendazole over several weeks, while the monensin bolus was designed to release monensin over 150 days. Slow-release rumen devices liberate drug by diffusion through a semipermeable membrane. The morantel tartrate sustained-release bolus uses this technology (Presson et al., 1984). Neither the monensin nor the morantel tartrate sustained-release bolus is currently marketed in the United States because the stainless-steel cylinders pose problems for the meat-packing industry.

Continuous-release rumen retention devices include the Laby device, laminated polymeric boluses (using the weighting method or expanding geometry method of rumen retention), and osmotic pumps. The Laby device has wings that spring open to prevent passage of the device out of the reticulorumen. The Laby device may be loaded with tablets that are exposed at one end of a cylinder to the rumen fluid. As the last tablet dissolves, the spring-loaded device ejects the remainder of the tablet into the rumen fluid. Tablet composition dictates the rate of drug delivery.

Both the morantel tartrate trilaminate bolus and the levamisole polymeric bolus utilize polymers impregnated with drug (Boettner et al., 1988). The morantel tartrate trilaminate has the drug-impregnated polymer sandwiched between two drug-impermeable polymeric films. The entire sandwich is rolled up using a gelatin band that dissolves when immersed in the rumen fluid, unfurling the trilaminate sandwich. On the other hand, the levamisole tube bolus is a levamisole-laden ethylene vinyl acetate copolymer with iron filings interspersed to adequately weight the bolus for rumen retention (Taylor et al., 1988).

Osmotic pumps consist of a flexible impermeable drug reservoir which contains a single portal for the exit of drug. The drug reservoir is encased by a saturated solution of an osmotic agent which is protected by a rigid water-permeable outer wall. This outer wall is the rate-controlling membrane for drug release. As water enters the rigid outer osmotically active reservoir, it collapses the flexible drug reservoir, thus "squeezing" drug from the reservoir at a constant rate. For example, ivermectin administration to cattle can be accomplished by use of an osmotic pump which is retained in the reticulorumen. Steady-state concentrations are achieved in 14 days, and constant amounts of drug are released for up to 90 days (Baggot, 1988; Pope, 1985).

Intermittent-release rumen boluses deliver drug to the animal from the rumen at discrete times after administration of the bolus. The Castex device delivers pulse doses of oxfendazole. Each pulse is produced by erosion of a central magnesium spindle that releases five successive plastic segments, each containing the discrete dose of oxfendazole. The interval between pulses is approximately 23 days, although pH differences increase the uncertainty of pulse intervals. The oxfendazole pulse bolus releases pulse doses of oxfendazole at approximately 3-week intervals (Jacobs et al., 1987; Rolands et al., 1988). It is significant that this timespan roughly coincides with the prepatent period of the major gastrointestinal tristrongylids of cattle. The electronic bolus releases three doses of albendazole spaced 31 days apart using custom circuitry powered by a watch-type battery. Every 31 days, the gas generator expels the medication and resets the timer for the second and third cycles. The precision is 15 min over 93 days (Delatour, 1987).

a. Advantages

1.   The maintenance of a relatively constant drug concentration in blood can reduce fluctuations in drug concentrations in tissues and at biological target sites, and thus may bring about a more uniform pharmacological response.

2.   The incidence and intensity of side effects that might be caused by excessively high peak plasma concentrations resulting from the administration of conventional dosage forms may be reduced.

3.   The number and frequency of doses are decreased; hence, there is a reduction in labor costs and trauma to the animal.

4.   Gradual release of a drug from a dosage form may reduce or prevent irritation to the gastrointestinal mucosa by drugs that are irritant to the tissue at high concentrations.

b. Disadvantages

1.   Controlled-release formulations are more costly to manufacture than conventional dosage forms. This may be largely offset by the reduction in labor costs and trauma to the animal, as previously mentioned.

2.   They do not permit termination during the period of drug release. In some instances—e.g., in ruminants, duration of release may be intended to be days, weeks, or even months. If the animal is intended for human consumption, drug residues will be present if the animal is slaughtered before depletion of the delivery device. In most instances, a withdrawal period is defined; hence, intelligent use should avoid problems with drug residues.

3.   Drugs with a narrow margin of safety may cause concern. If dose dumping (excessive instant release) occurs at any time due to a flaw in the formulation, toxic concentrations may be reached.

4.   A first-pass effect would occur to a variable extent, depending on the rate of release from the formulation. For example, Bevill et al. (1977) noted differences in relative absorption (total amount of drug and metabolites absorbed) and systemic availability of sulfamethazine administered to yearling cattle as a solution, rapid-release bolus, and sustained-release bolus. Because of the longer retention time in the rumen, and hence different extent of metabolism in the rumen, the sustained-release bolus yielded only 32% systemic availability, the fast-release bolus 63%, and the solution 81%.

11. Feed/Water/Lick Blocks

Feed blocks represent a form of controlled drug delivery. If carefully used, a feed block may overcome the problem of medication on a single occasion, since it is available over a period of days, during which time most animals have the opportunity to obtain an adequate intake. When considering feed block dosing, the variable intake of feed block, depending on the availability of grazing and other supplementary feeding, must be taken into account and animals should be acclimatized to block feeding. Extensive lush pastures or heavy concentrate feeding reduces block consumption. Intensive conditions with inadequate natural or concentrate feed, increases consumption. The medicament to be incorporated into a feed block should be nontoxic, stable, palatable, and preferably of low solubility so that no loss of the active principle occurs when the block is placed outside in the rain. Fenbendazole has been found to meet many of the above criteria and, incorporated into a feed block, was effective as an anthelmintic (Gaenssler et al., 1978; McBeath et al., 1979).

B.   Parenteral Administration

There are many different routes of parenteral administration. Intravenous, intramuscular, subcutaneous, inhalational, intratracheal, and sometimes transdermal administration of drugs can all be used to elicit a systemic response; intra-articular, epidural, intralesional, intrathecal, nebulization, and topical application of drugs all present drug to a localized portion of the body. The more important of these routes of administration will be discussed.

1. Intravenous Injection

Direct injection of a drug into the systemic circulation provides predictable concentrations of drug in the bloodstream and almost instantly elicits a pharmacologic response. Because the drug is administered directly into the bloodstream, there is by definition no absorption phase preceding the disposition of the drug. Therefore, a drug that is injected intravenously is completely bio-available, meaning that the entire dose of drug is available to exert its effect on the animal.

The intravenous route of administration has several unique advantages over other parenteral routes. First, because of the direct administration into the circulation, the response to an intravenously administered drug is more rapid than the response to the same drug by any other route. Second, because of this rapid response to the drug, an intravenous dose of drug can often be titrated to achieve only the desired response. For example, administration of thiopental for induction of anesthesia is titrated after administering the first two-thirds of the calculated dose (8 mg/kg) to achieve a plane of anesthesia that facilitates intubation and surgical preparation. This titration of the last one-third of the dose is often called "administering to effect." In this manner, animals that are particularly responsive to thiopental will be given less total drug than more stoic animals. Third, intravenous administration can provide transiently high plasma concentrations, which may be therapeutically important for drugs such as some bactericidal antibiotics (aminoglycosides). Fourth, the dose of a drug given intravenously to produce a desired systemic effect is generally lower than the dose of the same drug administered by any other route to produce the same effect.

The intravenous route is also potentially the most dangerous route of drug administration. Intravenous doses must be given slowly, because rapid administration of drug into the bloodstream may cause immediate collapse of the animal. Oxytetracycline administered as a rapid intravenous injection to horses will often result in collapse of the animal, which is thought to be due to local chelation of calcium in the heart. Injection volume may also be important to the safety of the animal. Obviously, injection of 100 ml of a drug intravenously will not be well tolerated by a cat weighing 4 kg. Apart from the drug, other components of the injection solution may cause untoward effects. For example, intravenous administration of large amounts of potassium penicillin G may increase plasma potassium to dangerously high concentrations. Additionally, any time the vascular system is perforated (injection), the possibility of iatrogenic septicemia exists. Furthermore, continued intravenous administration, either by venipuncture or by indwelling catheter, may cause localized thrombophlebitis.

2. Intramuscular Injection

The intramuscular injection of drugs has been a very popular route of administration used by veterinarians for many years. It is an easy way to give an injection, there are many different locations that can be used, absorption is in many instances rapid and complete, and local tissue reaction of the drug is often masked by the depth of the injection site. However, there are many pitfalls associated with intramuscular injection.

Absorption of drug from an intramuscular injection site is generally assumed to be a first-order rate process. However, first-order absorption from an intramuscular injection site is the exception rather than the rule. Furthermore, intramuscular injection of drugs does not always assure rapid or complete bioavailability. Prolonged-release injectable preparations may release drug so gradually that the onset of action is slower than after oral administration, and the drug concentrations at the site of injection of these preparations may remain high for long periods of time. Plasma/serum concentrations may be lower than those observed after oral or even topical administration of the drug.

Several factors can determine the absorption of drug from an intramuscular site. Although high lipid solubility of a drug promotes rapid diffusion into the capillaries, some degree of water solubility at physiological pH is required. Otherwise, the drug may precipitate in the interstitial fluid at the site of injection, precluding absorption into the capillary bed. This precipitation can also cause severe tissue reactions either by physical irritation or by the attraction of phagocytes which ingest the drug particles. Granuloma formation at the site of injection not only can be painful to the animal, but can be aesthetically displeasing in the meat of food-producing animals. Furthermore, precipitation of drug at the injection site can lead to incomplete absorption and hence ineffectiveness of the drug administered. Examples of commonly used drugs that are incompletely absorbed from intramuscular injection sites include ampicillin, cephaloridine, cephradine, chlordiazepoxide, phenytoin, dicloxacillin, digoxin, phenylbutazone, diazepam, and quinidine.

The concentration of a drug in the volume injected can affect absorption of the drug from the injection site. Atropine is absorbed more rapidly from a concentrated solution of the drug, whereas lidocaine is absorbed more slowly. When the osmolarity of the solution is increased by the addition of another compound, absorption may be delayed. Increasing the total surface area available for diffusion from the absorption site to the capillaries will enhance drug absorption. The surface area for absorption can be increased by massaging the area after the injection, by using high-pressure injection devices, or by dividing the volume to be injected into several aliquots which are administered at various locations on the animal. Blood flow is often the rate-limiting factor associated with absorption from an intramuscular injection site. Blood flow differs from one muscle to another, and the result is variation in absorption of drug from different sites. For example, blood flow is greater in the human deltoid muscle than in the vastus lateralis, and blood flow is least in the gluteal muscles. Accordingly, drug absorption from the deltoid is faster than from the vastus lateralis, and absorption from both of those locations is faster than from the gluteal muscles. Slow injection from the buttocks of women is often noted, primarily because the drug is more likely to be deposited in adipose tissue which is very poorly perfused. This situation can also occur when large beef animals are given intramuscular injections of drug. In many instances, the length of the needle is insufficient to fully penetrate the adipose tissue, and what was intended to be an intramuscular injection with a characteristic absorption pattern becomes an intra-adipose injection with quite different absorption kinetics. Absorption from an intramuscular injection site will be increased during exercise because of the increased blood flow to the skeletal muscles. When blood flow is decreased to skeletal muscle, such as in patients with circulatory shock, hypotension, congestive heart failure, myxedema, and other circulatory disturbances, absorption from the intramuscular injection site will be prolonged and decreased.

Nonlinear absorption from an intramuscular injection site is relatively common, since the factors that govern absorption often change as the drug is absorbed. For example, the drug concentration may change as the drug is absorbed, creating a different gradient for absorption. Many injectable drugs are hypertonic, and intramuscular injection of these drugs will attract extracellular fluid to the injection site, creating a constantly changing concentration gradient for absorption. Perhaps the most important factor to change over time after injection is the surface area for absorption. At first, the volume of drug is deposited in a small, spherical shape. Later, the volume spreads to cover a larger area of muscle. Furthermore, there is migration of the drug along fascial planes, between muscle masses, and along tendons. This dispersion increases the surface area for absorption, which is thereby enhanced. For this reason, injection into the musculature of the neck is more often intermuscular, compared with injection into the musculature of the rear legs or buttocks, which is more often true intramuscular. Finally, a drug may alter the blood flow to the injection site if it is a local vasoconstrictor or vasodilator such as many of the cardiovascular drugs currently used therapeutically. Concurrently administered drugs may also alter the rate of absorption, such as prolonged absorption of lidocaine when administered in combination with epinephrine hydrochloride.

Complications of intramuscular administration of drugs include pain at the injection site, tissue damage at the injection site with accompanying increase in serum creatine phosphokinase (an indicator of skeletal muscle damage), sciatic nerve damage after injection into the hind limb (particularly of cats), and residual drug within the muscle which is unacceptable for food-producing animals. Finally, if an adverse reaction develops shortly after intramuscular injection, further absorption of drug cannot be prevented. Because intramuscular injections are more frequently given by lay persons, proper emergency remedies for the injection reaction are unlikely to be implemented.

3. Subcutaneous Injection

Subcutaneous injection is another route of administration that has enjoyed much use in veterinary medicine. Subcutaneous administration of fluids to slightly dehydrated dogs and cats is often used except when the animal requires immediate vascular volume replacement. Subcutaneous injection has many of the same characteristics as intramuscular injection. However, absorption from subcutaneous sites is often slower and more erratic than from intramuscular sites because of the limited and variable blood flow to subcutaneous tissue. Precipitation, tissue damage, volume, and concentration of drug all affect subcutaneous absorption in a manner similar to intramuscular absorption. It appears that the tissue irritation of some formulations is more severe after subcutaneous injection than after intramuscular injection (Korsrud et al., 1993). Teleologically, this may be the result of the subcutaneous space as the second line of defense against cutaneous invasion by foreign material. Nevertheless, one major advantage of subcutaneous injection over intramuscule administration is that the subcutaneous site of injection is often trimmed away from the carcass of meat-producing animals. This reduces the problem of violative drug residues in the meat of these animals, and it also avoids damage to the muscle that intramuscular injection would cause.

Disadvantages of the subcutaneous route of administration include tissue reactions in visible locations of the body; erratic absorption, which is particularly obvious at extremes in ambient temperature; and the difficulty of administration perceived by many people, especially large-animal veterinarians and producers.

4. Parenteral Dosage Forms

Parenteral dosage forms include aqueous, aqueous organic, and oily solutions, emulsions, suspensions, and solid forms for implantation. Parenteral preparations must be sterile and pyrogen-free; they should, if possible, be buffered close to physiological pH and preferably be isotonic with the body fluids.

When considering whether a new drug would be suitable as a parenteral formulation that could be manufactured, and used clinically, the following advantages and disadvantages of parenteral therapy should be considered. The advantages and disadvantages of choosing a parenteral dosage form include:

a. Advantages

1. The time to onset of action can be controlled by the type of formulation and by the site of injection. The absorption of many drugs from intramuscular and subcutaneous sites of injection is rapid and often assumed to be complete. The release of drug from the parenteral preparation (the processes of particulate dissolution into aqueous media), rather than the absorption of dissolved drug, generally controls the rate of drug entry into the systemic circulation. While the sodium salt of ceftiofur is well absorbed after intramuscular and subcutaneous injection as a reconstituted solution (Brown et. al., 1991), the ceftiofur crystalline-free, acid-sterile oil suspension is more slowly absorbed (Brown, unpublished data). The greater solubility of the former in aqueous media

and the oil suspension of the latter probably account for this difference in the rate of absorption of ceftiofur from its dosage forms. The bioavailability of some drugs from their parenteral preparations is shown in Table 6.

Gentamicin and ketamine were rapidly and completely absorbed from intramuscular injection sites in dogs and cats, respectively. The data also show that drug absorption from some of the products administered was incomplete (ampicillin) or slow (tylosin, erythromycin). Drug absorption from extravascular parenteral sites may vary in rate and extent and is controlled mainly by release from the parenteral preparation. A drug is immediately available for absorption only when administered as an aqueous solution and when no precipitation occurs at the injection site. An oil vehicle delays absorption, and when the preparation is an aqueous or oily suspension, the absorption process has a number of stages (Table 7).

Each stage is controlled by numerous physicochemical factors (Wagner, 1961) including the volume of the injected formulation; the concentration of the drug in the vehicle; the presence or absence of enzymes such as hyaluronidase in the formulation; the surface area of the depot; the nature of the solvent or vehicle; the tonicity, viscosity, and intrinsic dissolution rate of the drug in the tissue fluid; the crystalline or polymorphic form of suspended drugs; average particle size and particle size distribution; the presence of any coating on the drug particles; the presence of pharmaceutical adjuvants such as suspending agents; the presence of vasoconstrictors; and the partition coefficient of the drug between the vehicle and tissue fluid which is, in turn, dependent on the chemical nature of the drug itself.

Two topics that warrant critical evaluation but which will only be commented on here are: First, is there variation among species in the rate of drug absorption from an intramuscular injection site? Second, is there a difference between intramuscular and subcutaneous administration in the rate and extent of absorption from these sites? By comparing the absorption half-lives of kanamycin from an intramuscular site in horses and dogs given single doses of kanamycin sulfate in aqueous solution (Table 8), it is evident that the drug was absorbed much more rapidly in dogs. While the data show that the rate of kanamycin absorption differed among the species, the different concentrations of the parenteral preparations for large and small animals may well have contributed to the difference in the rate of absorption. The vascularity of the injection site is the most important factor influencing drug absorption from an aqueous solution. Other factors include the degree of ionization and lipid solubility of drugs that are weak organic electrolytes, molecular size of lipid-insoluble substances, and the area over which the injected solution spreads (Schou, 1961; Sund and Schou, 1964).

It is interesting to compare the serum concentration-time curves (Fig. 3) for different species given 10% amoxycillin aqueous suspension by intramuscular

Table 6 Bioavailability of Drugs from Parenteral Dosage Forms

Drug product	Dose (mg/kg)	Species	Site of injection	Systemic availability (per cent)	Peak serum level Average time (min)	Peak serum level Average conc. (μg/ml)
Gentamicin sulfate	10	Dog	IM	>90	30	30
Ketamine hydrochloride	25	Cat	IM	92	10	12
Anhydrous ampicillin	10	Cat	IM	28	30	14
Ampicillin trihydrate	10	Cat	SC	56	60	15
Tylosin (in 50% propylene glycol)	12.5	Cow	IM	70–80	360 ± 120	0.85
Erythromycin	12.5	Cow	IM	70–80	600 ± 300	1.0

Table 7 Stages in Drug Absorption from Aqueous and Oily Suspensions

Aqueous suspensions	Oily suspensions
Drug particle in aqueous depot	Drug particle in oil depot
Drug dissolves in aqueous depot	Drug particle reaches oil/water boundary
	Drug particle becomes wetted

Drug dissolves in tissue fluid
Drug enters tissue fluid or capillaries
Drug passes into bloodstream

Source: Rasmussen and Svendsen (1976).

injection at 7 mg/kg body weight (with the exception of cats given 50 mg per cat, equivalent to perhaps 10 to 12 mg/kg). The trend is for smaller animals (piglets, dogs, cats) to show an early high peak concentration followed by a rapid decline, while larger animals (calves, horses) show a lower and relatively constant concentration of antibiotic in the serum (Marshall and Palmer, 1980). Ampicillin was absorbed at a similar rate from subcutaneous ($t_{1/2(a)}$ = 51 $\pm$ 16 min) and intramuscular ($t_{1/2(a)}$ = 57 $\pm$ 28 min) sites in cats (Mercer et al., 1977a,b). In this study different preparations of ampicillin were administered (Table 6). The drug's extent of absorption differed twofold, so the conclusion that the rate of absorption is similar might be applicable only to this particular case. A difference in the rate and extent of absorption may exist between different intramuscular regions, such as the gluteal compared with the neck in large animals. Consequently, in determining bioavailability data for an extravascular parenteral preparation it is important to specify the location of the injection site (see factors influencing bioavailability and Sec. IV). In many supposedly intramuscular injections, the parenteral preparation may have been deposited predominantly in an intermuscular location (Marshall and Palmer, 1980).

Table 8 Absorption of Kanamycin from Intramuscular Sites in Horses (n=6) and Dogs (n=6)[a]

Species	Concentration of drug in solution (mg/ml)	Systemic availability (%)	Absorption half-life (min)	Peak serum concentration (μg/ml)
Horses	200	>90	62.3 $\pm$ 13.1 (41.7–77.3)	30
Dogs	50	90	9.1 $\pm$ 1.1 (7.2–10.0)	28

[a]Each animal was given single doses (10 mg/kg) of kanamycin sulfate in aqueous solution by intravenous and intramuscular injection. The two phases of the bioavailability study were separated by an appropriate washout period for Ranamycin.

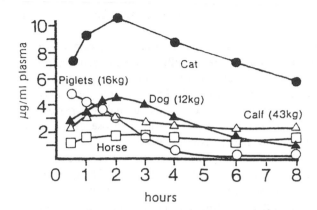

Figure 3 Effect of species/weight on bioavailability. Amoxycillin aqueous suspension (10%) was given by intramuscular injection to all species at the same dosage (7 mg/kg), except cats (50 mg/cat). (From Marshall AB, Palmer GH, 1980.)

2.   To avoid nausea and vomiting due to local gastrointestinal irritation, some drugs can be given parenterally.

3.   Many drugs are inactivated by acidic pH in the stomach, or metabolized by rumen or gastrointestinal enzymes or bacteria (Pope, 1975). Thus, for example, chloramphenicol is metabolized in the rumen (Theodorides et al., 1968), and blood concentrations were not detected after oral administration to goats (Davis et al., 1972). Insulin, parathyroid extracts, penicillin G, and certain other antibiotics are at least partly inactivated by gastrointestinal secretions.

4.   Onset of action is not only more rapid following parenteral administration, but blood concentrations are often more predictable because all of the drug is systemically available after intravenous administration and is likely to have relatively high bioavailability after intramuscular (IM), subcutaneous (SC) or intradermal (ID) administration, particularly compared with oral administration. In emergency situations, rapid onset of pharmacological effect is essential.

5.   When an animal is uncooperative or unconscious, parenteral administration is warranted. In fact, parenteral administration to an individual food-producing animal is much easier than oral administration of a dosage form that requires manual restraint and manipulation of the animal's head.

6.   Parenteral administration can be used when local effects are desired—e.g., in producing local or regional anesthesia or subconjunctival injection.

b. Disadvantages

1.   Since manufacture requires specialized facilities, excellent training of personnel, stringent adherence to good manufacturing practices (GMP), and a well-

planned quality control regimen, the cost per dose is greater than that of conventional preparations for administration by other routes.

2.   The dosage form has to be administered by trained personnel and requires adherence to aseptic technique. Clean needles and sanitized equipment are critical to minimize transfer of bacteria and viruses among animals.

3.   When a drug is administered parenterally (by injection), it cannot be removed from the body. This may be dangerous if the animal is experiencing adverse side effects or has been inadvertently given an overdose of the drug. Elimination is totally dependent on metabolism and excretion.

4.   Subcutaneous and intramuscular injections can produce severe local irritation and tissue damage. This may be due to the solvent used (Spiegel and Noseworthy, 1963), or, in some cases, the drug causes a reaction.

5.   Parenteral administration may be time-consuming unless one of the automatic syringes discussed in Chapter 5 is used. Mass medication of entire herds or flocks is difficult when the drug has to be administered parenterally.

6.   Special care must be taken in packaging parenteral dosage forms. Packaging usually entails use of an ampule (single dose), syringe (single dose), or vial (multiple doses). Multiple-dose containers have to contain a preservative, whereas the single-dose containers need not, as long as they are tamper-proof.

From a pragmatic standpoint, ampules are more difficult to handle than a multidose vial and often, when opened incorrectly, send a shower of glass particles into the solution. Furthermore, storage of any unused drug is inconvenient when an ampule is used. However, from a container closure integrity standpoint, ampules are the ideal container. Because the preparation is completely contained in type I glass, the best possible stability characteristics can be afforded, although problems may arise.

Multidose vials are usually preferred by veterinarians because of their convenience. However, the injection is in contact with the rubber closure, which can either leach out contaminants or adsorb the active drug or preservative, or both. In addition, penetration of the rubber septum with the needle may generate rubber particles. The in-use integrity of multidose vial closures is established to assure that the product purity is maintained. Typically, 10 to 20 insertions of the appropriate-diameter needle are tolerated without generating rubber particles or compromising sterility of the contents. Readers are directed to Chapter 2 for additional information.

5. Solutions

Although injectable solutions can be either aqueous or lipoidal, most solutions administered parenterally to animals are aqueous. In general, aqueous solutions are more rapidly absorbed and less irritating than comparable injected suspensions. Ideally, the solutions would be isotonic, at physiologic pH, and easily syringeable. Deviations from those criteria increase the likelihood that the injection will be irritating, although even solutions with those properties can be irritating to tissues due to the intrinsic properties of the drug. Injectable solutions can be formulated to be injected intravascularly, which can be a more forgiving route of administration for hypertonic and/or irritating solutions because the injected solution is quickly diluted and dispersed throughout the systemic circulation prior to the patient's response to the irritant. However, intravascular injectable formulations are expected to contain fewer particulates than extravascular injectable solutions.

6. Suspensions

Injectable suspensions are frequently administered to large animals. These injectable suspensions have a slower apparent absorption rate due to the additional steps of disintegration and dissolution that precede the true process of absorption of dissolved drug. Therefore, for the same compound and the same route of administration, suspensions are absorbed more slowly and possibly less completely than analogous solutions. As an example, ceftiofur sodium (when administered as the reconstituted solution) is absorbed somewhat more rapidly than ceftiofur hydrochloride sterile suspension, although the extent of absorption (i.e., systemic bioavailability) is the same (Brown et al., submitted). Because of the particular nature of suspensions, injection site reactions tend to be more severe than for comparable drug solutions inasmuch as the body defenses recognize particulate matter as foreign more readily than dissolved drug. Furthermore, suspensions are typically more viscous than solutions, decreasing their syringeability (particularly in cold environments) as compared with aqueous solutions.

C. Implants

1. Implantable Infusion Devices

Vapor-pressure-powered devices are dual-chamber, disk-shaped devices with an inexhaustible volatile liquid power source (Rohde and Buchwald, 1988). When drug is added into the bellows-shaped drug reservoir inside the vapor-liquid chamber by percutaneous needle injection through the septum, it compresses and condenses the charging fluid vapor and recharges the pump. Peristaltic pumps are the most popular implantable infusion devices available for human use today. These disk-shaped devices contain a flexible tube in a U-shaped chamber that is in contact with rollers that press against the tube, sending the fluid forward as the rollers rotate. It can be programmed by an external telemetric programmer, and it can signal completion of programming as well as low reservoir volumes and low battery power. Solenoid pumps utilize a pump that contains one-way check valves. These stepper motors pulse drug to the patient in 2-μl volumes from the 10-ml reservoir which can be refilled through a septum at the surface adjacent to the skin. Interrogation of the pump status can be made by a clinician. Programmable implantable medication systems (PIMS) administer and alter the basal drug delivery rate, turn the system on and off, and set limits on the medication usage (Fischell, 1988). The devices run on a battery-powered receiver and microprocessor, and can be programmed for individualized drug delivery, or reprogrammed, and reused after sterilization in another patient for delivery of another drug. Typically, these PIMS infuse drug into the patient at some basal rate, and superimpose repetitive pulse doses on top of the basal release rate. Osmotic pumps can also be implanted subcutaneously to provide sustained systemic drug delivery. Growth hormone-releasing factors have been given to steers and wethers using subcutaneously implanted osmotic pumps with significant changes in growth rate and feed efficiency (Wheaton et al., 1988).

2. Subdermal Implants

Diffusion reservoir systems are generally nonbiodegradable devices that release a constant amount of drug each day (zero-order release). If they leak, a toxic dose of the drug may be released into the animal (dose dumping). Diffusion matrix systems do not typically have a zero-order release rate, but with the appropriate combinations of device geometry, the release may be very close to a constant amount each day. Drug must be uniformly distributed in the polymeric device, and the device's geometry should be constant over the effective lifetime of the device. The polymer may or may not be very slowly degradable. Elanco's estradiol implants are an example of such a device. Some subdermal implants release drug by erosion of the polymeric matrix at the site rather than diffusion through the matrix. Irrespective of the degree of erosion, these implants must retain a relatively similar geometry to that of the original implant. Drug release from subdermal implants may be controlled by the swelling of the polymeric matrix. The rate of release is proportional to the surface area and rate of swelling of the outer surface of the polymer relative to the middle. If absorption of fluid from the environment is constant, so is the drug release. Magnetically controlled release of drug from subdermal implants is accomplished by interspersing drug and magnetic beads in the polymer matrix. In the presence of a magnetic field, the rate of release is increased.

D. Intramammary Administration

The selection of the method of treatment (systemic, intraparenchymal, or intramammary) is a primary consideration in the control of mastitis infections. Systemic treatment may be indicated where the parenchyma is intensely swollen, the milk duct system being either compressed or blocked by inflammatory secretions and cellular debris, thus preventing distribution of an intramammary infusion throughout the udder and access of antimicrobial drug to the site of infection.

Ziv (1980) discussed the pharmacokinetic properties that must be recognized in selecting a drug for systemic treatment. The drug should have a low minimum inhibitory concentration (MIC) against the majority of udder pathogens and should exhibit good availability from the intramuscular injection site. To ensure transport across the blood-milk barrier, it should have a low degree of protein binding, be nonionized in the blood, and be lipid-soluble. A long half-life would be desirable to ensure a suitable duration of concentrations above the MIC in the udder.

Some ion trapping of weakly basic antibiotics occurs in mastitic milk because the pH of mastitic milk is in the range 6.9 to 7.2. These weak bases would, however, accumulate more in normal milk than in mastitic milk, since normal milk has a lower pH (6.0 to 6.8) and hence could trap basic antibiotics more effectively. For instance, the milk:serum concentration ratio for acidic drugs is less than 1, whereas for weak organic bases the ratio is generally 1 or higher (Table 9). The closer the pK_a of the antibiotic to physiological pH, the greater the influence of milk pH on the milk:serum concentration ratio (Prescott and Baggot, 1988).

Parenchymal injection of drugs into mammary tissue by passing a needle through the skin into the body of the gland is not widely used. It is only occasionally recommended when the gland is so swollen that poor distribution of the intramammary infusion is likely. However, even in this situation, diffusion from a parenchymal injection may be greatly impeded.

If intramammary infusion is to be used, some of the criteria for optimizing effective formulations for use in dry and lactating cows are different. The criteria that are similar include:

1.   There should be minimal irritation to the udder. Specific target animal safety studies for udder irritation must be carried out and show that there is no irritation beyond the designated milk discard time. Udder irritation studies are classically done in cattle with normal quarters rather than mastitic quarters, and are evaluated by somatic cell counts at each milking after drug product administration. Irritation caused by dry cow formulations obviously has longer to resolve than lactating cow formulations. On the other hand, the dry cow formulation must be reasonably nonirritating to the udder because of the extended time it is expected to be in contact with the udder tissues. For example, although chlortetracycline is extensively used in the treatment of lactating cows, it should not be used in dry cows because of its tendency to cause chemical mastitis, especially when the udder is completely dry (Zinn, 1961).

2.   A short milk withholding period (discard time). This requirement poses the dilemma of deciding which is the more important parameter: efficacy or tissue residue. If emphasis is placed on efficacy and capacity to maintain an MIC for an extended time, veterinary surgeons and dairy farmers may be loath to use the formulation because of the long milk discard time. For a given anti-

Table 9 Comparison of Calculated and Experimentally Obtained Milk:Plasma Concentration Ratios for Antimicrobial Agents Under Equilibrium Conditions

Drug	pK$_a$	Milk pH	Concentration ratio (milk ultrafiltrate:plasma ultrafiltrate)	
			Theoretical	Experimental
Acids				
Penicillin G	2.7	6.8	0.25	0.13–0.26
Cloxacillin	2.7	6.8	0.25	0.25–0.30
Ampicillin	2.7, 7.2	6.8		0.24–0.30
Cephaloridine	3.4	6.8	0.25	0.24–0.28
Cephaloglycin	4.9	6.8	0.25	0.33
Sulfadimethoxine	6.1	6.6	0.20	0.23
Sulfadiazine	6.4	6.6	0.23	0.21
Sulfamethazine	7.4	6.6	0.58	0.59
Rifampin	7.9	6.8	0.82	0.90–1.28
Bases				
Tylosin	7.1	6.8	2.00	3.5
Lincomycin	7.6	6.8	2.83	3.1
Spiramycin	8.2	6.8	3.57	4.6
Erythromycin	8.8	6.8	3.87	8.7
Trimethoprim	7.6	6.8	2.83	2.9
Aminoglycosides	(7.8)	6.8	3.13	0.5
Spectinomycin	8.8	6.8	3.87	0.6
Polymyxin B	10.0	6.8	3.97	0.3
Amphoteric				
Oxytetracycline	—	6.5–6.8	—	0.75
Doxycycline	—	6.5–6.8	—	1.53

Source: Prescott and Baggot (1988).

biotic, if the formulation's desired milk discard time minimizes tissue residues, the therapeutic benefit of the product may be reduced. However, short milk discard times may not jeopardize the efficacy of a given antibiotic, since milk discard times are determined by safety of the residues for human consumption, whereas efficacy is determined by activity against the microorganism. One approach to the problem of dairy farmers inadvertently not withholding milk for a sufficiently long period is to include a marker dye such as brilliant blue in the antibiotic formulation. The excretion rate of the dye must be slower than or closely approach that of the antibiotic for it to be a reliable marker. Thus, each antibiotic must be considered individually, and the relationship between drug and dye excretion must be established (Bywater, 1977). Some countries require such a dye to be incorporated in approved intramammary formulations, despite the fact that the kinetics of the dye and the drug residues may not coincide.

3. The most important factor influencing the decrease in drug concentrations in the udder is the frequency of milking; the second largest factor is the efficiency of milkout (i.e., completeness of milkout at each milking). Thus, milk discard times can be altered by the efficiency and frequency of milking. This is one of the reasons why dry cow formulations can reside for a significantly longer period than lactating cow formulations, even if the formulations are reasonably similar.

4. With many countries now requiring sterile intramammary infusions, the method of sterilization and manufacture should be considered when considering the type of formulation. Because of the highly competitive nature of the intramammary infusion market, exact costs should be determined for a finished sterile tamper-proof formulation before development proceeds too far.

Compared with dry cow formulations, a lactating cow formulation may focus on the following:

1. A low degree of binding to milk and udder tissue proteins. This ensures a fast rate of drug distribution to various parts of the udder. Even for dry cow formulations, binding to milk and tissue proteins must be readily reversible and provide sufficient free drug to be effective against the target udder pathogens.

2. A vehicle that ensures fast and even distribution of the antimicrobial drug. Release may depend on the solubility and dissolution rate of the drug if it is suspended—e.g., in a nonaqueous medium. A water-soluble salt may be more suitable for treatment of lactating cow mastitis, whereas an insoluble salt may be more suitable for dry cow mastitis (Ziv et al., 1973). Other criteria, such as particle size and mechanical properties of formulations, have been discussed by Brander (1975). Particulate formulations may be more irritant to tissue (similar to parenteral injections) and may have an adverse effect on dispersion of drug in the gland, causing a lumping and/or aggregation of particles, which decreases the efficacy of the formulation by decreasing its dispersion in the system (Fulper, 1991). Furthermore, recovery of drug in milk over time is inversely proportional to formulation viscosity and interfacial tension, indicating less release of drug from the vehicle as formulation viscosity and interfacial tension increase (Fulper, 1991). However, formulation leakage through the teat canal increases as formulation viscosity decreases, making the formulation viscosity a two-edged sword.

Dry cow therapy has considerable merit because: 1. it prevents most new infections during the dry period; 2. milk is not contaminated; 3. damaged tissue is allowed to regenerate, and 4. it results in a high proportion of uninfected cows at the time of highest milk yield (Dodd et al., 1969). When considering development of an intramammary infusion for use in the dry cow, the following should be considered:

1. In contrast to formulations for the lactating cow, the selection of a drug with a high degree of binding to the secretions and udder tissue proteins is desirable for dry cow formulations. Although binding diminishes antibiotic activity in vitro, it is not a measure of its activity in vivo (Kunin, 1967; Meyer and Guttman, 1968). An equilibrium exists between the bound and unbound antibiotic so that the bound fraction decreases as the unbound antibiotic is removed by diffusion or metabolism. Therefore, binding can serve as a depot for some antibiotics, provided the concentration of active free drug is higher than the MIC.

2. The dry cow formulation should ideally exhibit stability of antimicrobial activity throughout the entire dry period, the release from the formulation preferably being of a zero-order type. The formulation in the dry cow thus determines the antibiotic concentration time profile to a larger extent than in the udder of the lactating cow. Hence, efforts to prepare sustained-release formulations have incorporated the drug molecules into a particular matrix by adsorption onto insoluble inert compounds, by microencapsulation, by suitable choice of vehicles and thixotropic agents, and by desolubilizing the drug by conjugation with degradable desolubilizing moieties (Ziv, 1978; Dowrick and Marsden, 1975; Wilson et al., 1972).

E. Topical Application

The topical route of administration is appealing to food-animal (livestock) producers because it is less labor-intensive than many of the conventional routes of administration. When a drug is applied topically, an exact dose can be administered, the first-pass effect of the liver associated with oral administration is bypassed, there is less trauma and tissue damage than might occur after injection of the drug intramuscularly or subcutaneously, and there is no drug residue at a site of injection. Drugs are much more likely to traverse the skin of cattle and sheep by way of adnexa than through the stratum corneum simply because of the high density of appendageal structures in their skin. For example, whereas a square centimeter of human skin contains an average of 40 to 70 hair follicles and 200 to 250 sweat ducts, the same area of cattle skin contains approximately 2000 hair follicles with the associated sweat and sebaceous glands, and sheep skin can contain up to 10,000 follicles per square centimeter. Dogs have approximately 100 to 600 hair bundles per square centimeter, and cats have 800 to 1600 hair groups per square centimeter.

For local therapeutic and systemic effects, the following should be considered in selecting the dose delivery formulation of choice. There are three potential rate-determining barriers to percutaneous penetration: the rate of dissolution of the drug in the vehicle; the rate of diffusion of solubilized drug through the vehicle to the skin; and the rate of permeation of the drug through the stratum corneum. Formulation overcomes the problems of drug dissolution in the vehicle. Either the vehicle can be changed or the chemical properties of the active drug suitably altered—e.g., salt formation.

The vehicle also governs the rate of diffusion and release of the drug to the stratum corneum. Thus, the vehicle is exceedingly important in determining topical bioavailability (Malone et al., 1974; Paulsen et al., 1968; Oishi et al., 1976; Ostrenga et al., 1971a,b; Coldman et al., 1969). Nevertheless, vehicle design is often ignored in the development of a suitable delivery system.

The penetration of levamisole through cattle and sheep skin is not affected by removal of the stratum corneum to the same degree that would be predicted by denuding human skin of the keratinized layers. This indicates that percutaneous absorption of the only moderately lipophilic levamisole in cattle and sheep occurs through the appendageal pathway. The hair and wool of animals consist of modified keratin which contains chemically reactive groups such as thio, amino, and carboxyl groups and hydrophobic regions. These chemically reactive groups can alter (often reduce) the thermodynamic activity of drugs through the process of binding. Hair is coated with an emulsion of sebum and sweat, particularly in animals that are not dipped regularly. This emulsion rapidly dissolves many topically applied drugs. Seasonal changes in the composition of the emulsion result in variable absorption at different times of the year. For example, penetration of levamisole through the skin of cattle was 10 times faster when administered during the summer than when administered during the winter (Forsyth et al., 1983). This can substantially decrease the efficacy of topically applied drugs during the winter months. Dissolution in the emulsion therefore controls diffusion through the skin.

Extremely lipid-soluble drugs penetrate into the stratum corneum quite effectively but need some degree of water solubility to pass through the epidermis and dermis. Because of this characteristic absorption pattern, highly lipid-soluble drugs accumulate in the corneal layer of the skin without being absorbed (Kydonieus, 1987a). Intercellular regions of the stratum corneum are filled with lipid-rich amorphous material. Intercellular volume in dry stratum corneum is approximately 5%, compared with approximately 1% in fully hydrated stratum corneum. This volume is an order of magnitude larger than that estimated for the intraappendageal pathway. Therefore, the intercellular diffusion volume is probably the most important determinant of diffusion.

The hydration of the stratum corneum dramatically affects the percutaneous absorption of drugs. As the degree of hydration increases, so does the absorption of drug across the stratum corneum. Hydration of the stratum corneum can be increased by soaking the skin with water or an aqueous solution, or by occluding the topical site with plastic sheeting or other impermeable film. Occlusion causes an accumulation of sweat and condensed water vapor underneath the dressing, which hydrates the outer keratinized layers of the skin. This results in an increased porosity between cells of the stratum corneum and a decreased density of cells per unit volume of skin by imbibing the dead cells with water. Altering these properties of the stratum corneum can increase the penetration of drugs as much as eightfold (Kydonieus, 1987a).

When the drug has penetrated the stratum corneum, it must then pass through the viable epidermis and the dermis. As previously mentioned, the dermis and living epidermal layers are much more easily penetrated by drug molecules than the cornified epithelium (see Fig. 2). Having traversed these barriers, the drug must then be absorbed into the vascular bed of the skin. The vascular surface for absorption in the skin is only approximately 1 to 2 cm^2/cm^2 skin, much less than the corresponding ratio encountered in muscle (Kydonieus, 1987a). Furthermore, ambient temperature can greatly alter the blood flow through the capillary beds of the skin. Thus, the percutaneous absorption of a drug in the same animal may be much different, depending on whether the animal is housed at a constant ambient temperature or is outdoors, where the ambient temperature can vary widely. The absorption of acetylsalicylic acid (aspirin) and glucocorticoids increases 10-fold when the environmental temperature is raised from 10°C to 37°C (Kydonieus, 1987a). However, except for small, lipid-soluble compounds the resistance to absorption of drugs caused by perfusion is very small compared with the resistance of the stratum corneum. For those compounds that penetrate the stratum corneum more rapidly than n-octanol, or for drugs that damage the stratum corneum, perfusion would be the limiting factor for absorption (Ohshima et al., 1985). For most drugs, however, the stratum corneum is the rate-limiting barrier.

The stratum corneum may serve as a reservoir for many classes of drugs. The retention of drugs increases as the concentration in the applied solution increases and may occur for as long as 72 h after topical administration of some drugs (Kydonieus, 1987a). There are two populations of molecules that comprise the drug reservoir. The first is the mobile molecules, which can diffuse anywhere in the stratum corneum or may leave the stratum corneum by diffusion. Diffusion of drugs within the stratum corneum is mainly transcellular rather than intercellular, although both routes are thought to play a role in the movement of drugs within the stratum corneum. The second group of drug molecules in the drug reservoir of the stratum corneum comprises those that are reversibly bound to a fixed number of binding sites in the protein component of the stratum corneum. As such, these bound molecules are not free to diffuse through the stratum corneum. However, they may become unbound as the concentration of mobile, free drug molecules diminishes. The analogous situation most veterinarians are familiar with is the reversible binding of drug molecules to plasma proteins.

Occlusion of the surface to which the drug is topically applied will also reduce the bound fraction of the drug in the stratum corneum. Retention of drug in the stratum corneum, also known as the reservoir effect, is most notable with corticosteroids and sex steroids, whereas alcohols and related drugs have very little residue binding within the stratum corneum. In addition to the bound portion of drug within the stratum corneum, there is evidence that drug may be recovered from the surface of the keratinized epithelium long after the drug has been applied.

Bulk transport of neutral molecules with low molecular weights occurs largely via skin appendages rather than the transcellular pathway, which predominates in human skin penetration. The rate and extent of drug absorption are substantially influenced by the composition and physical properties of the sebum-sweat emulsion.

1. Topical Preparations

The skin is one of the most easily accessible organs. Topical formulations, in this context, are intended for local therapy and not for transdermal treatment of systemic infections. However, cutaneous administration is meant to mean a site of administration of formulations that may be intended for local (topical) treatments or for systemic treatment of generalized diseases. Topical veterinary preparations may be used for local protective or therapeutic reasons (dusting powders, solutions, suspensions, lotions, liniments, creams, ointments, aerosols), or for systemic activity (pour-ons).

These topically applied products offer the same physicochemical advantages and disadvantages of solutions, suspensions, emulsions, and solids, as already discussed. The important considerations in the formulation of a dermatological preparation are: whether it is to be applied to a broken wound or an abrasion; whether it is to be rubbed into the affected area; whether it has to exhibit adhesiveness upon application to the skin; and whether it will deliver the active ingredient to the site required.

Dusting powders, lotions, and aerosols are recommended formulations for application to abraded sites; lotions, liniments, creams, and ointments are suitable for unabraded sites.

In the design of a suitable delivery system, the following should also be considered:

1. Permeability of the stratum corneum may be increased if it is hydrated by a suitable vehicle (dimethyl sulfoxide can result in superhydration [Stoughton and Fritsch, 1964; Maibach and Feldmann, 1967]) or by occlusive dressings or vehicles. Transport in some species, however, may be via skin appendages; agents promoting increased stratum corneum

permeability may thus have no or even a negative effect (Pitman and Rostas, 1981).

2. The thermodynamic activity of the drug in the vehicle should be higher than in the stratum corneum (Higuchi, 1960).

3. The formulation should be buffered to a pH such that the active ingredient is in the nonionized state. Levamisole (pK_a 7.94) penetrated sheep skin eight times faster when buffered at pH 8.90 (90% nonionized) than when buffered at pH 5.95 (1% nonionized) (Ponting and Pitman, 1979).

4. Keratolytics, lipid and polar solvents (acetone and alcohol), surfactants after protracted use, and some vehicles may cause damage to the stratum corneum, thus increasing penetrability.

5. Viscosity of the medium is inversely related to flux.

The above principles also apply to pour-on preparations where systemic activity of the active drug is required. For example, the pour-on may contain an organophosphorus insecticide which, when poured evenly along the animal's back, kills cattle grubs (genus *Hypoderma*) and lice (order Mallophage or Anaplura) in all parts of the body (Pope, 1978). Levamisole in a pour-on preparation has even proved successful as an anthelmintic (Brooker and Goose, 1975; Curr, 1977).

Further studies are required to clarify the relationship between formulation and skin penetration in animals, but because of the many advantages of this method of application, this type of formulation should be seriously considered for a wide range of drugs.

2. Creams/Ointments

Creams and ointments are generally considered for local topical therapy of cutaneous infections or inflammation. Creams are aqueous-based products, whereas ointments are generally oil-based formulations. Creams, because of their aqueous vehicle, tend to dissipate from the site of application, leave little greasy residue, and are shorter-lived at the site of action. Ointments tend to remain at the site of application for a longer time, trapping moisture between the skin and the ointment film, thereby hydrating the skin and rendering it a less effective barrier to drug absorption. Ointments also provide a better contact between the skin and the formulation. Neither type of formulation is typically used to administer products that are applied cutaneously for treatment of systemic diseases.

Because cats are constant groomers, any drug substance applied topically is likely to be ingested. Even disinfectants and other chemicals applied to cages, boxes, and floors are picked up on the cat's paws and eventually ingested. Therefore, a disinfectant considered safe for use in kennels may be detrimental if used in an area accessible to cats (Spinelli and Enos, 1978).

Behavioral differences of a breed within a species may also present problems in drug selection. Flea collars (see Chap. 2) often cause severe local reactions in dogs when wet. Thus, water-loving breeds, such as the labrador retriever, may show the problem more often or more severely than other breeds (Spinelli and Enos, 1978).

3. Pour-ons

Pour-on formulations are liquid solutions, frequently containing anthelmintics and/or ectoparasiticides, that are poured onto the dorsal skin (back) of the animal. These formulations typically contain sufficiently lipid-soluble compounds to allow them to be absorbed and exert their desired effects systemically. Nevertheless, a high concentration of drug may remain at the surface with residual drug in the stratum corneum and stratum germinativum that provides a prolonged effect at the surface. To be effective as a pour-on, there must be sufficient penetration into the stratum corneum and other layers of the epidermis to allow for systemic absorption and/or prevent removal of the drug by environmental conditions (e.g., rain) or mechanical removal (e.g., rubbing).

4. Transdermal Patches

The usual rate-limiting step for transdermal drug absorption, which is penetration of drug through the stratum corneum, presents an absorption barrier that may not be readily penetrated although penetration is highly variable between individuals. A transdermal delivery device must release the drug to be absorbed at a slower rate than the rate of penetration through the stratum corneum, yet absorption must be rapid and sufficiently complete to provide therapeutic plasma concentrations (Govil, 1988; Kydonieus, 1987b). The imposed rate-limiting factor is the release of drug from the device, which can be more precisely controlled.

Drug release from the delivery device is dependent on diffusion of the drug within the device and the relative affinity of the drug for the components of the device (e.g., polymers) relative to the skin. The amount of drug absorbed is proportional to the amount of drug in the device and the surface area of skin in contact with the device. Occlusion of the area over an applied drug can increase absorption of the drug.

a. *Monoliths.* When a drug is mixed to the point of saturation into a single polymeric matrix which is then polymerized, the result is a block of substance with a uniform dispersion of drug. The rate of drug diffusion is the rate of release of drug from the monolith. The major problem with monolithic patches is that, if the skin is unusually permeable, the skin may provide such an inefficient barrier to absorption that the drug may be absorbed rapidly at first, resulting in an overdose (toxic plasma concentrations) of drug.

b. Membrane-Limited Release.   Drug may be incorporated into a reservoir, and a rate-limiting membrane is then placed between the reservoir and the skin. This membrane can be adhesive; the adhesive can be added as an additional layer to the device, or the adhesive can be incorporated only on the perimeter of the patch. The zero-order rate of release from these membrane-limited devices is proportional to the permeability of the membrane to the drug.

c. Microreservoir Device.   The microreservoir device consists of a polymer matrix which has microscopic cavitations containing dissolved drug. The dissolved drug must diffuse from its location inside the polymer through the polymer to the surface, where it can then be absorbed through the skin. This type of delivery system can be used when the drug is not very easily dispersed in the polymer.

d. Laminated Polymeric Systems.   The release from these polymers is similar to that of devices that use a rate-limiting membrane to ensure zero-order release. These polymeric patches can be made extremely thin, resulting in better adhesion and improved patient compliance.

e. Iontophoresis.   The application of an electric current to the skin increases the transdermal absorption of ionic drugs and small peptides (e.g., insulin). Penetration of ionic drugs through the skin principally by transfollicular or transappendageal route is proportional to the strength of the current (Tyle and Kari, 1988). The amperage that can be applied is limited by the pain and tissue damage produced, although sufficient current can be provided to enhance transdermal absorption with minimal to no sensation or tingling. In addition, increasing the ionic strength of the drug solution increases the iontophoretic transfer across the skin. Drugs for which iontophoresis has shown benefit include antihistamines, antibiotics, insulin, vasopressin, LHRS analogs (e.g., leuprolide), and steroids. The drug must be appreciably ionized and capable of carrying a current; molecules larger than 10,000 MW have uncertain current-carrying capacity.

F. Body Cavity

1. Rectal

This route is infrequently used in veterinary practice for any other purpose than local action on the rectum and lower colon. Glucose, digested proteins, and anesthetics are occasionally administered by high colonic irrigation to obtain systemic effect.

2. Vaginal

Vaginal tablets are ovoid or pear-shaped and prepared by granulation and compression. They can be formulated to exhibit two types of release mechanism: first, a slow-release dissolution which retains the tablet's original shape. This tablet is similar to a lozenge in structure and is ideal for drugs requiring low concentrations in the cavity for long periods. Second, effervescent and disintegrating tablets release drug quickly and ensure rapid distribution of the active drug for total local effect throughout the cavity. Both forms may often contain a buffer to maintain or change the vaginal pH to that required for normal physiologic vaginal flora. Vaginal pessaries require specialized manufacturing equipment. The mass is usually a glycogelatin, theobroma oil, or synthetic base solid at room temperature but which dissolves or is liquid at body temperature. Pessaries pose special stability problems and should be considered only if the vaginal tablet does not give satisfactory drug efficacy.

Intravaginal delivery of reproductive hormones (e.g., progestins) has been used as a means to conveniently administer and easily remove the formulation and thereby synchronize estrus. Drug-laden intrauterine sponges have served this purpose and are marketed in several countries. Intravaginal and intrauterine irrigations are frequently used in cows as antiseptic infusions, and occasionally as irritants to stimulate uterine activity. Most of the problems associated with formulating solutions apply.

3. Otic

Otic dosage forms are intended for administration either on the outer ear or into the auditory canal. They include a number of dosage forms: solutions, suspensions, ointments, otic cones, and powders. Their primary use is either to remove ear wax or supply local drug delivery. The type of formulation and its intended use are extremely important. For instance, if otic powders are used misguidedly for the treatment of canker, in unskilled hands the powder may adhere to wax in the ear canal, resulting in blockage. More powder then completes the impaction, and pressure is built up against the tympanum, producing vertigo and incoordination of the hind limbs. In extreme cases, the pressure may cause the tympanum to rupture.

4. Intranasal

Inhalation of drugs often results in onset of action comparable to administration by intravenous injection while avoiding many of the potential problems associated with intravenous administration. Most drugs are absorbed by this route; however, only certain drugs can be administered because the pulmonary tract is extremely sensitive to foreign bodies. If irritation occurs, the respiratory passages will constrict, leading to impaired absorption and interference with oxygen and carbon dioxide exchange.

Vaccines and drugs can be administered intranasally as solutions or powders to one or a number of animals sequentially or simultaneously (Chap. 2). The effectiveness of action depends on both the formulation and the method of delivery. Spray and mist dispensers may be used for mass inoculation. The depth of penetration into the pulmonary tree is highly dependent on the particle size distribution in the inhaled materials. The larger the particle, the greater is its tendency to impact and be retained in the upper respiratory tract. Very fine particles are inspired to deeper regions, but small particles may be exhaled, and the total amount of drug retained would be less than for larger particles. Hence, the use of the nasal route for drug delivery requires extremely precise manufacture and packaging to effect reproducible formulation presentation. Problems with use must also be considered. For example, with mass inhalation inoculation, the size of the air space to be filled and the stability of the mist should be considered.

5. Ophthalmic

Ophthalmic preparations are sterile aqueous or oily solutions, suspensions, emulsions, or ointments for topical administration by instillation. The use of topical medication should be considered for treatment of diseases involving the superficial cornea, conjunctiva, third eyelid, and nasolacrimal drainage apparatus, as well as inflammation of the anterior chamber of the eye. Conditions involving deeper ocular structures do not lend themselves to topical therapy alone; they require systemic treatment with a combination of topical and systemic therapy (Magrane, 1977).

Ophthalmic solutions are usually isotonic and buffered, to minimize irritation to the eye. All multidose eye preparations must contain a bacteriostatic agent. However, these cannot be used in the injured eye or during surgical intervention in the anterior chamber, because of possible irritation.

Solutions exhibit a fast drug pulse delivery: they produce an initial high concentration that rapidly declines to very low concentrations. Because the normal tear volume is quite small, eye drops overflow the lacrimal lake and eyelid margin (Pavan-Langston, 1976), sometimes losing more than 80% of the solution instilled. Thus, the instilled medicinal drop is more or less immediately lost from the preocular film. Most solutions also increase tear flow because they cause a stinging sensation when instilled. Such reflex tearing actually hastens removal of the solution from the eye (Fraunfelder and Hanna, 1977). In the irritated eye, where tearing is already increased, washout occurs even faster than in the normal eye. Thus, to achieve adequate corneal levels, solutions should be administered every 30 min to 1 h. The interval for administration may be increased if suitable viscosity-inducing compounds (e.g., methylcellulose or polyvinyl alcohol) are added to the formulation (Magrane, 1977; Trueblood et al., 1975).

Ophthalmic suspensions should preferably consist of an aqueous vehicle containing a drug of low solubility. The duration of action produced by a suspension is more prolonged than that from an aqueous solution. Their disadvantage is the possibility of irritation due to suspended crystals or particles. Solutions and suspensions offer a number of advantages, however. They are easily instilled, cause no interference with vision, cause few skin reactions, and do not interfere with mitoses of the corneal epithelium (Campbell, 1979).

Ointments remain in the eye for a longer time than solutions, both in the precorneal tear film and the conjunctival fornices, thereby increasing absorption of active ingredients (Campbell, 1979). If the lid margin contacts the ointment, as the ointment on the lid gradually melts, drug is released into the precorneal film, thereby prolonging therapeutic concentrations of the drug up to 6 h (Fraunfelder and Hanna, 1977; Campbell, 1979). Other advantages offered by ointments include comfort upon initial instillation, less lacrimonasal transfer, and enhanced stability, particularly for some antimicrobial agents. Disadvantages include formation of a film which partially obstructs vision and the potential for slowing the healing of superficial abrasions due to possible interference with epithelialization of the cornea (Heerema and Freedenwald, 1950).

In some situations, ointment should not be used; solutions are preferred. Since the commonly used ointment bases may be toxic to the interior of the eye, they should not be used when the cornea has been penetrated. Invasion of the ointment base into the internal chambers of the eye causes toxic endothelial damage, corneal edema, vascularization, and scarring (Campbell, 1979).

Ocular inserts, such as hydrophilic polymers and contact lenses, have been investigated for the long-term delivery of therapeutic agents including antibiotics in cattle. In fact, polymers as simple as gelatin have been successfully used as methods for antibiotic delivery into bovine eyes (Punch et al., 1985, 1987; Slatter et al., 1982).

IV. PHARMACOKINETICS

A. Drug Absorption

The term *bioavailability* refers to both the rate and extent of drug absorption. The formulation of the dosage form and route of administration affect the bioavailability of a drug. This is particularly the case when a solid dosage form is administered in the feed or given to an animal after feeding. Because of differences in anatomical arrangement of the gastrointestinal tract and digestive physiology among domestic animal species, the bioavailability of drugs administered as oral dosage forms can vary widely, particularly between ruminant and monogastric animals.

An indication of the rate of drug absorption can be obtained from the peak (maximum) plasma concentration (C_{max}) and the time to reach peak concentration (t_{max}), based on the measured plasma concentration-time data. A pilot study is useful for selecting sample collection times that will enable the plasma concentration-time curve to be well-defined. Since C_{max} and t_{max} are features of the peak plasma concentration, the utility of these parameters for estimating the rate of drug absorption depends on how well the peak is defined. When absorption occurs rapidly, a distinct peak is usually evident, but when absorption is prolonged, the plasma concentration-time curve may show an elevated plateau.

In addition to the estimated rate of absorption provided by C_{max} and t_{max} (observations), the "method of residuals" can be used to determine whether absorption is an apparent first-order process and to obtain the absorption rate constant (Gibaldi and Perrier, 1982). This allows calculation of the absorption half-life of the drug. Either a one- or two-compartment pharmacokinetic model with apparent first-order absorption adequately describes the plasma concentration-time data for most drugs that are rapidly absorbed (i.e., $t_{max} \leq$ 2-3h).

In bioavailability studies it is preferable to use a two-way crossover design in which the drug is administered intravenously and orally (or by a nonvascular parenteral route) to the same animals with an appropriate intervening washout period. Riegelman and Collier (1980) have applied statistical moment theory to the gastrointestinal absorption of a drug after oral administration of a solid dosage form. Their analysis permits the estimation of a mean dissolution time of a drug from its dosage form.

The extent of absorption or rather the fraction of dose that reaches the systemic circulation unchanged (systemic availability) is generally of greater interest than the rate of drug absorption, especially following oral administration of a conventional dosage form. It is only when a drug is administered intravenously that complete systemic availability (F = 100%) can be assumed. The usual method for estimating the systemic availability of a drug employs the method of corresponding areas, with correction for dose (when required):

$$\text{Systemic availability (F)} = \frac{AUC_{PO}}{AUC_{IV}} \times \frac{Dose_{IV}}{Dose_{PO}}$$

where AUC is the total area under the plasma concentration-time curve after drug administration by the intravenous (IV) or oral (PO) route, through infinite time. Area under the curve (AUC) is calculated by numerical integration using the trapezoidal rule (from zero time to the last measured plasma concentration) with extrapolation to infinite time. The areas under the extrapolated portion of the curves are estimated by:

$$\frac{C_{p(last)}}{\beta}$$

where $C_{p(last)}$ is the last measured plasma concentration and β is the overall elimination rate constant of the drug (Fig. 4). This method for estimating systemic availability (sometimes used synonomously with bioavailability) involves the

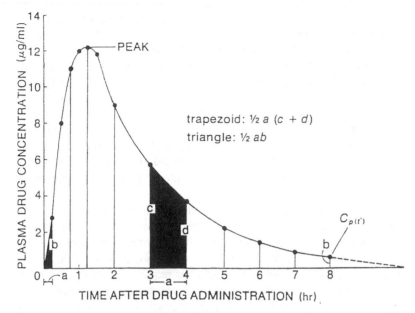

Figure 4 The usual form of the plasma drug concentration vs. time profile that follows the oral administration or nonintravascular injection of a drug. The area under the curve may be calculated by the trapezoidal rule. (From Baggot JD, 1977.)

assumption that systemic (body) clearance of the drug is not changed by the route of administration. The necessity to make this assumption could be avoided by simultaneously administering an intravenous, stable isotope-labeled formulation and the oral dosage form of the drug (Strong et al., 1975).

1. Relative Bioavailability

If an intravenous preparation of the drug is not available, an oral reference standard (usually an aqueous solution or an elixir) may be used for comparison, in which case the relative rather than absolute bioavailability is obtained. Relative bioavailability has a number of important applications, including bioequivalence assessment, in veterinary pharmacology. Bioequivalence refers to the comparison made between a generic formulation of a drug or a product in which a change has been made in one or more of the ingredients or in the manufacturing process, and a reference (standard) dosage form of the drug. This comparison is based on an estimate of the relative bioavailability together with a measure of the uncertainty (variance) of the estimate. The statistical evaluation of bioavailability/bioequivalence studies should be based on confidence interval estimation rather than hypothesis testing (Metzler, 1988; Westlake, 1988). The confidence interval approach, using $1-2\alpha$ or 90%, should be applied to the individual parameters of interest (i.e., the parameters that estimate the rate and extent of drug absorption). Although the observed peak (maximum) plasma concentration provides an estimate of the rate of drug absorption, it is also affected by the extent of absorption. It has been suggested (Tozer, 1994) that $C_{max}/AUC_{0\text{-}LOQ}$ where $AUC_{0\text{-}LOQ}$ is AUC from zero time to the limit of quantification of the analytical method, may provide a more reliable measure of the rate of drug absorption, except when multiexponential decline is extensive. In the design of bioequivalence studies, blood sampling times should be selected to characterize the peak (C_{max}, t_{max}), and sample collection should extend for a period corresponding to at least four apparent half-lives of the drug beyond the expected time of the peak plasma concentration.

AUCs are based on the measured concentrations from zero time to the limit of quantification of the acceptable analytical method employed. The duration of sampling is important since extrapolation to infinite time is not involved and the area under what would be the extrapolated (terminal) portion of the curve should be less than 10% of the total area. Another requirement of the design is that a suitable washout period be allowed to elapse between the phases of a crossover study. For more information regarding bioequivalence studies and their requirements, refer to Chapter 7.

Other applications of relative bioavailability are to determine the effect of different routes or IM sites of administration and the influence of feeding relative to the time of oral dosing on the rate and extent of absorption of a drug from the same dosage form.

2. Urinary Excretion Data

The systemic availability of a drug can be estimated by comparing the cumulative urinary excretion of the unchanged (parent) drug after oral administration with that following intravenous injection of the drug. This method provides an alternative to comparing areas under the plasma concentration-time curve, but is cumbersome to apply since total collection of the urine voided during the excretion period for the drug (at least four times the half-life) is necessary to measure urine volume. Use of cumulative urinary excretion data to determine the relative bioavailability of different dosage forms of a drug assumes that the ratio of the total amount excreted unchanged in the urine to the amount absorbed remains constant. Urinary excretion rate data cannot be relied on to estimate the rate of absorption of a drug; the rate of drug absorption can be obtained only from plasma concentration-time data.

3. Multiple-Dose Approach

When single-dose studies are considered to be unreliable for determining the relative bioavailability of a drug, a multiple-dose approach is warranted. This is based on comparing the areas under the plasma concentration-time curve during a dosage interval at steady-state, which implies that a fixed dose is administered repeatedly (at least five times) at a constant interval. A practical advantage of multiple-dose studies is that the dosage form may be crossed over in the same animals without an intervening washout period. It is necessary to apply the same dosing rate in both phases of the multiple-dose study and desirable to administer six doses in each phase to ensure that steady-state plasma concentrations are attained.

The circumstances under which a multiple-dose approach should be used to determine the relative bioavailability of a drug from different oral dosage forms include:

1. There is a difference in the rate, but not in the extent, of absorption
2. There is excessive variation in bioavailability from animal to animal
3. The concentration of the active drug/therapeutic moiety in plasma following a single dose is too low for accurate quantification by the analytical method
4. The drug product is a controlled-release dosage form

4. Factors Influencing Bioavailability

A number of factors may influence the bioavailability of a drug. They include the physicochemical properties of the drug, formulation of the dosage form and its route of administration, the temporal relationship between feeding and oral dosing, and the species of animal. When determining the relative bioavailability of a drug from oral dosage forms, a standardized schedule of feeding should be applied throughout the study. With regard to parenteral dosage forms for intramuscular administration, the plasma concentration profile and bioavailability of a drug can vary widely with the formulation of the drug product. Intramuscular injection, in the lateral neck region of ruminant calves, of five different parenteral preparations of ampicillin at a similar dose level (7.7 $\pm$ 1.0 mg/kg) yielded plasma concentration profiles for ampicillin that differed widely (Fig. 5). Location of the intramuscular injection site can influence bioavailability of a drug from a parenteral dosage form, particularly prolonged-release formulations. This was shown for penicillin G administered as an aqueous suspension of procaine penicillin G to horses (Fig. 6). The peak plasma concentration and systemic availability of penicillin G, in descending order, were: M. serratus ventralis cervicis > M. biceps > M. pectoralis > M. gluteus or subcutaneously in the cranial part of the pectoral area (Firth et al., 1986).

Some parenteral preparations are incompletely absorbed from IM injection sites. This may be attributed to low solubility of the drug at the pH of muscle tissue or to a damaging effect (tissue irritation) caused by the drug product at the site of injection. The sensitivity of horse muscle tissue to injection site irritation precludes the use in horses of some parenteral preparations developed for intramuscular administration to cattle.

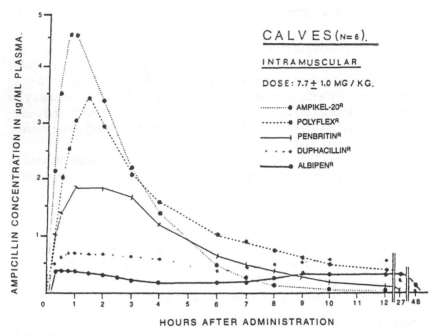

Figure 5 Mean plasma ampicillin concentration in ruminant calves after intramuscular injection of five parenteral ampicillin preparations at similar dose levels (7.7 ± 1.0 mg/kg). (From Nouws et al., 1982.)

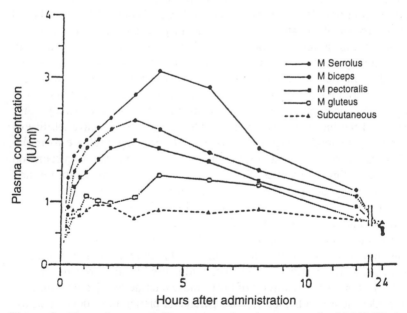

Figure 6 Mean plasma penicillin concentration–time curves after 20,000 IU of procaine penicillin G/kg was administered to five animals (4 horses and 1 pony) at five different sites. (From Firth et al., 1986.)

5. Oral Administration

Because of its large surface area and rich blood supply, the small intestine is the principal site of absorption of orally administered drugs. Knowledge of the morphology of the small intestine is essential for understanding the mechanism by which drugs are absorbed. The intestinal mucosa comprises a uniform single layer of columnar epithelial cells. These cells are formed in the crypts of Lieberkuhn and migrate as they mature to the tips of the villi, where they become senescent and slough into the intestinal lumen. The average life of mucosal cells is 2 to 5 days, depending on the species of animal. The absorptive capacity of the immature cells in the crypts is less than that of the mature cells midway up the villi. As the cells approach the tips of the villi, where they become senescent, their absorptive capacity is reduced. Therefore, the major site of drug absorption from the small intestine is the central portion of the villi. The contribution of other cells in the mucosa, such as goblet cells and granulated cells, to drug absorption is negligible, due to their small numbers.

The intestinal mucosal cells are joined at their apical surface by tight junctions. Below these junctions, the intercellular space becomes larger. Passive cation fluxes are almost entirely via the tight junctions, and movement of water is thought to occur in part through the tight junctions and intercellular spaces (extracellular pathway). Dissolved drugs, particularly large molecules, which could not penetrate the epithelial cells, could be carried in the water through the intercellular spaces. The extracellular pathway may also be important for passage of lipophilic nonelectrolytes. Intestinal absorption of drugs that cannot use the extracellular pathway occurs by passive diffusion through the mucosal epithelial cells.

The aqueous stagnant layer is a recognized component of the barrier to drug penetration through the mucosa, especially for drugs that are rapidly absorbed. Segmental contractions of the intestine decrease the thickness of the stagnant layer. However, the layer of mucus adjacent to the apical surface of the mucosal cells is not affected by these contractions and therefore represents the minimum thickness of the stagnant layer. The mucin layer, in addition to protecting the gastrointestinal mucosa from the acidic environment, appears to reduce the permeability of the mucosa to positively charged drugs. This effect is presumably due to binding of the positively charged drugs, such as tetracycline and quaternary amines, by the negatively charged mucin.

The apical membrane is the next component of the barrier to drug absorption from the small intestine. This membrane is in the form of microvilli which contribute to the extensive surface area for absorption. The cores of the microvilli contain longitudinal protein strands (microtubules), which come together to form the terminal web just beneath the apical surface. The microvilli are covered by a thin, loose covering of mucopolysaccharide filaments which constitute the fuzzy coat, the mechanism by which the mucous layer is anchored into place. The pH at the apical membrane has been reported to be about 5.5,

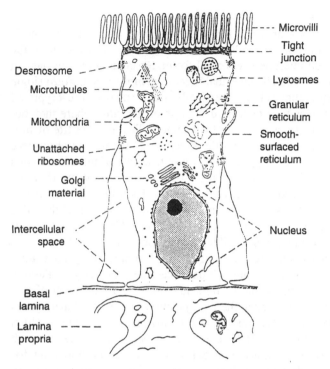

Figure 7 Diagram of mucosal cells from human small intestine. Note microvilli, tight connections of cells at mucosal edge (desmosomes), and space between cells at base (intercellular space). (Reproduced with permission, from Trier, 1967.)

which may partly explain the discrepancy between experimental and theoretical rates of penetration of weak organic electrolytes.

Once the drug molecule has traversed the mucosal layer and the apical membrane, it must then diffuse through the highly structured epithelial cell contents (Fig. 7). Within the cell, the nucleus is located near the serosal edge, whereas the endoplasmic reticulum forms a continuous network distributed throughout the cytoplasm. The Golgi apparatus, which serves to segregate, store, and modify materials that are absorbed and synthesized in the cell, is located near the nucleus. The mitochondria and lysosomes are dispersed throughout the cell. The cellular contents may therefore present conduits or barriers for diffusion of drug from the mucosal to the serosal edge of the cell. Furthermore, because of the structured contents of the cell, large concentration gradients may exist from one edge to the other.

The basal cell membrane constitutes the next component of the barrier to drug absorption. Immediately beneath this is the basement membrane, which is located within 0.5 µm of the villous capillaries. Drug passage across the basal cell and basement membranes has the same requirement of lipid solubility as for diffusion through any biological membrane. Penetration of the capillary wall can be either transcellular or via mucopolysaccharide-lined fenestrae. Capillary endothelial fenestrae, 2 to 4 nm in diameter, account for approximately 10% of the total surface area of the capillary wall and are responsible for absorption of many of the water-soluble drugs administered orally.

Blood flow to intestinal villi can also be a rate-limiting factor in drug absorption. Villous blood flow can vary widely, depending on how recently food was ingested, the presence of disease, or the effects produced by concurrent drug treatment.

Wide variation in the bioavailability of a drug is more likely to follow oral than intramuscular administration. This is due to the wide variety of dosage forms that may be available for oral administration, the physiological factors that influence absorption of a drug from the gastrointestinal tract, the effect of feeding prior to or in conjunction with drug administration (particularly of powders and granules), and, importantly, the influence of the first-pass effect on systemic availability of the drug. When a drug is given orally in solid dosage form, it must first be released from the preparation, then traverse the gastrointestinal mucosal barrier to enter the portal venous blood in which it is conveyed to the liver, and, finally, pass through the liver to reach the systemic circulation (Fig. 8). Each of these events has the potential to decrease the amount of drug that reaches the systemic circulation unchanged; the net effect is reflected in the bioavailability profile, which is characterized by peak plasma concentration (C_{max}), time of peak (t_{max}), area under the curve (AUC), and apparent half-life ($t_{1/2}$) or mean residence time (MRT). Wide individual variation in bioavailability is an unavoidable consequence of oral drug administration, particularly for drugs that are well absorbed. Metronidazole (20 mg/kg), administered as an aqueous suspension of crushed tablets (250 mg) by nasogastric tube, showed wide variation in both the rate (C_{max}: 16.7 to 24.3 µg/ml; t_{max}: 0.75 to 4 h) and extent (F: 58.4% to 91.5%) of absorption in six Quarterhorse mares that were fasted for 12 h prior to dosing (Baggot et al., 1988).

Oral dosage forms range from liquid preparations of soluble drugs or salts to solid preparations, some of which are of poorly soluble drugs. When drugs are administered as solid dosage forms, release (which includes disintegration and dissolution) from the dosage form is frequently rate-limiting for the overall absorption process (Levy, 1968). The physiological factors (gastric emptying, intestinal transit rate, and pH of the gastrointestinal contents) influence dissolution, stability, and, mainly, the rate of absorption of the drug. The effect of feeding is variable but generally decreases or at least delays drug access to the site of absorption. The first-pass effect decreases the systemic availability of a drug to an extent that depends on the rate of metabolism of the drug

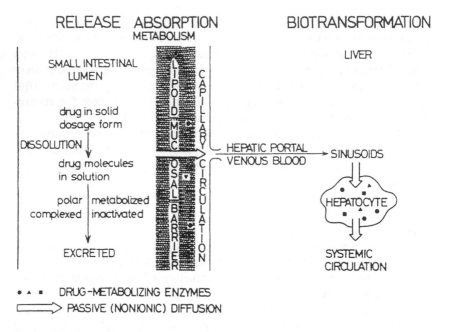

RELEASE ABSORPTION   BIOTRANSFORMATION
METABOLISM

Figure 8 Events that may decrease the systemic availability of a drug given orally as a solid dosage form.

by the liver. High hepatic clearance (by metabolism) is a characteristic feature of drugs (such as diazepam, propranolol, and lidocaine) that show a significant first-pass effect. A combined perfusion-compartmental pharmacokinetic model (Fig. 9) has been developed to describe the time course of drugs that are subject to preabsorptive, intestinal epithelial, and hepatic first-pass metabolism (Colburn and Gibaldi, 1978; Colburn, 1979). By using the equations associated with this model it might be possible to estimate the fraction of the administered dose that is metabolized at each of the three sites.

Some drugs that show incomplete systemic availability following oral administration to dogs are listed in Table 10. For the majority of these drugs presystemic metabolism is responsible. The liver is assumed to be the site of metabolism in most cases. In the human, the bioavailability of digoxin was enhanced when the drug was administered as an aqueous alcoholic solution in a gelatin capsule, rather than in tablet form, even though the tablets had a satisfactory dissolution rate (Mallis et al., 1975; Johnson et al., 1976; Lindenbaum, 1977). Similarly, increased bioavailability of flufenamic acid was achieved when the drug was administered in soft gelatin capsules to dogs (Angelucci et al., 1976). In both instances the increased bioavailability was attributed to physi-

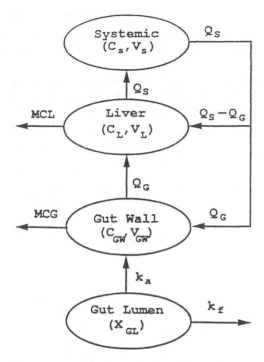

Figure 9 Perfusion-compartmental pharmacokinetic model used to describe gut lumen, gut epithelial, and hepatic "first-pass" metabolism. MCG and MCL represent intrinsic metabolic clearance in the gut and liver, respectively, whereas k_a and k_f are the apparent first-order absorption and gut floral metabolic rate constants. X, C, V, and Q terms denote amount, concentrations, compartment volumes, and blood flows respectively. (From Colburn, 1979.)

Table 10 Systemic Availability of Drugs After Oral Administration to Dogs

Drug (dosage form)	Dose (mg/kg)	Systemic availability (%)	Contributing factors
Diazepam (tablet)	2	1–3	Metabolism (intestinal wall and liver)
Digoxin (Lanoxin tablet)	1 mg total	80	Dissolution
Propranolol (tablet)	80 mg total	2–17	Hepatic metabolism
Lidocaine (solution)	10	15	Hepatic metabolism
Salicylamide (solution)	30	22	Metabolism (intestinal wall and liver)
Levodopa (solid in gelatin capsule)	25	44	Metabolism (GI lumen and/ or intestinal wall)
Sulfadimethoxine (suspension)	55	50	(Dissolution and hepatic metabolism)?

cochemical factors, which were brought into play by adjuvants in the soft gelatin capsules.

When a drug is incompletely available systemically from an oral dosage form, either a higher dose might compensate for the deficit or, preferably, an improved formulation should be developed. The influence that formulation, or rather the use of an alternative oral dosage form, can have on the systemic availability of a drug was shown in horses given racemic ketoprofen (2.2 mg/kg). The systemic availabilities of the $S(+)$ and $R(-)$ enantiomers were 54.2% and 50.5% following administration of micronized racemic ketoprofen powder in hard gelatin capsules to horses with restricted access to feed. When powder from the same batch was administered as an oil-based paste, systemic availabilities of the $S(+)$ and $R(-)$ enantiomers were 5.75% and 2.7%, respectively, regardless of the feeding schedule (Landoni and Lees, 1995). There was no enantioselective difference in bioavailability.

6. Species Differences

The species of domestic animals differ markedly in anatomical arrangement of the gastrointestinal tract and digestive physiology. The greatest difference is between ruminant species (cattle, sheep, and goats) and monogastric species (pigs, dogs, and cats). The horse is unique in that it is a monogastric herbivore in which essential bacterial fermentation processes take place in the large intestine (cecum and colon). In ruminant species, drugs are slowly absorbed from the reticulorumen, subject to metabolism (hydrolytic and reductive reactions) by ruminal microorganisms, and gradually pass into the abomasum, where they could be degraded by the acidic environment (pH 3.0), and small intestine. The fraction of an oral dose that reaches the small intestine may be rapidly absorbed. In monogastric species, like humans, some drugs are degraded in the stomach but most pass unchanged into the small intestine which is the principal site of absorption. In the horse, degradation of drugs in the stomach is variable since the pH reaction of gastric contents can vary widely (1.13 to 6.8). Drug absorption takes place from the small intestine, as in other monogastric species, provided feeding has not occurred shortly prior to or concurrently with drug administration. Horse feed may hinder access of a drug to the mucosal lining of the small intestine and convey the drug to the large intestine, where absorption takes place, following its release through digestion of the fibrous feed. The plasma concentration profile may show two peaks—an early peak corresponding to absorption from the small intestine, and a second peak several hours later corresponding to large-intestinal absorption. Hence drug absorption in the horse may be described as variable in rate and may occur in two phases, depending on the temporal relationship between feeding and oral dosing, and the dosage form administered.

The first-pass effect is an important source of species variation in the systemic availability of orally administered drugs. Because of the generally higher capacity of the liver of herbivorous species (horses and ruminant animals) to metabolize lipid-soluble drugs by microsomal metabolic pathways (oxidative reactions and glucuronide synthesis), the first-pass effect is likely to decrease systemic availability of rapidly metabolized drugs to a greater extent in herbivorous than nonherbivorous species.

The combination of gastrointestinal physiology on drug stability and absorption and the first-pass effect on systemic availability largely contributes to the wide species differences in the bioavailability of drugs from oral dosage forms.

7. Effect of Food/Feeding

Any effect of the feeding schedule on the bioavailability of drugs administered orally is generally applicable only to monogastric species. The physiological factors that greatly influence dissolution are gastric emptying and intestinal transit rate. Since the small intestine is the principal site of drug absorption, factors that accelerate gastric emptying are likely to increase the rate of drug absorption, while factors that delay gastric emptying would probably have the opposite effect, regardless of the physicochemical properties of the drug. Release from the dosage form is a prerequisite for absorption of a drug and could be a shortcoming of some enteric-coated preparations. The administration of solid dosage forms of sparingly soluble drugs (such as griseofulvin), enteric-coated preparations, and modified-release products in conjunction with feeding may increase their dissolution by prolonging their residence time in the stomach.

The slower rate of absorption of ampicillin administered as the trihydrate compared with the sodium or potassium salt in Beagle dogs can be attributed to the difference in the rate of dissolution of the oral dosage forms (Cabana et al., 1969). Systemic availability of ampicillin did not differ significantly between the dosage forms. Ampicillin trihydrate capsules and anhydrous ampicillin tablets were reported to be bioequivalent in Beagles (Bywater et al., 1977). The effect of feeding on the bioavailability of a drug would be expected to vary with the oral dosage form. Feeding, by promoting gastric acid secretion and delaying emptying, would decrease the bioavailability of ampicillin in dogs, particularly when the antibiotic is administered as an oral suspension except when given in esterified form (pivampicillin). The systemic availability of amoxicillin, which is acid-stable, is not affected by feeding.

Based on AUC the systemic availability of mitotane (an adrenocortical suppressant) was approximately 16 times higher when the drug was administered as an emulsion than as tablets to fasted dogs. When mitotane was administered as ground tablets in oil added to the food, systemic availability was significantly increased (25-fold) compared with tablets administered to fasted dogs (Watson et al., 1987).

Composition of the food may affect both the absorption and biotransformation processes that influence the systemic availability of lipid-soluble drugs. The

repeated feeding of a protein-rich diet to dogs may enhance, while a carbohy-
drate-rich diet may reduce, the rate of hepatic microsomal oxidative reactions.
This would affect the systemic availability of drugs that are extensively metabolized by this metabolic pathway, by changing the influence of the first-pass effect.

In horses the time of feeding relative to oral dosing has been shown to affect the systemic availability and/or absorption pattern of some drugs. When
an oral paste preparation of trimethoprim-sulfadiazine (30 mg/kg of the combination) was administered to fed and unfed horses, feeding was shown to decrease the peak plasma concentration and systemic availability of trimethoprim,
while the absorption of sulfadiazine was not affected (Bogan et al., 1984). The
administration of rifampin (5 mg/kg) as an aqueous suspension by nasogastric
tube to horses 1 h after feeding yielded systemic availability of 25.6% compared with 67.6% when the drug was administered 1 h before feeding (Fig. 10).
Absorption of rifampin was preceded by a short lag time (ca. 0.25 h), and the
rate of absorption was not influenced by the time of feeding relative to drug
administration (Baggot, 1992).

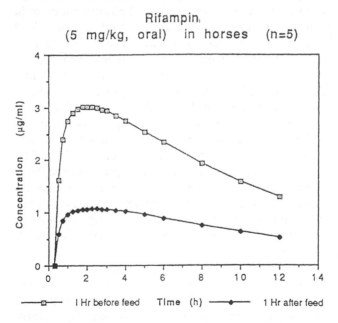

Figure 10   Mean plasma rifampin concentration–time curves in horses (n = 5) after
oral administration of the drug (5 mg/kg) 1 h before or 1 h after feeding. Area under
the curves, indicative of systemic availability, differed significantly while the rate of
absorption was not influenced by time of feeding relative to oral dosing. (Baggot JD,
1992.)

The administration of phenylbutazone (4.4 mg/kg) as an aqueous suspension of a granular (powder) formulation to Welsh Mountain ponies yielded marked variations in the absorption pattern (plasma concentration profile) with different feeding schedules (Maitho et al., 1986). When access to hay was permitted before and after feeding, the time to peak plasma concentration (t_{max}) was 13.2 $\pm$ 2.9 h, and double peaks in the plasma concentration-time curve were common. Double peaks were also found when phenylbutazone was given to ponies deprived of food prior to, but allowed access to hay immediately after, dosing. In this circumstance, the time to peak concentration was much shorter and generally similar to that found (5.9 $\pm$ 4.4 h) when phenylbutazone administration was preceded and followed by moderate periods of fasting. Absorption was more regular and double peaks were less apparent when food was withheld both before and after dosing. It was tentatively postulated that, while some of the administered dose may be absorbed in the small intestine, some may become adsorbed/bound to the ingested hay and be subsequently released by fermentative digestion for absorption in the large intestine. Subsequent in vitro and in vivo studies of phenylbutazone have confirmed binding of the drug to hay and equine digesta. Delayed absorption of phenylbutazone in ponies given access to hay was not accompanied by a significant reduction in the extent of absorption; the systemic availability of the drug was estimated to be 69% in fed and 78% in unfed ponies. Oral administration of multiple doses of phenylbutazone in an oil-based paste to horses deprived of feed around the time of dosing showed an episodic absorption pattern with two or three peaks in the plasma concentration profile after the administration of each dose at 24-h intervals (Lees et al., 1986). Dissolution, which controls the pattern of absorption, occurs in a phasic manner in an oil-based paste and differs from drug dissolution in a granular formulation. The feeding schedule did not significantly affect the low systemic availability of the S(+) and R(−) enantiomers of ketoprofen administered to horses as an oil-based paste of micronized racemic ketoprofen powder (Landoni and Lees, 1995).

Diet—hay and concentrate feeding compared with grazing on pasture—was shown to affect the C_{max} and AUCs of the active moieties of rafoxanide and triclabendazole administered as oral suspensions of the flukicides to 4- to 5-month-old parasite-free lambs (Taylor et al., 1993). Systemic availability of both drugs was higher in the housed lambs fed hay and concentrates than in the grazing lambs due to slower passage of the digesta through the gastrointestinal tract. This afforded more time for absorption of the drugs.

In sheep, reduction in the level of feed intake decreases the rate of onward passage of digesta from the rumen to the abomasum and small intestine. Hennessy et al. (1995) showed that orally administered oxfendazole associated extensively with particulate digest in the rumen. Following passage of this material into the abomasum, the drug is released and subsequently absorbed from the small intestine. When the rate of passage of digesta from the rumen was decreased, by temporarily reducing feed intake, the systemic availability of oxfendazole was increased due to the extended residence time at the site of absorption in the digestive tract (Ali and Hennessy, 1995).

B. Drug Distribution

Drug distribution deals with the movement and relative concentrations of drugs in the blood and the various tissues of the body. After a drug is absorbed or injected into the bloodstream, it may be distributed into interstitial and cellular fluids. The initial phase of drug distribution depends on the cardiac output and regional blood flow. Because the heart, liver, kidneys, brain, and other highly perfused organs receive a high proportion of blood flow, most of the drug during the first few minutes after absorption reaches these organs and tissues. Delivery of drug to muscle, viscera, skin, and fat takes much longer as a result of the disparity of blood flow to these organs compared with the highly perfused organs. Therefore, equilibration between the blood and these poorly perfused tissues may take hours to occur.

Most drugs exert their effects by interaction with a very small number of receptors in a very limited number of tissue cells. Because of this, only a very small proportion of the total amount of drug in the body at any time is producing the pharmacological effect. Most of the drug remains in other tissues and body fluids, which do not respond to the drug. In light of that, the extent of drug distribution largely determines the total amount of drug that must be administered to achieve the appropriate concentration at the responsive tissue. As a corollary, if the distribution of the drug is known or can be determined, the concentration of drug in any body fluid can be related to the concentration of drug at the site of action and, hence, a particular effect.

Diffusion, lipid solubility, ionic character, and molecular weight of the drug have similar effects on distribution and absorption. Weakly acidic drugs tend to accumulate in more alkaline areas of the body, due to the ion-tapping phenomenon, whereas weakly basic drugs accumulate in the more acidic regions of the body, such as the gastric juice and rumen. For example, sodium dicloxacillin, which has a pK_a of 4.4, diffuses appreciably only into tissues in its nonionized form. The pH of the normal mammary gland can be as low as 6.4. The ratio of ionized to nonionized dicloxacillin can be predicted by the Henderson-Hasselbalch equation:

$$pK_a - pH = \log \frac{[\text{protonated form}]}{[\text{nonprotonated form}]}$$

Thus, in the milk the equation becomes:

$$4.4 - 6.4 = \log\frac{[HA]}{[A^-]}; \quad \text{antilog}(-2)$$

The actual ratio of nonionized to ionized dicloxacillin is 1:100. In the plasma, where the pH is approximately 7.4, the ratio of nonionized to ionized dicloxacillin can be calculated from the Henderson-Hasselbalch equation:

$$4.4 - 7.4 = \log\frac{[HA]}{[A^-]}; \quad \text{antilog}(-3)$$

Therefore, the ratio of nonionized to ionized dicloxacillin is 1:1000. When the mammary gland becomes infected, the pH of the milk becomes more alkaline (pH of approximately 7.4). Therefore, mastitic milk has the same ratio of nonionized dicloxacillin to ionized dicloxacillin as plasma. The distribution of dicloxacillin into mastitic quarters would be expected to be greater than into normal quarters, simply because more dicloxacillin exists in the ionized form in the mastitic gland, and the ionized dicloxacillin is trapped in the gland.

Differential blood flow to different regions of the body and the different physicochemical properties of each tissue can substantially affect the disposition of drug in various types of tissue. Drugs will distribute more rapidly to highly perfused tissues, but the extent of distribution of drug into each tissue is governed by the solubility of drug in each tissue. Highly lipid-soluble drugs will distribute extensively into the adipose tissue of the body, although distribution to adipose tissue will be relatively slow because of the limited blood flow to it.

An offshoot of this differential disposition is the concept of redistribution, whereby the drug redistributes to a part of the body where it does not exert its effect. That portion of the body can act as a reservoir for drug, effectively soaking up drug from the other tissues (including the site of action), thereby decreasing the concentration of drug at the site of action. The decline in the concentration of drug at the active site will terminate the activity of the drug without truly eliminating the drug from the body. In general, redistribution of drug occurs from high-flow to low-flow areas, and requires that the drug be more soluble in the tissues that are less highly perfused but which may constitute a substantially large reservoir to terminate a drug's activity. The ultra-short-acting thiobarbiturates (e.g., thiopental) represent a group of drugs that rely on the redistribution phenomenon to terminate their anesthetic action in the brain. Initially, concentrations of thiobarbiturates in the plasma and brain are quite high, causing the rapid onset of anesthesia.

However, these drugs very rapidly partition from the plasma and brain into two major reservoirs. One reservoir is the adipose tissue, into which the thiobarbiturates migrate because of the high degree of lipid solubility that they possess. The second tissue reservoir into which thiobarbiturates redistribute is the skeletal muscle. Redistribution into the skeletal muscle is a result of the large mass of tissue that skeletal muscle represents. Therefore, although the thiobarbiturates partition much more easily into adipose tissue than skeletal muscle, the large fraction of the body that skeletal muscle comprises makes it perhaps an even larger reservoir for redistribution than adipose tissue. Irrespective of which tissue takes up the largest proportion of the thiobarbiturates, redistribution into these tissues prevents the thiobarbiturates from being present in the central nervous system for a prolonged period of time (Wertz et al., 1988). If multiple doses of the same thiobarbiturate are administered, or if the drug is given by continuous infusion, equilibration will occur between the brain and the reservoirs for redistribution. Under these conditions, the concentrations of the thiobarbiturate will remain high in the brain, resulting in an extremely long duration of action.

1. Plasma Protein Binding

Drugs bind to varying degrees to plasma proteins (Table 11). When a drug molecule is bound to plasma protein, it is restricted in its distribution and elimination because it cannot leave the bloodstream. Likewise, bound drug cannot exert its effect in the body because it cannot reach the location of action (unless that location of action is within the bloodstream and can be accessed by the drug·protein complex).

Acidic drugs commonly bind to albumin, which is abundant in the plasma. Basic drugs often bind to α_1-acid glycoproteins and lipoproteins, whereas many endogenous compounds (e. g., vitamins and minerals) bind primarily to globulins. Binding to any of these proteins is reversible; the drug can bind to and dissociate from the protein. The rates of association and dissociation are usually quite rapid (milliseconds). As a result, both the bound and the unbound

Table 11 Extent of Protein Binding of Drugs at Therapeutic Plasma Concentrations

Drug	Concentration (µg/ml)	% Protein-bound (therapeutic concentration range)
Digoxin	0.0015	27
Digitoxin	0.005	89
Morphine	1	12
Warfarin	6.75	97
Penicillin G	Not reported	50
Gentamicin	1–10	20

forms of the drug can be assumed to be in equilibrium with each other, obeying the law of mass action:

$$\text{Free drug} + \text{Protein} \underset{k_2}{\overset{k_1}{\rightleftarrows}} \text{Drug} \cdot \text{Protein}$$

In this reaction, k_1 and k_2 are the rate constants of the association and dissociation processes, respectively. The half-times of these rate processes is in the order of a few milliseconds. Therefore, equilibrium is achieved almost instantaneously. The affinity between a drug and its binding sites can be expressed as a concentration ratio of the drug in its bound form to the product of the free drug and the binding protein.

The unit of K_a is liters per mole. By itself, affinity is more commonly expressed in terms of the dissociation equilibrium constant (K_d), the reciprocal of K_a, and is therefore expressed in units of moles per liter. However, because it is a direct measure of the affinity of the protein for the drug, K_a is often used in equations to determine the bound and the unbound fractions of drug in the body. For example, the unbound fraction of drug (f_u) multiplied by the total drug concentration in the body yields the free concentration of drug ($f_u \cdot C$). The bound concentration of drug can be represented as $(1 - f_u)C$. It follows that:

$$f_u = \frac{1}{1 + K_a \cdot [\text{protein}]} = \frac{1}{1 + \dfrac{[\text{drug} \cdot \text{protein}]}{[\text{drug}]}}$$

From this equation, it can be seen that the unbound fraction depends not only on the affinity of the protein for the drug, but also on the concentration of unbound protein in the circulation. In turn, the concentration of unbound protein depends on the total concentration of protein and the concentration of the drug·protein complex. Usually, the fraction of the total concentration of protein accounted for by the protein bound to drug is minuscule; therefore, the fraction of unbound protein is relatively constant and independent of drug concentration. Alterations in either the apparent affinity of a protein for the drug or the plasma protein concentration can affect the fraction of unbound drug. Decreases in the apparent affinity of the binding protein for the drug can occur during uremia, fever, alterations in the pH of the plasma, and in the presence of other drugs which compete for the same binding site on the protein (Table 12). Lipid components of lipemic blood can also compete with some drugs for plasma protein-binding sites, thereby decreasing the apparent affinity of the protein for the drug. Decreased plasma protein concentrations can be observed in several instances. For example, decreases in the α_1 acid glycopro-

Table 12 Drugs that Compete for Binding Sites on
Albumin

Site 1	Site 2
Iophenoxic acid	Ethacrynic acid
Phenylbutazone	Flufenamic acid
Oxyphenbutazone	p-Chlorophenoxyisobutyric acid
Sulfadimethoxine	
Sulfinpyrazone	
Warfarin	

teins can occur in patients with nephrotic syndrome, and decreased lipoproteins
are observed in instances of hyperthyroidism and injury. Decreases in the albumin concentration in the plasma can be seen with age, hepatic cirrhosis, gastrointestinal disease, nephrotic syndrome, acute pancreatitis, and renal failure.
Increases in albumin concentration can be seen in patients with benign tumors
and hypothyroidism. Increased α_1-acid glycoproteins can be observed due to
age, myocardial infarction, renal failure, and stress; increased lipoproteins are
often seen as a result of diabetes, hypothyroidism, and the nephrotic syndrome.

Only the unbound drug fraction (f_u) can diffuse into tissues. Drug molecules
that are bound to plasma proteins are limited to the plasma. They may not be
metabolized, filtered, or distributed into cells. However, the drug·protein complexes do act as a circulating drug reservoir, which provides free drug as the
concentration of free drug in the plasma declines owing to distribution, metabolism, and excretion. The rate of distribution of lipophilic drugs is dependent on
the fraction of the drug that is unbound. In the absence of a diffusion limitation, as with lipophilic compounds, the initial rate of uptake of drug is equal
to the rate of presentation, irrespective of the extent of protein binding. Thus,
as the bound fraction of a drug increases, the fraction of the drug available for
immediate distribution decreases, and the initial rate of uptake is faster. It follows also that bound drug cannot exert its action because it is not available to
bind to its active site and produce effects.

Changes in protein binding have a substantial effect on the response of the
animal to the drug only if the bound form of the drug comprises more than 85%
of the drug in the body. If that is the case, the distribution of the majority of
the drug from the vascular compartment is restricted. However, the free drug
may be very widely distributed and will be masked by the drug that is protein
bound. For example, warfarin is 97% protein-bound; only 3% of it remains
unbound. That 3% is responsible for all of the therapeutic effect. If the plasma
protein binding decreases to 94%, by decreased affinity for the protein or com-

petition between warfarin and another acidic drug (e.g., phenylbutazone) for the binding site on albumin, or if there is a decrease in albumin concentration in the patient's plasma, then what was once 3% of the administered warfarin exerting an effect is now 6%, or double the amount of active drug. Because of the slight change in the bound fraction of warfarin, there is now twice as much active drug present in the patient. When this scenario occurs, the patient will exhibit bleeding tendencies because of the overload of active warfarin, in spite of the administration of a normally tolerated dose of warfarin.

Distribution of unbound drug can be limited either by perfusion to the tissues or by diffusion of drug across membranes. The perfusion rate limitation exists primarily for nonpolar, highly lipid-soluble drugs. This concept is analogous to the perfusion rate limitation to the absorption of highly lipid-soluble drugs. When drug disposition is limited by perfusion, distribution equilibrium may take longer to achieve when perfusion is low and when the permeability coefficient (partition coefficient) is high. When the permeability coefficient is high, more drug must be presented to the tissue before equilibrium takes place. On the other hand, diffusion rate-limited distribution occurs primarily for polar drugs. For these drugs, the rate of entry of drug into a tissue is a function of the permeability coefficient of the drug, the concentration gradient, the distance over which the drug travels, the surface area for diffusion, the temperature, and the molecular weight of the drug.

The factors that increase the rate of diffusion will also increase the extent of distribution of drug within the body since diffusion is the primary mechanism for drug distribution. Therefore, increases in the lipid:water partition coefficient, decreases in the degree of ionization, increased perfusion, and increases in the permeability of the diffusion barriers will all contribute to increased distribution of drug in the body.

2. Barriers to Drug Distribution

For a drug to freely diffuse from the bloodstream into many tissues, it must traverse several barriers. Some of the barriers to drug distribution are quite impermeable to most drugs, whereas other sites in the body present virtually no barrier to drug distribution.

a. Blood-Brain Barrier. Perhaps the most impermeable of the sites in the body is the central nervous system. Although blood flow is high to the brain, the permeability of the capillaries to polar compounds is very limited. In fact, the blood-brain barrier acts as a very strict lipid barrier, because the capillaries of the brain contain no fenestrations, as do other capillary beds in the body. Furthermore, the endothelial cells of the central nervous system are joined to each other by continuous tight intercellular junctions. In addition, certain cells within the brain connective tissue (the astrocytes) have long processes which form sheaths that surround most of the capillaries of the CNS. Because of these anatomic differences from normal capillaries, only drugs that are lipid-soluble or are actively transported into the CNS will be able to penetrate these barriers and distribute appreciably into the CNS. However, because of the high cerebral blood flow (approximately 16% of cardiac output), drugs that are highly lipid-soluble readily diffuse into the CNS. It is for this reason that the inhalational anesthetics diffuse rapidly into the CNS and exert their anesthetic effects.

Cerebrospinal fluid (CSF) is formed by active secretion of fluid, primarily by the choroid plexus of the third, fourth, and lateral ventricles. The CSF flows from the ventricles across the surfaces of the brain and spinal cord, then into the venous blood sinuses through an absorptive network of channels and valves in the arachnoid villi. Unlike the other capillary networks in the CNS, the capillaries of the choroid plexus have open junctions and are therefore more porous than other CNS capillary beds. However, the choroidal epithelial cells are joined at their apical surface by tight junctions. As a result, the same functional barrier is present in the choroid plexus as in other capillary beds of the CNS. This "blood-CSF barrier" prevents all but the most lipid-soluble drugs from entering the CSF, in a manner similar to the degree of penetration of drugs into the extracellular fluid of the brain.

However, the concentration of drug in the CNS tissue (which is highly lipoidal) may be significantly different from the concurrent concentration in the aqueous CSF. This is evidenced by the fact that clindamycin CSF concentrations were substantially lower than brain clindamycin concentrations after oral dosing in cats (Brown et al., 1990).

Drugs leave the brain by diffusion, whereas they leave the CSF by bulk flow into the venous sinuses and subsequent diffusion into the capillaries. There is an active organic acid transport system in the choroidal epithelial cells, oriented such as to pump organic acids out of the CSF back into the general circulation. Penicillin, probenecid, and other organic acids are examples of drugs that are actively pumped out of the CSF, in the rare event that these compounds get into the CNS.

b. Placental Barrier. Although the blood flow in the maternal-fetal unit is discontinuous from the mother to the fetus, the passage of many xenobiotics is relatively efficient, especially for drugs that are not extremely polar or ionized. The placental barrier does not restrict drug movement as much as the blood-brain barrier.

Tissue binding can occur in addition to or without concurrent plasma protein binding. Drug is sequestered outside the bloodstream in tissue(s) impossible to predict from serum concentrations. For example, basic amines tend to concentrate in the lung. High concentrations of chlorphentermine, propranolol, imipramine, serotonin, and norepinephrine are often found in pulmonary tissue. On the other hand, acridines (e.g., quinacrine) bind to nucleic acids in the liver. Because of this extensive tissue binding outside the bloodstream, the volume of distribution of acridines can be over 100 L/kg, making the volume of distribution of acridines one of the largest of any therapeutic drug administered nowadays. Even drugs that normally do not enter cells can be bound to certain tissues. Aminoglycosides become trapped inside renal tubular epithelium and inner-ear tissue. Renal concentrations of these aminoglycosides can be up to 1000-fold higher than concurrent serum concentrations.

Drug bound in the tissues at sites other than the site of action is no more active than bound drug in the bloodstream. Only if the drug is bound to the receptor at which it exerts its action will bound drug be active. Lipophilic drugs may also be sequestered in fat depots, effectively reducing the concentration of drug at the active site (thiobarbiturates).

3. Experimental Methods for Measuring Tissue Concentrations

When evaluating tissue concentrations of a drug, the clinician must remember that it is not just the averaged tissue concentrations that are important, but rather the concentrations at the site of action. Knowledge of whether the target is an extracellular bacterium, a surface-bound receptor, or an intracellular receptor is critical to understanding which experimental methods of assessing drug delivery to the site of action are clinically relevant. Several methods for determining "tissue concentrations" and "tissue fluid" concentrations have been developed experimentally. The techniques range from total tissue homogenization, to chambers implanted in a tissue, to threads and microfibers implanted in the tissue of interest. All of these are fraught with interpretation difficulties with regard to efficacy.

The simplest method of determining tissue concentrations is to obtain a specimen of tissue from the animal (either antemortem or postmortem), grind and homogenize the tissue in an aqueous buffer, and assay the supernatant from the tissue homogenate. This technique liberates drug that is intracellular and extracellular, and the resultant concentration is a weighted average of the concentration of drug in the "total tissue water" for that tissue. Depending on the drug and the extraction techniques employed, recovery of drug bound to cellular structures may vary. Nevertheless, it is difficult to determine whether the measured concentration reflects primarily extracellular concentrations, intracellularly accumulated drug, or a balanced average of the two.

Devices implanted into the tissue allow for sampling and analysis of a fluid from the tissue, and this fluid is typically termed "tissue fluid." However, the device in most instances has altered the local environment of the tissue, most of the time by creating an inflamed and fibrotic mass of connective tissue around the implant. Furthermore, diffusion kinetics must be considered for implanted hollow devices, remembering that for devices, such as tissue chambers, the surface area:volume ratio is much different from what is present in the tissue itself. These tissue chambers more closely represent a tissue abscess model (either sterile or infected, depending on the presence of infectious bacteria in the chamber) rather than true interstitial fluid. The implanted microfibers provide a geometry much closer to true tissue structure than implanted tissue chambers, although at the expense of sampling ease and sampling volume.

C. Pharmacokinetic Studies

The design of pharmacokinetic studies must be such that the plasma drug concentration-time curve is well defined. The completeness of the curve depends on the sample collection times, the frequency and duration of sampling, and the sensitivity of the analytical procedure used to measure concentrations of the drug in plasma. Well-designed pharmacokinetic studies performed in the target animal species provide the information on which dosage calculations can be based and predictions made on preslaughter withdrawal time (Baggot, 1977, 1983). The application of pharmacokinetics to design of dosage regimens rests largely on the premise that the therapeutic range of plasma concentrations can be defined for a drug. It is generally assumed that the range of therapeutic concentrations is the same in domestic animal species as in humans. This assumption is usually but not invariably valid. While the parameters obtained from single-dose studies can be used to predict the steady-state plasma concentrations that multiple dosing regimens will produce, the accuracy of this prediction should be verified. The magnitude of the pharmacological effect at steady state will indicate the suitability of the dosage regimen. For antimicrobial drugs different criteria have to be used for assessment of dosage.

A particular requirement of the experimental design in drug residue studies is that the pharmacokinetic behavior of the drug be compared at both ends of the range of recommended dose levels (mg/kg). A disproportionate increase in concentrations of drug with increase in dose is evidence that the drug shows nonlinear pharmacokinetic behavior. It is important to stress that insight into the drug residue "profile" can be obtained only by linking fixed-dose pharmacokinetic studies with determination of the amount of drug in selected tissues and organs of the target animal species. Economy is an important advantage of this approach in that a smaller number of animals would be required for residue studies.

1. Pharmacokinetic Modeling

The roles and relative merits of compartmental and noncompartmental modeling in pharmacokinetics were reviewed by Gillespie (1991). In domestic animal species, as in humans, the disposition kinetics of most drugs can be analyzed in terms of a two-(generally) or three-compartment open model (Baggot, 1977). An example of a disposition curve that can be mathematically described

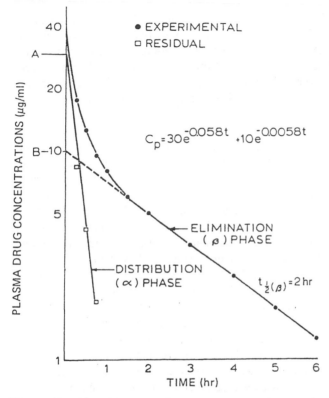

Figure 11 Disposition curve for a drug administered as a single intravenous dose (10 mg/kg). The biexponential equation describing the curve is shown. The distribution phase was obtained by the method of residuals. (From Baggot JD, 1977.)

by a biexponential equation (inset) and analyzed according to a two-compartment open model is shown (Fig. 11). The pharmacokinetic model selected must relate to the plasma concentration-time data obtained in an individual animal. In other words, the same model might not apply to all the animals in a group. Another criterion of model selection is the proposed application of the parameters, whether for design of dosage regimen, calculation of infusion rate, or prediction of withdrawal time.

The AUC is a pivotal variable which is best calculated by the trapezoidal rule with extrapolation to infinite time (last measured plasma concentration/terminal exponent). Perhaps the most common error in compartmental pharmacokinetic analysis is overestimation of the terminal exponent, on which half-life is based and may be used in estimating the area under the terminal (extrapolated) portion of the plasma concentration-time curve. This error occurs either from sampling for too short a time or when the analytical method is insufficiently sensitive to quantify plasma drug concentrations during the true elimination phase. A significant consequence of this error is that the plasma concentration at steady state produced by multiple-dose therapy will differ from that predicted. Also, the predicted preslaughter withdrawal time would be erroneous (too short).

When a conventional dosage form of a drug is administered orally or by a parenteral route other than intravenously, the absorption and disposition kinetics can generally be described by a one- (or in some individual animals, a two-compartment model with apparent first-order absorption. Even though the compartmental model gives an estimate of the absorption rate constant, the time to reach peak plasma concentration (t_{max}) and the peak height (C_{max}) provide a more useful indication of the rate of drug absorption. Too few data points during the absorption phase is a common feature of pharmacokinetic studies. The extent of absorption (F, systemic availability) can be calculated by the method of corresponding areas, with correction for dose (when required). Comparison of AUC following oral (or intramuscular) administration with that following intravenous bolus administration gives the absolute bioavailability of the drug; comparison of AUCs following administration by the same route of two dosage forms of a drug gives the relative bioavailability. Use of this method involves the assumption that clearance of the drug is not changed by the route of administration or dosage form. In bioavailability studies it is advisable to use a crossover design with a suitable washout period between the phases of the study. When the half-life of a drug in the same animals differs between the phases of an absolute bioavailability study, correction may be made for the change in half-life assuming that it reflects solely a change in clearance of the drug and apparent volume of distribution remains unchanged (Gibaldi and Perrier, 1982). Half-life correction must never be used when a change in half-life reflects more persistent or prolonged absorption of drug from one dosage form than another (relative bioavailability).

A crossover study is not always feasible in food-producing animal species, particularly in young animals that are rapidly gaining in body weight. In this situation a parallel-groups design, using animals of similar weight and maintained under the same conditions, may be justifiable.

2. Pharmacokinetic Parameters
The basic pharmacokinetic parameters that describe the disposition (i.e., distribution and elimination) of a drug are systemic (body) clearance (Cl_B), which measures the ability of the body to eliminate the drug, and volume of distribution (V_d) which denotes the apparent space available in the body (both systemic

circulation and tissues in which the drug is distributed) to contain the drug. The half-life ($t_{1/2}$) and mean residence time (MRT), which is the statistical moment analogy to half-life, indicate the overall rate of elimination of a drug and the average time molecules of the drug reside in the body, respectively. When a drug is administered orally or by a nonvascular parenteral route (such as intramuscular or subcutaneous), the systemic availability (F), which represents the fraction of the dose that reaches the systemic circulation unchanged, becomes an essential parameter. *Bioavailability* refers to both the rate and extent of drug absorption, but this term is often used synonymously with systemic availability.

a. Clearance. Clearance indicates the volume of biological fluid such as blood or plasma that would have to be completely cleared of drug per unit of time to account for its elimination. It is expressed in units of volume per unit of time (ml/min). The systemic (body) clearance of a drug represents the sum of the clearances by the various organs that contribute to elimination of the drug. It can be calculated by dividing the systemically available dose by the total AUC (from zero time to infinity):

$$Cl_B = \frac{F \times Dose}{AUC}$$

It is only when the drug is administered intravenously that the dose can be assumed to be completely available systemically. The concept of clearance is extremely useful in clinical pharmacokinetics since the clearance of a given drug is usually constant over the range of concentrations encountered clinically. This is because the elimination of most drugs follows first-order kinetics whereby a constant *fraction* of drug is eliminated per unit of time. For drugs that exhibit saturable or dose-dependent elimination, clearance will vary with the plasma concentrations that are attained. Elimination of these drugs obeys zero-order kinetics which implies that a constant *amount* of drug is eliminated per unit of time. Examples of drugs that are eliminated by zero-order kinetics include salicylate in cats (Yeary and Swanson, 1973), phenylbutazone in dogs (Daytona et al.,1967) and horses (Piperno et al., 1968), and phenytoin in humans (Houghton and Richens, 1974).

Systemic clearance is an important parameter to consider in designing dosage regimens with the objective of maintaining plasma drug concentrations within the therapeutic range. Under multiple dosing conditions (i.e., the repeated administration of a fixed dose at a constant dosage interval), the dosing rate required to produce a given average steady-state concentration ($C_{p(avg)}$) of the drug is:

$$\frac{F \times Dose}{Dosage\ interval} = C_{p(avg)} \times Cl_B$$

Dosing rate is defined as the systemically available dose divided by the dosage interval. Assuming knowledge of the average steady-state concentration of a drug, the dosing rate is determined by systemic clearance of the drug.

When a drug is administered by continuous (constant rate) intravenous infusion, the infusion rate (R_o) required to produce a desired steady-state plasma concentration is given by:

$$R_o = C_{p(ss)} \times Cl_B$$

Based on this equation it follows that the mean concentration at steady-state is directly proportional to the rate of infusion and inversely proportional to the systemic clearance of the drug. The time required to reach within 90% of the desired steady-state concentration or to change from one steady-state concentration to another depends solely on the half-life of the drug, being approximately four times the half-life.

When clearance is based on blood concentrations, the maximum possible clearance is equal to the sum of blood flow to the various organs of elimination (liver, kidney, lung, and other tissues in which drug elimination occurs). Clearance based on plasma concentrations can have values that are not "physiologic"; conversion is accomplished by experimentally determining the blood:plasma concentration ratio, which is a function of the hematocrit (HcT) and the binding of drug to both plasma proteins and blood cell components. Estimation of the extraction ratio (blood clearance/blood flow; E) of a drug across an organ of elimination requires the use of blood clearance. When a drug is eliminated by more than one organ (such as liver and kidneys), blood clearance of the drug is additive for the organs of elimination.

Organ clearance of a drug is determined by the product of blood flow to the organ and extraction ratio of the drug across the organ of elimination. The liver is the principal organ of elimination for lipid-soluble drugs. The hepatic clearance (Cl_H) of a drug, with respect to blood concentration, is:

$$Cl_H = Q_H \times E_H$$

where Q_H is the blood flow to the liver and E_H the hepatic extraction ratio. The processes of hepatic elimination are metabolism of the drug and/or carrier-mediated excretion of the unchanged drug in the bile. Several factors may affect the ability of the liver to extract a drug from the blood for elimination by these processes. The combination of factors is presented in the following equation:

$$Cl_H = \frac{Q_H \times f_u \times Cl_{int}}{Q_H + (f_u \times Cl_{int})}$$

where Q_H is liver blood flow, f_u is the fraction of drug unbound to plasma proteins, and Cl_{int} is the hepatic intrinsic clearance, which is a measure of the

Table 13 Pharmacokinetic Aspects of Hepatic Drug Elimination

Pharmacokinetic variable	Human	Dog	Rat
Valproate			
Half-life (h)	15.9	1.7	0.82
Body clearance (ml/min·kg)	0.11	3.03	4.17
Plasma protein binding (%), f_b	94.8	78.5	63.4
Blood clearance (ml/kg·min)	0.20	4.33	5.96
Hepatic extraction ratio, E	0.009	0.10	0.17
Fraction of free drug $(1-f_b)$	0.052	0.21	0.37
Diazepam			
Half-life (h)	32.9	7.6	1.1
Body clearance (ml/min·kg)	0.35	18.9	81.6
Plasma protein binding (%), f_b	96.8	96.0	86.3
Blood clearance (ml/kg·min)	0.64	35.0	214.7
Hepatic extraction ratio, E	0.029	0.81	6.31
Fraction of free drug, $(1-f_b)$	0.032	0.04	0.14

Sources: Valproate (Loscher, 1978); Diazepam (Klotz et al., 1976).

maximal ability of the liver to eliminate (by metabolism and/or biliary excretion) a drug from blood in the absence of limitations imposed by blood flow and plasma protein binding of the drug.

A species comparison of the influence of plasma protein binding and the significance of hepatic extraction ratio on the elimination of two drugs valproate and diazepam, which are eliminated entirely by hepatic metabolism, is made in Table 13. For valproate, the hepatic extraction ratios are low (less than 0.3) in the species compared and less than the fraction of the unbound (free) drug, indicating that valproate elimination is independent of liver blood flow and that only the unbound drug is available for metabolism (i.e., restrictive elimination). Therefore, protein binding of valproate plays a role in its elimination and may be largely responsible for the species difference in the half-life of the drug (Loscher, 1978).

In the case of diazepam, there is a contrast between humans and the two species of animals in the relationship between clearance of the drug and blood flow to the liver. Clearance of diazepam is independent of liver blood flow in the human, but is limited by protein binding of the drug (i.e., restrictive elimination). In the dog and the rat, clearance is dependent on liver blood flow; hepatic extraction ratio is high (> 0.6). Therefore, plasma protein binding appears to play a major role in the availability of diazepam for metabolism only in humans (Klotz et al., 1976). This approach to interpreting pharmacokinetic data for drugs that undergo extensive hepatic metabolism provides a further dimension in explaining species variations in drug elimination.

Renal clearance of a drug results in the excretion of unchanged (parent) drug in the urine. The processes involved include glomerular filtration, carrier-mediated tubular secretion, and pH-dependent passive tubular reabsorption. The rate of filtration of a drug depends on the volume of plasma that is filtered at the glomerulus and the unbound concentration of drug in plasma, since drug bound to macromolecules (plasma proteins) or blood cells is unable to pass across the glomerular membranes. The renal clearance of exogenous creatinine or inulin, an exogenous polysaccharide, provides a close measure of GFR. Creatinine clearance is not a reliable measure of GFR in ruminant species. The rate of secretion of a drug by the kidney (proximal tubule) depends on the binding of drug to the proteins (carriers) involved in active transport relative to binding to plasma proteins, the degree of saturation of these carriers, and the rates of delivery of the drug to the proximal tubules and its transfer across the tubular membrane.

Separate mechanisms exist for secreting organic acids (anions) and bases (cations), including quaternary ammonium compounds, from the plasma into the tubular lumen. These carrier-mediated transport processes are subject to competitive inhibition. Renal tubular secretion is inferred when the rate of excretion exceeds the rate of glomerular filtration of a drug. The renal clearance of para-aminohippurate, which is filtered at the glomerulus and secreted by the proximal renal tubule but not reabsorbed, provides an estimate of the effective renal plasma flow (ERPF). Reabsorption of the nonionized form of weak organic electrolytes from tubular fluid decreases their excretion in the urine. Tubular reabsorption takes place by passive diffusion, and the extent to which it occurs depends on the physicochemical properties of the drug and is influenced by urinary pH and urine flow rate.

The contribution of an organ of elimination (kidney or liver) to the systemic clearance of a drug can be estimated if the fraction of the intravenous dose that is eliminated by the particular organ is known. In the case of the kidney,

$$Cl_R = f_{ex} \times Cl_B$$

where Cl_R represents renal clearance, f_{ex} is the fraction of dose excreted unchanged in the urine, and Cl_B is the systemic clearance of the drug. The difference between the systemic and renal clearances ($Cl_B - Cl_R$) indicates the nonrenal clearance, which represents the sum of the clearances by other organs and tissues.

b. Volume of Distribution. The apparent volume of distribution (V_d), which relates the amount of drug in the body to the concentration in the plasma, provides an estimate of the extent of distribution of a drug. It reflects the apparent space, in both the systemic circulation and the tissues of distribution, available to contain the drug. This space, however, does not necessarily refer to an identifiable physiological volume. Even though the volume of distribution of a drug may be numerically similar to the volume of a body fluid (physiological) compartment, it cannot be concluded that distribution of the drug is limited to that body fluid. The distribution pattern of a drug can only be described by measuring the levels/amounts of drug in the various organs and tissues of the body.

The volume of distribution (area method) can be calculated from the equation:

$$V_{d(area)} = \frac{F \times Dose}{AUC \times \beta}$$

where AUC is the total area under the plasma concentration-time curve and β is the overall elimination rate constant of the drug, obtained from the linear terminal (elimination) phase of the semilogarithmic disposition curve. This implies that the drug was administered as an intravenous bolus dose. When the drug is administered by any other route, correction must be made for systemic availability, and the apparent first-order elimination rate constant must be substituted for β.

By definition, the systemic clearance is the product of the apparent volume of distribution (area method) and the overall elimination rate constant:

$$Cl_B = V_{d(area)} \times \beta$$

If the systemic availability (F) of a drug has not been determined, the term Cl_B/F should be used. This situation arises when the drug cannot be administered intravenously.

The volume of distribution of a drug is determined by the chemical nature and physicochemical properties (in particular pK_a and lipid solubility) of the drug and by the degree of binding to plasma proteins and extravascular tissue constituents. This parameter (volume term) can vary widely among domestic animal species due to differences mainly in body composition (Table 14) and partly in plasma protein binding. Anatomical features of the gastrointestinal tract that distinguish ruminant from monogastric species may contribute to the species variation in volume of distribution.

Knowledge of the volume of distribution allows estimation of the amount of drug in the body at any time during the elimination phase of its disposition. It is on this basis that volume of distribution is used in calculating the dose (mg/kg) required to produce a plasma concentration within the therapeutic range:

$$Dose_{iv} = C_{p(ther)} \times V_{d(area)}$$

Drug administration by a route other than intravenous may require upward adjustment of the dose to compensate for incomplete systemic availability (extent of absorption) of the drug.

Table 14 Body Composition of Various Species (percent of live weight)

Organ/tissue	Horse	Dog	Goat	Ox	Human
Blood	8.6	–	–	4.7	7.9
Brain	0.21	0.51	0.29	0.06	2.0
Heart	0.66	0.82	0.48	0.37	0.47
Lung	0.89	0.89	0.88	0.71	1.4
Liver	1.30	2.32	1.95	1.22	2.6
Spleen	1.11	0.26	0.25	0.16	0.26
Kidney	0.36	0.61	0.35	0.24	0.44
Gastrointestinal tract	5.8	3.9	6.4	3.8	1.7
Gastrointestinal contents	12.7	0.72	13.9	18.4	1.4
Skin	7.4	9.3	9.2	8.3	3.7
Muscle	40.1	54.5	45.5	38.5	40.0
Bone	14.6	8.7	6.3	12.7	14.0
Tendon	1.71	–	–	–	2.0
Adipose	5.1	–	–	18.9	18.1
Total weight (kg)	308	16	39	620	70
Reference[a]	a	b	b	c	d

[a]*References*: a. Webb and Weaver (1979).
           b. Neff-Davis et al (1975).
           c. Matthews et al. (1975).
           d. International Commission on Radiological Protection (1975).

Because of the wide range of body weights of domestic animal species (cats compared with mature cattle and horses), dosage can be satisfactorily expressed only on a unit/weight basis. When specifying the dose in mg/kg body weight, the term *dose level* rather than *dose rate* might be preferable, since the latter implies a component of time. The term *dosing rate* can be applied when stating the rate of an intravenous infusion (mg/kg · h) or the dosage regimen for a drug (mg/kg at x-h intervals). The latter is equivalent to dose level/dosage interval. The dose unit for antineoplastic drugs is an exception, since it is based on body surface area (mg/m^2) rather than body weight.

The distribution of drugs is likely to be altered in disease states, such as congestive heart failure, fever, and uremia, as well as in physiological conditions (e.g., pregnancy, the neonatal period, dehydration). When interpreting the influence of disease states or physiological conditions on the disposition of a drug, the volume of distribution at steady state ($V_{d(ss)}$) is the preferred pharmacokinetic volume term. It can be calculated by the use of areas (Benet and Galeazzi, 1979):

$$V_{d(ss)} = \frac{Dose_{IV} \times AUMC}{(AUC)^2}$$

where AUC is the total area under the curve (zero moment) and AUMC is the area under the first moment curve—i.e., the area under the curve of the product of time and plasma concentration from time zero to infinity. When using trapezoidal summation to estimate AUMC, the area under the extrapolated portion of the curve is estimated by:

$$\frac{t^* \times C_{p(last)}}{\beta} + \frac{C_{p(last)}}{\beta^2}$$

where β is the overall elimination rate constant of the drug and t^* is the time of the last measured plasma concentration of the drug. It is desirable that the areas under the extrapolated portion of the curves be less than 10% for AUC and 20% for AUMC. This method of calculation does not involve the assumption of a compartmental pharmacokinetic model, nor does it require a curve-fitting procedure. However, $V_{d(ss)}$ is valid only when the dose is administered as an intravenous bolus.

Although $V_{d(area)}$ is an easily calculated parameter and may be determined following intravenous and nonvascular routes of administration, it varies when the elimination rate constant for a drug changes, even when there has been no change in the distribution space. The volume of distribution at steady state ($V_{d(ss)}$) is not subject to this disadvantage. As the rate constant for drug elimination decreases, $V_{d(area)}$ approaches $V_{d(ss)}$.

Analog computer-generated curves, based on the microconstants associated with the compartmental pharmacokinetic model describing the disposition of a drug and the calculated zero-time plasma concentration, can be informative (Baggot, 1977). The curves show the levels of drug (expressed as fraction of the intravenous dose) in the central and peripheral compartments of the model and the fraction eliminated as a function of time. For a drug that is eliminated entirely by renal excretion (unchanged in the urine), the elimination curve would represent cumulative urinary excretion of the drug. Simulated curves depicting penicillin levels in the central (serum) and peripheral (tissue) compartments of a two-compartment open model and derived from disposition studies in normal and febrile dogs are shown in Figure 12. The tissue level curves indicate that the peak level in the febrile dogs represented 21% of the dose, compared with 7% in normal dogs. This could be interpreted as increased distribution of penicillin in the febrile state. How the distribution pattern of penicillin is altered in the febrile state cannot be deduced. The peak in the tissue level curve corresponds to the time at which pseudo-distribution equilibrium is attained and the overall tissue:serum level ratio is in agreement with the value of k_{12}/k_{21} (the first-

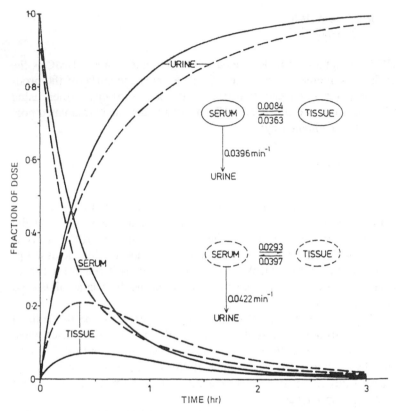

Figure 12 Analog–computer-generated curves showing penicillin G levels (as fraction of the intravenous dose) in the central and peripheral compartments of the two-compartment open model and the cumulative amount excreted in the urine as a function of time. The curves are based on the first-order rate constants associated with the model used to describe the disposition kinetics of the drug in normal (——) and febrile (– – –) dogs. (From Baggot JD, 1980.)

order transfer rate constants between the central and peripheral compartments of the two-compartment pharmacokinetic model). This approach may contribute to explaining disease-induced change in the magnitude of response to a fixed dose of a drug.

Although the apparent volume of distribution provides an estimate of the extent of distribution of a drug, it does not distinguish between widespread distribution and high-affinity (selective) binding with restricted distribution. When the disposition kinetics of a drug can be adequately described by a two-compartment pharmacokinetic model, the volume of distribution represents the sum of the apparent volumes in the central (plasma) and peripheral (tissue) compartments of the model. Since it is only the unbound (free) fraction of drug in the plasma that is available to distribute to the tissues, an estimate of the volume of the tissue compartment is given by:

$$V_t = \frac{f_{ut}(V_d - V_p)}{f_u}$$

where V_p is the volume of the plasma (central) compartment, V_t is the volume of the tissue (peripheral) compartment, and f_u and f_{ut} are the unbound fractions of drug in the plasma and tissues, respectively.

Application of this equation requires a knowledge of the extent of plasma protein binding of the drug and assumes that selective binding to a component of tissue does not take place. An indication of the validity of this assumption can be obtained by measuring the decline in plasma concentrations of the drug for a prolonged period. A decrease in the rate of decline at low (often subtherapeutic) plasma concentrations is evidence that selective tissue binding or sequestration of the drug in a component of tissues is a feature of distribution. This implies that the number of compartments, which corresponds to the number of exponents in the equation describing the disposition curve, in the pharmacokinetic model must be increased to accommodate the prolonged terminal elimination phase. In addition to the disposition study based on measurement of plasma drug concentrations, the amount of drug excreted unchanged and as metabolites in both urine and faeces for a period exceeding six times the terminal half-life of the drug should be quantified. Generally, the recovery of >95% of the administered dose is evidence that the disposition kinetics and fate of the drug are adequately described. Notwithstanding the utility of this information, particularly for clinical applications, the pattern of distribution of a drug can only be revealed by quantifying the amounts (levels) of drug in the various organs and tissues of the body (Baggot, 1983). This requires that samples/specimens be collected from the organs and tissues, which must be weighed, of the target species of animal at predetermined times after the administration of the drug at three dose levels (mg/kg) in the final dosage form and by the intended route.

    c. Half-Life. The half-life of a drug expresses the time required for the plasma concentration, as well as the amount of drug in the body, to decrease by 50% through the process of elimination. Half-life measures the rate of decline in plasma drug concentrations during the elimination phase of the disposition curve, and is obtained from the expression:

$$t_{1/2} = \frac{0.693}{\beta}$$

where β is the overall elimination rate constant of the drug which, numerically, is the negative value of the slope of the linear terminal phase of the semilogarithmic disposition curve; 0.693 is ln 2. At therapeutic doses elimination of the majority of drugs follows first-order kinetics, which implies that a constant fraction is eliminated per unit of time—e.g., 50% each half-life. Elimination of a small number of drugs is dose-dependent (i.e., obeys zero-order kinetics), whereby a constant amount of drug is eliminated per unit of time, in certain species. This can generally be attributed to saturation of a major metabolic pathway for the drug at therapeutic dosage.

With greater analytical sensitivity, a multiexponential pattern of decline in plasma concentrations may be obtained for some drugs. It follows that the half-life would relate to the exponent associated with the decline phase considered to represent "elimination" of the drug. The decline (elimination) phase selected would depend on the plasma concentration range of interest and the intended application of the half-life, whether for selection of the dosage interval or prediction of withdrawal time. A prolonged terminal phase could indicate selective tissue binding and provides a cautionary warning in predicting withdrawal time. The significance of the phase selected for determining the half-lives of aminoglycoside antibiotics has been reviewed (Brown and Riviere, 1991). Gentamicin half-lives of 3.5 h and 142 h were obtained in the horse (Bowman et al., 1986), and of 2.4 h and 30.4 h in sheep (Brown et al., 1985). In another study, the half-life of the prolonged terminal (elimination) phase of gentamicin in sheep was 88.9 h (Brown et al., 1986). The relevant application of a particular half-life may be defined in terms of the fraction of the clearance and volume of distribution that is related to each half-life and whether plasma concentrations or amounts of drug in the body are best related to measures of response (Benet, 1984). This statement could be expanded to include drug withdrawal times in food-producing animals where residual amount of drug is the primary concern.

Half-life is a derived parameter that changes as a function of both systemic clearance and volume of distribution. The relationship between the clinically relevant half-life and the basic parameters is given by:

$$t_{1/2} = \frac{0.693 \times V_{d(area)}}{Cl_B}$$

It follows that alteration in either volume of distribution or clearance may result in a change in the clinically relevant half-life. The disposition of a drug may be altered in the presence of disease states or in certain physiological conditions, or pharmacokinetic drug interactions could alter disposition. Altered disposition is of particular concern for drugs with a narrow margin of safety (e.g., digoxin). Half-life cannot be used as the sole pharmacokinetic parameter to interpret the underlying changes associated with altered disposition of a drug.

Half-life does have certain useful applications. It provides a good indication of the time required to reach steady state during constant intravenous infusion or after the initiation of a dosage regimen (the repeated administration of a fixed dose at constant intervals) and is the basis for selection of an appropriate dosage interval. The rate at which a drug administered by constant infusion or multiple dosing approaches a steady-state concentration is determined solely by the half-life of the drug; a period of four times the half-life is required to attain plasma concentrations within 90% of the eventual steady-state concentration. Assuming that the half-life relating to drug removal from tissues that bind the drug has been determined, this parameter can be used to predict withdrawal time. When a drug is administered by a route other than intravenous, the apparent half-life of the drug will vary with the route of administration and the formulation of the dosage form.

d. Mean Residence Time. The mean residence time (MRT) represents the average time the molecules of a drug reside in the body. The calculation of MRT is based on total areas under the plasma concentration-time curves, which are estimated by numerical integration using the trapezoidal rule (from time zero to the last measured plasma concentration) with extrapolation to infinite time:

$$MRT = \frac{AUMC}{AUC}$$

where AUC is area under the curve (zero moment) and AUMC is area under the (first) moment curve obtained from the product of plasma concentration and time versus time from time zero to infinity. The areas under the extrapolated portion of the curves are estimated by:

$$\frac{C_{p(last)}}{\beta}$$

for AUC and

$$\frac{t^* \cdot C_{p(last)}}{\beta} + \frac{C_{p(last)}}{\beta^2}$$

for AUMC, where β is the overall elimination rate constant of the drug and t^* is the time of the last measured plasma drug concentration ($C_{p(last)}$). It is desirable that the areas under the extrapolated portion of the curves be less than 10% of the total AUC and less than 20% of total AUMC.

Alternatively, the total AUCs can be calculated from the coefficients and exponents of the equation describing the disposition curve associated with compartmental pharmacokinetic analysis of the plasma concentration-time data following intravenous administration of a bolus dose:

$$AUC_{IV} = \frac{A}{\alpha} + \frac{B}{\beta}$$

$$AUMC_{IV} = \frac{A}{\alpha^2} + \frac{B}{\beta^2}$$

Regardless of whether a noncompartmental or compartmental analysis of the data is used, the limit of quantification of the analytical method and the duration of blood sampling are important features of the design of the pharmacokinetic study.

The advantages of using noncompartmental methods for calculating pharmacokinetic parameters, such as MRT, systemic clearance (Cl_B), volume of distribution ($V_{d(area)}$), and systemic availability (F), are that they can be applied to any route of administration and do not require the selection of a compartmental pharmacokinetic model. The only assumption made is that the absorption and disposition processes for the drug obey first-order (linear) pharmacokinetics. The volume of distribution at steady state ($V_{d(ss)}$) is valid only following the intravenous (bolus dose) administration of a drug.

After intravenous administration of a bolus dose of drug, the following relationship applies:

$$MRT_{IV} = \frac{1}{\overline{K}}$$

where $\overline{K}$ is a first-order rate constant equal to the ratio of systemic clearance (Cl_B) to volume of distribution at steady state ($V_{d(ss)}$). It follows that:

$$V_{d(ss)} = Cl_B \cdot MRT_{IV}$$

It may be appropriate in most instances to consider the product of 0.693 and MRT_{IV} as the "effective" half-life of a drug when the disposition kinetics is described by a two- (or more) compartment pharmacokinetic model (Gibaldi and Perrier, 1982). The effective half-life ($t_{1/2(eff)}$) is best calculated from plasma concentration-time data obtained after both single and steady-state dosing. Under these conditions, it is a direct and accurate parameter reflecting accumulation from the dosage form and dosage regimen (Boxenbaum and Battle, 1995)

MRT_{iv}, represents the time for 63.2% of an intravenous bolus dose of a drug to be eliminated. When the drug is administered by another route (oral or intramuscular) or mode, such as short-term, constant-rate intravenous infusion,

MRT_{ni} (noninstantaneous) is obtained. The difference in mean residence times following drug administration in a noninstantaneous manner (MRT_{ni}) and as an intravenous bolus dose (MRT_{IV}) provides an estimate of the rate of absorption:

$$MAT = MRT_{ni} - MRT_{IV}$$

where MAT is the mean absorption time.
When drug absorption can be described by a single first-order process:

$$MAT = \frac{1}{k_a}$$

where k_a is the apparent first-order absorption rate constant. Under these conditions, the absorption half-life ($t_{1/2(a)}$) is given by:

$$t_{1/2(a)} = 0.693 \cdot MAT$$

The statistical moment method of calculating the absorption rate constant (and absorption half-life) of a drug is generally more accurate than the use of a compartmental pharmacokinetic model. Too few data points during the absorption phase is a common shortcoming of the design of bioavailability studies.

When drug absorption is a zero-order process:

$$MAT = \frac{T}{2}$$

where T is the time over which absorption takes place.

D. Application of Pharmacokinetic Parameters to Dosage

The design of dosage regimens, particularly for drugs that produce pharmacological effects (therapeutic drugs), represents an important application of pharmacokinetics. However, a safe and effective dosage regimen can be designed only when the range of therapeutic plasma concentrations for the drug is known. It is generally assumed that the range of therapeutic concentrations developed in humans can be applied to domestic animal species. The validity of this assumption depends largely on whether pharmacological activity is associated entirely with the parent drug, since the rate and extent of formation of an active metabolite would likely vary between species (e.g., N-acetylprocainamide). The determination of a therapeutic range is based on establishing the relationship between plasma drug concentrations (generally total rather than free) and the principal pharmacological effect produced by the drug, while recognizing that due consideration has to be given to the other effects produced and the suitability of the drug for use in a particular species of animal.

Values of the pharmacokinetic parameters that are used in designing a dosage regimen are obtained, at least initially, in healthy adult animals of the target species. Disease or physiological states or drug interactions may alter the disposition of a drug by changing the values of some of these pharmacokinetic parameters. It is important to realize that the plasma drug concentration-effect relationships are meaningful only at steady state or when the ratio of the drug concentration at the site of action to that in the plasma can be expected to remain constant over time—i.e., during the elimination log-linear phase of the plasma concentration-time curve. During multiple-dose therapy, close observation of the pharmacological effects associated with the steady-state plasma concentrations that have been achieved may indicate that the dosage regimen should be adjusted or treatment with the drug terminated.

1. Dose Regimen Design

A dosage regimen entails the administration of a series of maintenance (fixed) doses at a constant dosage interval. The formulation of the available dosage forms largely determines the route and mode (manner) of administration, as well as the suitability for the animal species. Pharmacokinetic parameters are involved in calculating the dosing rate, in predicting the average steady-state plasma concentration and the degree of fluctuation in plasma concentrations during a dosage interval, and, together with the range of therapeutic plasma concentrations, in selecting the dosage interval. The margin of safety or therapeutic ratio (i.e., the ratio of the toxic to effective plasma concentrations) and the overall rate of elimination (half-life) of a drug are the factors which limit the size of the dose and the duration of action, respectively. The response to treatment is the principal criterion to determine the duration of therapy. When the response is a quantifiable physiological variable, such as heart rate or disturbance of rhythm, the objective of therapy should be predefined and the response monitored to determine if and when the objective has been achieved.

2. Average Steady-State Plasma Concentration

When designing a dosage regimen the objective may be to maintain either an average steady-state plasma concentration (C_{ss} or $C_{p(avg)}$) or plasma concentrations within the limits of the therapeutic range ($C_{p,ss(min)}$ to $C_{p,ss(max)}$). The therapeutic range specifies the minimum effective ($C_{p,ss(min)}$) and maximum desirable ($C_{p,ss(max)}$) plasma concentrations of a drug at steady state. The width of the range indicates the acceptable degree of fluctuation in steady-state plasma concentrations, while the units of plasma concentration (e.g., ng/ml for digoxin, µg/ml for theophylline) reflect the pharmacological activity of the drug.

Clearance, which relates the rate of elimination to the plasma concentration of a drug, is the most important pharmacokinetic parameter in the development of a dosage regimen. It determines the relationship between the dosing rate and the average steady-state plasma concentration. When a drug is administered at a constant rate (e.g., as an intravenous infusion), the clearance of the drug determines the rate of infusion (R_o) that would be required to produce a desired steady-state plasma concentration:

$$R_o = Cl_{(B)} \cdot C_{ss}$$

A similar situation should apply when a controlled-release dosage form delivers drug to the systemic circulation at a constant rate (e.g., the ivermectin ruminal bolus for cattle).

When a drug is administered as maintenance doses at a constant dosage interval, the dosing rate can be calculated:

$$\text{Dosing rate} = Cl_B \cdot C_{p(avg)}$$

Dosing rate is defined as the systemically available dose divided by the dosage interval;

$$\text{Dosing rate} = \frac{F \cdot \text{Dose}}{\text{Dosage interval}}$$

The steady-state (plateau) plasma concentration achieved by continuous infusion or the average steady-state plasma concentration following multiple dosing depends only on the clearance of the drug. Prediction of the average plasma concentration at steady state that will be achieved by a dosage regimen may be based on:

$$C_{p(avg)} = \frac{F \cdot \text{Dose}}{Cl_B \cdot \tau}$$

Since the area under the plasma drug concentration-time curve during a dosage interval at steady state is equal to the total AUC after a single intravenous dose, the average plasma concentration at steady state can be estimated from:

$$C_{p(avg)} = \frac{AUC}{\tau}$$

where AUC is the total area under the curve after a single dose and τ is the dosage interval.

The average plasma concentration of a drug at steady-state ($C_{p(avg)}$) is neither the arithmetic nor the geometric mean of the maximum desirable ($C_{p,ss(max)}$) and minimum effective ($C_{p,ss(min)}$) concentrations. Rather, it is a plasma concentration within the therapeutic range which when multiplied by the dosage interval (τ) equals the area under the plasma concentration-time curve during a dosage interval at steady state. A disadvantage associated with basing the design of a dosage regimen on the average plasma concentration at steady state is that it gives no information regarding the degree of fluctuation in the steady-state concentrations during the dosage interval. This limits its application to drugs that have a relatively wide margin of safety and the selection of a dosage interval approximately equal to the half-life.

The time to reach steady state is solely determined by the half-life of the drug. From a practical point of view, a plasma concentration within 95% of the eventual average steady-state concentration will be attained after five half-lives of the drug, regardless of whether it is administered by intravenous infusion or as multiple doses at constant intervals (equal to the half-life).

For drugs that have long half-lives, the time to reach steady-state is appreciable. The desired average steady-state plasma concentration can be promptly attained by administering a loading dose. The volume of distribution ($V_{d(area)}$) is the proportionality factor that relates the plasma concentration to the total amount of drug in the body. Consequently, this parameter is used to calculate the size of the loading dose:

$$\text{Loading dose} = C_{p(avg)} \cdot V_{d(area)}$$

The loading dose is slowly administered intravenously, either as a single entity or in increments at short intervals to avoid excessively high plasma concentrations during the distribution phase—i.e., before pseudo-distribution equilibrium has been attained. Quinidine and lignocaine (lidocaine) (antiarrhythmic drugs) are administered in increments of the calculated loading dose. The administration of loading doses of these drugs as an entity would likely produce toxic effects and could be fatal.

3.   Fluctuation in Steady-State Plasma Concentrations

A dosage regimen designed to produce an average steady-state plasma concentration could show wide variation in the steady-state peak ($C_{p,ss(max)}$) and trough ($C_{p,ss(min)}$) concentrations, depending on the dosage interval. A large degree of fluctuation in steady-state concentrations during the dosage interval is unacceptable for drugs that have a narrow range of therapeutic plasma concentrations. At least for these drugs, the maintenance dose should be precisely calculated and a dosage interval should be selected with the objective of limiting the degree of fluctuation to a level consistent with maintaining steady-state concentrations within the therapeutic range.

The steady-state trough (minimum effective) plasma concentration obtained after the administration of a maintenance dose is based on the volume of distribution, the fraction of dose remaining ($f_r = e^{-\beta\tau}$), and the fraction of dose eliminated ($f_{el} = 1-e^{-\beta\tau}$) during the dosage interval. Based on a desired trough plasma concentration ($C_{p,ss(min)}$) at the end of the dosage interval, the maintenance dose that would be required may be calculated:

$$\text{Dose} = C_{p,ss(min)} \cdot V_d \, (e^{+\beta\tau} - 1)$$

Table 15 Relationship Between the Relative Dosage Interval ($\varepsilon = \tau/t_{1/2}$) and the Extent of Drug Accumulation (R_A) on Multiple-Dosing

Relative dosage interval (ε)	Extent of accumulation (R_A)
0.1	14.9
0.5	3.41
1.0	2.00
2.0	1.33
3.0	1.14
4.0	1.07
5.0	1.03

Selection of the dosage interval (τ) is based on the half-life ($t_{1/2(\beta)}$) and the acceptable degree of fluctuation in the plasma drug concentration-time profile. The longer the dosage interval relative to the half-life, which is called the relative dosage interval (ε), the greater the fluctuation. As the relative dosage interval increases, the extent of drug accumulation on multiple dosing decreases (Table 15). The extent of accumulation (R_A), which is reflected by the plasma concentrations at steady state relative to those occurring after the first dose, is dependent on the fraction of the dose eliminated during the dosage interval (f_{el}):

$$R_A = \frac{1}{f_{el}} = \frac{1}{(1 - e^{-\beta\tau})}$$

where R_A is the accumulation factor (extent of accumulation), β is the overall elimination rate constant of the drug, and τ is the dosage interval (Baggot, 1977). When the dosage interval is equal to the half-life of the drug (i.e., $\varepsilon = 1$), the accumulation factor is 2.00. The administration of maintenance doses at intervals longer than the half-life ($\varepsilon > 1$) gives lower values of the accumulation factor. For example, when the dosage interval is twice the half-life ($\varepsilon = 2$), the accumulation factor is 1.33.

When designing a dosage regimen to maintain steady-state plasma concentrations within a specified range, the dosage interval corresponding to the length of time for plasma drug concentration to decline from the maximum desirable ($C_{p,ss(max)}$) to the minimum effective ($C_{p,ss(min)}$) concentrations can be calculated:

$$\text{Dosage interval} = \frac{\ln(C_{p,ss(max)}/C_{p,ss(min)})}{\beta}$$

where β is the overall elimination rate constant of the drug. A dosage interval close to the calculated value should be selected. The wider the range of therapeutic plasma concentrations, the larger the acceptable degree of fluctuation in steady-state plasma concentrations and the longer the dosage interval relative to the rate of elimination (half-life) of the drug. The relationship between the length of the dosage interval and the acceptable degree of fluctuation in steady-state plasma concentrations may be stated in the following way. If the ratio of $C_{p,ss(max)}$ to $C_{p,ss(min)}$ (upper and lower limits of the therapeutic range) for a drug is 2^{ε} (where ε is the relative dosage interval), then the dosage interval should be $\varepsilon \times t_{1/2}$ of the drug. For most drugs the ratio of the maximum desirable to minimum effective plasma concentrations is less than 8:1. This means that dosage intervals less than three times the half-life have to be used to maintain steady-state plasma concentrations within the therapeutic range. Sustained-release dosage forms will, in many cases, allow less frequent dosing and cause a lower degree of fluctuation in steady-state plasma concentrations.

When the time to reach steady state by administering maintenance doses is long, approximately five times the half-life of the drug, it may be desirable to initiate therapy with a loading dose. For this type of design, calculation of the loading dose is based on achieving steady-state trough concentrations at the end of the first dosage interval:

$$\text{Loading dose} = C_{p,ss(min)} \cdot V_d \, (e^{+\beta\Upsilon})$$

—that is,

$$\text{Loading dose} = \frac{\text{Maintenance dose}}{(1 - e^{-\beta\Upsilon})}$$

A designed dosage regimen allows prediction to be made of the maximum desirable and minimum effective steady-state plasma drug concentrations:

$$C_{p,ss(max)} = \frac{F \cdot \text{Dose}}{V_d \cdot f_{el}}$$

where dose refers to the maintenance dose and f_{el}, the fraction of dose eliminated during a dosage interval, is $(1 - e^{-\beta\Upsilon})$.

$$C_{p,ss(min)} = \frac{F \cdot \text{Dose} \cdot e^{-\beta\Upsilon}}{V_d(1 - e^{-\beta\Upsilon})}$$

—that is,

$$C_{p,ss(min)} = C_{p,ss(max)} \cdot f_r$$

where f_r, the fraction of dose remaining at the end of the dosage interval, is $e^{-\beta\Upsilon}$.

A dosage regimen consisting of a loading (or priming) dose equal to twice the maintenance dose and a dosage interval of one half-life is satisfactory for drugs with half-lives between 8 and 24 h. Far fewer drugs have half-lives in this range in domestic animals, particularly ruminant species, compared with human beings.

Unlike the estimates of dosing rates and average steady-state plasma concentrations, which may be determined independently of any pharmacokinetic model in that systemic clearance is the only pharmacokinetic parameter used, the prediction of peak and trough steady-state concentrations requires pharmacokinetic compartmental model assumptions. It is assumed that drug disposition can be adequately described by a one-compartment pharmacokinetic model, that disposition is independent of dose (i.e., linear pharmacokinetics apply), and that the absorption rate is much faster than the rate of elimination of the drug, which is always valid when the drug is administered intravenously. For clinical applications, these assumptions are reasonable.

The administration of a rapidly absorbed drug at intervals exceeding five times the half-life practically constitutes single dosing, since a relative dosage interval (ε) above 5.0 has an accumulation factor (R_A) of less than 1.03. Assuming first-order elimination, the duration of therapeutic plasma concentrations ($t_{Cp(ther)}$) produced by a single dose depends on the size of the administered dose (D_o) relative to the minimum effective dose (D_{min}) and the half-life of the drug:

$$t_{Cp(ther)} = \frac{\ln(D_o/D_{min})}{\beta}$$
$$= \ln(D_o/D_{min}) \cdot 1.44 \cdot t_{1/2}$$

The relationship between these variables is such that geometric increases in the dose produce linear increases in the duration of therapeutic plasma concentrations. This implies that if twice the minimum effective dose produces therapeutic plasma concentrations for a length of time equal to one half-life of the drug, eight times the minimum effective dose would have to be administered to extend the duration of therapeutic plasma concentrations to three half-lives. The margin of safety of a drug primarily limits the size of dose that can be administered without producing toxic effects. For most pharmacological agents and antimicrobial agents (with the notable exception of penicillins), a dose exceeding five times the minimum effective dose would be likely to produce toxic effects. Moreover the bulk or volume of prepared dosage forms may make the administration of high doses cumbersome, particularly in large animals. On the other hand, the concentration of drug in parenteral dosage forms must be given special attention when estimating the total volume for administration to small animals, particularly cats and toy breeds of dog, or neonatal animals (such as piglets).

E. Therapeutic Concentrations

With regard to the basis of therapeutic plasma concentrations, a clear distinction must be made between drugs that combine with receptor sites in the body to produce pharmacologic effects (pharmacologic agents) and drugs that act on microorganisms (antimicrobial agents) or helminth parasites (anthelmintics).

Most pharmacologic agents have a defined range of therapeutic plasma concentrations, which relate to the magnitude of their principal effects. (Some examples are given in Table 16.) Even though the therapeutic range may be almost invariant for a drug, the pharmacokinetic parameters that influence the half-life—i.e., systemic clearance and volume of distribution—may change in the presence of disease states or certain physiological conditions. A change in the half-life may require adjustment of the dosage interval or size of the maintenance dose to maintain plasma concentrations within the therapeutic range. The satisfactoriness of the adjusted dosing rate can only be assessed when the new steady-state concentration has been attained, which is following the administration of at least four doses. In the absence of drug concentration monitoring, which is the usual situation with animals, the pharmacologic effect produced by the drug is the sole criterion for assessment of dosage.

For specific therapy with antimicrobial agents, both drug selection and dosage estimation are based on the quantitative susceptibility, measured in vitro and expressed as minimum inhibitory concentration (MIC), of the pathogenic microorganisms isolated from the infected animal to an appropriate range of drugs. The need to determine quantitative susceptibility will largely depend on the

Table 16 Principal Effect and Therapeutic Range of Plasma Concentrations of Some Drugs

Drug	Principal effect	Plasma concentrations	
		Therapeutic	Toxic
Quinidine	antiarrhythmic	1–4 µg/ml	>6 µg/ml
Lidocaine	antiarrhythmic	1.5–5 µg/ml	>8 µg/ml
Digoxin	+ inotropic	0.6–2.4 ng/ml	>2.5 ng/ml
Theophylline	bronchodilation	6–16 µg/ml	>20 µg/ml
Phenobarbital	anticonvulsant	12–25 µg/ml	>30 µg/ml
Phenytoin	anticonvulsant	9–18 µg/ml	>20 µg/ml
Valproic acid	anticonvulsant	30–100 µg/ml	>150 µg/ml
Meperidine	analgesic	0.4–0.7 µg/ml	
Morphine	analgesic	0.065 µg/ml	
Cimetidine	50% inhibition of	0.8 µg\ml	
Ranitidine	gastric acid secretion	0.1 µg/ml	

microorganism isolated. While the susceptibility of some pathogenic bacteria can be anticipated, that of others, especially coagulase-positive staphylococci and enteric microorganisms *(Escherichia coli, Klebsiella* sp., *Proteus* sp., and *Salmonella),* is unpredictable. Since quantitative susceptibility of bacteria to an antimicrobial drug varies and infections occur at different locations in the body, the specification of a defined range of plasma/serum concentrations is not appropriate, particularly for drugs that exert a bactericidal action and yield a postantibiotic effect (e.g., aminoglycosides and the newer fluoroquinolones). However, a somewhat flexible range of effective concentrations is applied and used for estimation of a corresponding dose range. The bioavailability and pharmacokinetic parameters that describe disposition of the drug in domestic animal species are known. The peak plasma concentration ($C_{p,ss(max)}$) relates to the margin of safety of the drug, while the trough concentration ($C_{p,ss(min)}$) is generally based on the average MIC of the majority of pathogenic microorganisms susceptible to the drug. It is highly desirable for bactericidal drugs and essential for bacteriostatic drugs to maintain plasma/serum concentrations above the minimum effective concentration for the duration of therapy. It follows that the selection and dosage of an antimicrobial drug should take into account both the microbiologic (mechanism of action, quantitative susceptibility) and pharmacologic (toxic and residue potential, pharmacokinetic) properties of the drug. The range of effective plasma/serum concentrations for systemic therapy with an antimicrobial agent is based on these properties.

Traditionally, little thought was given to determining whether a relationship existed between the activity of anthelmintic drugs and the plasma concentration profiles following their administration. The margin of safety or adverse effects produced were, however, considered in determining dose levels (mg/kg) for the various species of animals. It is only in more recent years that a linkage between the activity (antiparasitic effectiveness) and pharmacokinetic properties of some anthelmintic drugs (e.g., ivermectin, benzimidazoles, probenzimidazoles, levamisole) has been established (Bogan et al., 1982; Bogan and Marriner, 1983; Prichard et al., 1985, 1991; Lo et al., 1985; Lanusse et al., 1990).

The active moiety of an anthelmintic may be the parent (unchanged) drug and/or active metabolite formed in the liver (mainly oxidative reactions). The peak plasma concentration (C_{max}) and time of the peak after drug administration(t_{max}) relate to the rate of absorption or formation of active metabolite and the margin of safety. Area under the curve (AUC) relates to the extent of absorption or formation of active metabolite and the apparent half-life ($t_{1/2}$) or mean residence time (MRT) relates to the duration of action and residue potential. A knowledge of the systemic availability (F) of the active moiety would be required to determine the systemic clearance (Cl_B) and volume of distribution (V_d). The value of F is generally not known owing to the unavailability of intravenous dosage forms. Relative bioavailability is useful for comparing extent of absorption of either different oral dosage forms of an anthelmintic or formulations administered by different routes to the same animals (Baggot and McKellar, 1994). What has not been established is the relationship between plasma concentration of the active moiety, including importantly the minimum effective concentration and anthelmintic activity. There does not appear to be an in vitro method, analogous to the broth dilution method for determining quantitative susceptibility of bacteria to antimicrobial drugs, for relating active moiety concentration to anthelmintic activity in parasites. Since anthelmintics are generally administered, at least for prophylactic purposes, as a single dose, their duration of action is largely dependent on release from the dosage form to maintain effective concentrations in the plasma and at the site of action (in the parasite). This is presumably related to AUC above some minimum effective concentration. So long as assessment of clinical efficacy, which is semi-quantitative, remains the criterion of anthelmintic activity, the duration of action of anthelmintics on various parasites will be poorly defined. Without knowledge of the minimum effective concentration of the active moiety, the therapeutic range of concentrations cannot be defined for an anthelmintic. A further complication of anthelmintic usage is that 100% clinical efficacy is not necessarily desirable, since it would reduce the development of relative immunity to the parasite in the host animal.

F. Tissue Residues

The application of pharmacokinetic modeling to the prediction of tissue residues of drugs in food-producing animals was comprehensively reviewed by Dittert (1977), and an experimental protocol that might be adopted in the early phase of drug development using the target species was presented by Mercer et al. (1977a,b). One objective of pharmacokinetic modeling in food-producing animals is to predict when and if an animal meat will be contaminated without having to slaughter the animal. Such models are useful in determining withdrawal times for new drugs, dosage forms, or routes of administration. Although such a paradigm is theoretically attractive, pragmatic limitations such as those outlined below make the use of pharmacokinetic prediction of tissue residues extraordinarily difficult and perilous.

In developing a pharmacokinetic model on which predictions of tissue residues can be based, excretion data must be correlated with the pharmacokinetic variables derived from the disposition curve. The predictions of the model are then compared with tissue residues (levels) determined experimentally in a small number of animals at various times after dosing. The drug must attain distribution equilibrium rapidly, and all tissue residues must decline in parallel with the plasma concentrations for the model to be valid. The validity of the predictions, however, depends on the distribution and elimination processes obeying linear (first-order) kinetics. Saturable metabolism and protein binding are nonlinear processes that would cause predictions to be erroneous, but well-designed pharmacokinetic studies can detect these deviations from linear kinetics. The binding of dihydrostreptomycin to certain subcellular components of the kidneys of rats and swine is an example of this type of nonlinear tissue distribution (Bevill, 1972). If the model predictions match the experimental data, it can be assumed that the time required for the drug level in a particular tissue to fall to any predicted level can be determined by mathematical extrapolation from the model. Based on those requirements, pharmaceutical companies may argue that, since the tissue residue data are required to "validate" the model, no economic advantage is gained beyond doing the residue study. Furthermore, the regulatory authorities will likely be reluctant to make decisions regarding human food safety based on anything but an extremely well-validated predictive model system.

In addition, pharmacokinetic models are relatively insensitive to processes that contribute only a small fraction to the observed area under the plasma concentration-time curve, even though if those processes affect only one specific tissue much different tissue residue depletion can occur. As an example, plasma pharmacokinetics are unable to reliably distinguish between a formulation that is 100% bioavailable from the injection site and one that is 98% bioavailable. However, a formulation that gives 2% residual drug at the injection site may translate into very large residues at the injection site for a prolonged time. Finally, the validity of a pharmacokinetic model as a predictive tool requires not only that it be adequately descriptive for historical data but that the pharmacokinetic model continue to correctly predict future studies when perturbations of the system are imposed.

In essence, the studies required to sufficiently validate a pharmacokinetic model as a predictive system for tissue residues is likely to be more costly and undoubtedly less certain than simply conducting the tissue residue study. Perhaps the most defensible use of pharmacokinetics in the prediction of tissue residues is one in which plasma pharmacokinetics can be well described beyond the withdrawal time required for tissue residues to be considered safe. In this situation, it is much more likely that components of the pharmacokinetic model that directly impact upon tissue residue depletion are recognized and identified. Otherwise, it is entirely possible that the pharmacokinetic process that governs tissue residue depletion remains undefined by the pharmacokinetic model employed.

Nonetheless, studies of the pharmacokinetic and metabolic behavior of a drug are important for understanding tissue residues. It can only be concluded that pharmacokinetic studies are most useful for elucidating the general behavior of the drug (and its metabolites, if measured) in the animal.

G. Injection Site Irritation

Irritation at the site of injection can be attributed to several factors, all of which may interact with one another to be either additive or synergistic regarding irritation. The following is a nonexhaustive list of formulation factors that may affect injection site irritation.

1. Injection Volume

There are many formulations in which the degree of injection site irritation is proportional to injection volume. Unfortunately, very few studies have been published that deconvolute the components of a formulation to know whether it is the increased drug at the site, increased vehicle, or increased excipients (all of which are associated with increased volume of a formulation injected at a site) that contributes to the injection site irritation. Nevertheless, from a practical standpoint, it is often the case that decreasing the volume of an irritating formulation injected into a site will decrease the degree of tissue irritation at that site. Although based on little data but rather on usual veterinary practice, many injectable veterinary dosage formulations suggest that no more than 20 ml of formulation be administered at each injection site. Dividing a given volume of drug into multiple injections at several sites may decrease the irritation at each site, but it will create multiple sites of irritation. Injection of a larger volume of a more dilute formulation of the active drug will often result in much less discomfort than a smaller volume of more concentrated active drug due to osmotic effects. The decision must often be left to a veterinarian's clinical judgment. However, the veterinarian must also realize that absorption of a formulation from multiple sites may be different than absorption of the same total volume of drug formulation injected at one site.

2. Injection Vehicle

The most innocuous vehicles for drugs are aqueous solutions that are nearly isotonic, are isosmotic, and have similar electrolyte composition. Therefore, aqueous solutions of drugs are usually less irritating than nonaqueous solutions. Other vehicles often used are propylene glycol, polyethylene glycol, other glycol derivatives, and various oils. These vehicles often provide better pharmaceutical properties (e.g., physical and/or chemical stability) for certain drugs than aqueous vehicles. These nonaqueous vehicles are often more tissue-irritating than aqueous vehicles. After intramuscular injection of physiological saline or sterile water, little or no tissue reaction was observed, while vehicles containing glycerol formal or propylene glycol caused severe damage at the injection site in swine, hens, and rabbits (Rasmussen and Svendsen, 1976; Blom and Rasmussen, 1976).

3. Injected particles

The body reacts to foreign bodies very profoundly compared with low-molecular-weight compounds in solution. Phagocytic cells may actively engulf particulates, and/or opsonization may occur more readily against particulates than solubilized molecules. As a result, injected suspensions are generally more irritating than injected solutions of comparable material. Given that injectable sustained-release formulations are generally developed by devising and injecting a formulation of a sparingly water-soluble salt of a drug that has dissolution-dependent absorption, most sustained-release injectable formulations are more tissue-irritating than their rapid-release counterpart. Pharmaceutical formulations must therefore balance sustained-release with tissue irritation for an optimal formulation of a drug.

Formulation techniques can often be used to overcome irritation and pain at the injection site. For example, if the crystalline drug is administered in the form of a suspension or an implant, the crystal shape will play an important role. Crystals with sharp edges, long habit, needles, etc., will cause pain and discomfort (Speiser, 1966).

4. Injected Drug Properties

Depending on a compound's physicochemical properties and/or its biological properties, the compound itself may be a tissue irritant. As a general rule, water-soluble salts of β-lactam antibiotics are relatively nonirritating, whereas water-soluble fluoroquinolone antibacterials tend to be tissue-irritating. Drug formulators may attempt to investigate structure-irritancy relationships to find a less irritating compound, but those investigations must be secondary to structure-activity relationships.

The injection of preparations containing various sulfonamides and others with a sulfonamide and trimethoprim combined caused macroscopic lesions, described as areas of necrotic tissue surrounded by a hemorrhagic zone, at the injection site. The preparation containing sulfadoxine and trimethoprim in glycerol formal caused pronounced necrotic areas at injection sites in cows, horses, and swine (Rasmussen, 1980). The intramuscular injection of oxytetracycline and tetracycline in water caused severe tissue reactions, consisting of necrosis with a peripheral zone of hemorrhage, edema, and fibrosis, which were seen 6 days after the injections in cows, swine, and hens. Similar tissue-damaging effects were seen in swine injected with neomycin in water, erythromycin, and tylosin in propylene glycol (Rasmussen and Hogh, 1971). Thirty days after the intramuscular injection of preparations that caused severe local damage, the injured muscle tissue had been replaced by scar tissue with small remnants of necrotic tissue. Aqueous preparations containing salts of penicillin G and procaine penicillin G-streptomycin combined caused only slight local tissue damage, as seen at 6 days after injection in cows and swine (Rasmussen and Hogh, 1971).

5. *Species Variations*

In hens, streptomycin caused necrosis at the injection site, which was seen microscopically as a necrotic area surrounded by a demarcating zone with pronounced vascular and fibroblastic proliferation (Blom and Rasmussen, 1976). This local reaction appears to represent a species variation in response to the injection of a parenteral preparation but might be attributed to the concentration of the drug or the volume of the solution injected.

Rasmussen (1980) has studied and reviewed the tissue-damaging effect at the site of intramuscular injection of various preparations of antibiotics, other chemotherapeutic agents (sulfonamides, trimethoprim), certain drugs (lidocaine, diazepam, digoxin), and some vehicles. Pain is also likely to occur after intramuscular injection. There is always a risk that an irritant preparation, when given by this route, will cause damage to a nerve trunk or be inadvertently injected into a blood vessel or into the fascial planes between muscle masses. In small animals, these adverse effects may be minimized if injection is made slowly into the quadriceps muscle mass (Cobb, 1979).

Local anesthetics have been included in formulations of some products to overcome pain and irritation. Fetisov (1977) has reported that the addition of 1% to 2% procaine to injectable solutions of levamisole and tetramisole reduced the frequency of swelling at the subcutaneous site to 10% to 15%.

V. COMPARATIVE PHARMACOKINETICS

A. Dose-Effect Relationship: Species Variations

The physiological and biochemical processes that determine the absorption and disposition, which refers to distribution and elimination, of drugs are qualitatively similar in humans and domestic animal species. These processes include passage of drug molecules across biological membranes (cellular barriers), binding to plasma proteins, blood flow to organs and tissues, pathways of biotransformation, and mechanisms of excretion. Quantitative differences in the contribution made by these processes largely determine species variations in drug response. Such variations can generally be accommodated by designing dosage regimens appropriate for the species. This constitutes the pharmacokinetic component of the dose-effect relationship.

Drugs interact with, to activate or inhibit, the same receptor types and have the same biochemical mechanisms of action in mammalian species. However, the distribution of receptors, particularly receptor subtypes, in tissues can differ among species. Differences in receptor distribution may largely account for species variations in sensitivity, reflected by magnitude of response, to a fixed dose (mg/kg) of a drug. This source of variation constitutes the pharmacodynamic component of the dose-effect relationship.

An understanding of the dose-effect relationship in an animal species can generally be obtained by linking the pharmacokinetic behavior with the pharmacodynamic activity of a drug in the species. The therapeutic range of plasma concentrations is central in linking the pharmacokinetic and pharmacodynamic components of the dose-effect relationship. In designing dosage regimens an important assumption made is that the therapeutic range of plasma concentrations of a drug is the same in domestic animal species as in humans. Interspecies extrapolation of dosage can be applied only when either the pharmacokinetic or the pharmacodynamic components of the dose-effect relationship do not vary widely among the species.

Some examples of species differences in drug dosage are presented in Table 17. The 10-fold difference in the dose of morphine between dogs and cats and in the dose of xylazine between horses and cattle is attributed to species variations in receptor sensitivity (pharmacodynamic activity) to these drugs. This explanation is based on clinical observation rather than measurement of pharmacodynamic properties (affinity and intrinsic activity/maximal efficacy) of the drugs in the different species. The wide variation among species in the dose of succinylcholine required to produce an equivalent degree of skeletal muscle

Table 17 Species Variations in Drug Dosage

Drug (route of administration)	Animal species	Dose (mg/kg)	Dosage interval (h)
Single dose			
Xylazine hydrochloride (IV)	Dog	1.0	
	Cat	1.0	
	Horse	0.75	
	Cattle	0.075	
Morphine sulfate (IM)	Dog	1.0	
	Cat	0.1	
Succinylcholine chloride (IV)	Dog	0.3	
	Cat	1.0	
	Horse	0.1	
	Cattle	0.02	
Multiple doses			
Aspirin (PO)	Dog	10	8–(12)
	Cat	10	24–(48)
	Cattle	100	12
Aminophylline (PO)	Dog	10	8
	Cat	5	12
	Horse	5	12

relaxation has been attributed to differences in activity of plasma pseudocholinesterase.

Species differences in the dosing rate (dose/dosage interval) can be anticipated for drugs that are extensively metabolized, particularly by hepatic microsomal oxidative reactions and glucuronide synthesis. Since these differences have
a pharmacokinetic origin, they can generally be accommodated by adjustment
of the dosing rate. This assumes that the therapeutic range of plasma concentrations of a drug is the same in different species. For example, the dosage
interval for aspirin in cats is 24 h, compared with 8 h in dogs. The longer dosage interval for aspirin in cats allows for the relative deficiency in microsomal
glucuronyl transferase activity in this species. Acetaminophen is toxic in cats
since an alternative metabolic pathway to glucuronide synthesis yields a hepa-
totoxic metabolite. Adjustment of both the dose and dosage interval is required
for conventional aminophylline tablets to produce a sustained bronchodilator
effect, based on maintaining an average-steady state plasma theophylline concentration of 10 µg/ml, in horses and dogs. The systemic availability of theophylline following the administration of this oral dosage form of the drug is
approximately 90% in both species. The short half-life of most anticonvulsant
drugs, apart from phenobarbital, in dogs compared with humans makes their
therapeutic use impractical in dogs. This difficulty could presumably be overcome by developing sustained-release oral dosage forms of these drugs that
would provide effective concentrations for 24 h in dogs.

When drugs are administered orally, even in solution, marked differences
in bioavailability between ruminant (cattle, sheep, and goats) and monogastric
(horses, dogs, and cats) species are usual. This is because of the large volume
of reticuloruminal contents (100 to 225 L in cattle; 10 to 25 L in sheep and
goats), possibly metabolism (hydrolytic and reductive reactions) by ruminal
microorganisms, and the high capacity of the liver to metabolize lipid-soluble
drugs (first-pass effect) by a variety of oxidative reactions. Collectively these
factors decrease the systemic availability of orally administered drugs in ruminant species.

Breed differences in clinical response to certain drugs have been observed.
Thiobarbiturates (thiopental and thiamylal) produce longer periods of struggling
and relapses into sleep during recovery from anesthesia in Greyhounds than in
mixed-breed dogs (Robinson et al., 1986). This observation correlates with significantly higher plasma concentrations of the drugs, which could be attributed
to decreased tissue distribution or metabolic clearance in Greyhounds due to a
difference in plasma protein binding (Sams et al., 1985).

Comparison of the disposition kinetics of propofol, a highly lipophilic hypnotic drug, in mixed-breed dogs and Greyhounds showed significant differences
in blood concentrations during recovery from anesthesia and in the times for
return to the sternal and standing positions (Zoran et al., 1993). The return to

sternal position in individual dogs corresponded to the time when the blood propofol concentration was < 1.0 µg/ml in mixed-breed dogs and slightly > 1.0 µg/ml in Greyhounds. Interestingly, the return of consciousness was evident in humans when blood propofol concentration was < 1.0 µg/ml (Cockshott et al., 1987; Adam et al., 1982). With regard to pharmacokinetic parameters, the apparent volume of distribution at steady state was significantly larger and systemic (body) clearance of propofol was significantly higher in the mixed-breed dogs than in Greyhounds. The differences in these basic parameters could be attributed to variations in body fat content and in the rate of hepatic metabolism of propofol.

The unusual sensitivity, manifested by neurologic signs, of a subpopulation of Collies to ivermectin (100 µg/kg administered orally) is not associated with increased bioavailability or decreased clearance (pharmacokinetic properties) of the anthelmintic, but may be due to variation in permeability of the blood-brain barrier and/or in the release of gamma-aminobutyric acid in the central nervous system (Tranquilli et al., 1989).

No explanation has been offered as to why Brahman cattle are supposedly more sensitive than other breeds to dioxathion (an organophosphorus compound) and Isle of Rhum red deer, off the west coast of Scotland, are more sensitive to xylazine (based on the immobilizing dose) than red deer on the mainland.

B. Species Differences in Pharmacokinetic Behavior

When a drug is administered orally, the bioavailability (i.e., rate and extent of absorption) of the drug can differ widely among domestic animal species. Absorption generally occurs more slowly in ruminant than in monogastric species, although drug absorption from the gastrointestinal tract of the horse may occur in two phases which are separated by 8 to 12 h, particularly when a solid dosage form is administered in conjunction with or shortly after feeding. Species differences in the extent of absorption (systemic availability) are largely due to the influence of the first-pass effect, which is generally greater in herbivorous (horses and ruminant animals) than in carnivorous (dogs and cats) species. In pigs (omnivore) and humans, the pattern of absorption is reasonably similar to that in dogs.

Species differences in the disposition (i.e., distribution and elimination) of drugs can generally be attributed to the rate of elimination, particularly biotransformation. Although the activities of oxidative, reductive, and hydrolytic reactions vary unpredictably among domestic animal species, some synthetic reactions are either defective or absent in certain species (Table 5). Hepatic microsomal oxidative reactions and glucuronide synthesis take place slowly in neonatal animals.

The half-lives of drugs that undergo extensive hepatic metabolism vary widely among the species of domestic animals and humans (Table 18). In general, the

Table 18   Species Variations in the Half-life (h) of Drugs Eliminated Mainly by Hepatic
Metabolism

Drug	Cattle	Horse	Dog	Human
Antipyrine	2.5	1.75	2.8	10.3–12.7
Pentobarbital	0.8	1.5	4.5	22.3
Salicylate	0.8	1.0	8.6	12.0[a]
Phenylbutazone	55.1	4.1–4.7	2.5–6.0	72.0
Theophylline	6.9	14.8	5.7	9.0
Metronidazole	2.8	3.9	4.5	8.5
Trimethoprim	1.25	3.2	4.6	10.6
Sulfadiazine	4.1	4.6	5.6	9.9
Sulfadimethoxine	7.9–8.6	11.0	13.2	40.0

[a]$t_{1/2}$ at usual anti-inflammatory doses.

herbivorous species metabolize lipid-soluble drugs more rapidly than do carnivorous species. However, there are notable exceptions to this trend, such as
theophylline in horses and phenylbutazone in cattle, that defy explanation. On
the basis of half-life comparison, humans metabolize many drugs more slowly
than do domestic animals. It has been suggested (Boxenbaum, 1982) that the
lesser quantitative ability of humans to metabolize drugs (especially by hepatic
microsomal oxidative reactions) may be correlated with their enhanced longevity
(maximum life span potential).

   Although the domestic ruminant species are often considered as a group, the
rate of metabolism (and half-life) of many drugs differ among cattle, sheep, and
goats. Sheep have a significantly higher hepatic microsomal protein content
(expressed as mg/g liver) than cattle or goats, while the concentrations of hepatic cytochrome P-450 (which catalyzes oxidative reactions) are similar in cattle
and goats but significantly lower in sheep (Dalvi et al., 1987). Cattle and goats
appear to have a higher capacity than sheep for metabolizing (oxidative reactions) fenbendazole, oxfendazole (fenbendazole sulfoxide), and albendazole to
the sulfoxide and sulfone metabolites. This may account for the lower anthelmintic activity, and consequently higher dosage requirement, of these drugs in
cattle and goats. Marked differences in the evolution of the enantiomers of
albendazole sulfoxide have been shown in sheep, goats, and cattle, with the S(+)
enantiomer representing 86%, 80%, and 91% of the total albendazole sulfoxide AUC, respectively (Delatour et al., 1991). These differences could cause
variation in anthelmintic efficacy among the species, since enantiomers often
differ in quantitative biological activity (Ariens, 1984). The mean half-life of
chlorsulon (7 mg/kg, IV) was shorter in goats (14.37 ± 8.30 h) than in sheep
(18.16 ± 8.39 h), and the AUC in goats was 64% of that in sheep; systemic
(body) clearance of chlorsulon in goats was 1.56 times that in sheep (Sundlof

and Whitlock, 1992). The more rapid elimination of chlorsulon in goats than in sheep is consistent with the lower efficacy of the anthelmintic in goats.

The half-lives of at least some drugs that are eliminated mainly by hepatic metabolism are shorter in goats than in cattle. Phenylbutazone, for example, has a half-life (mean $\pm$ SEM) of 15.9 $\pm$ 1.5 h in goats (Eltom et al., 1993), compared with 55.1 $\pm$ 5.6 h in cows (DeBacker et al., 1980). Pygmy (dwarf-like) goats metabolize antipyrine (microsomal oxidation), sulfonamides (hydroxylation), and chloramphenicol (glucuronide synthesis) more rapidly than other breeds of goat. This observation is significant because of the notion that pygmy goats could serve as an animal model representative of domestic ruminant species. The mean half-life of antipyrine varies among the ruminant species: sheep (3.25 h), cattle (2.5 h), goats (2.25 h), and pygmy goats (0.75 h). Although antipyrine half-life is useful for comparing hepatic microsomal oxidative activity in different species, it does not reflect the activity of all hepatic microsomal metabolic pathways (Vesell et al. 1973). Hepatic intrinsic clearance of antipyrine (Cl_{Uint}) has been estimated to be 0.22 L/min/kg liver weight in a wide range of mammalian species, with the notable exception of humans (Boxenbaum, 1980). Hepatic blood flow is directly proportional to liver weight which, in turn, is heterogonically related to body weight:

$$L = 0.037B^{0.849}$$

where L and B are liver and body weights expressed in kilograms.

Species variations in the half-lives of drugs that are eliminated both by hepatic metabolism and renal excretion could be attributed to differences in the activity of metabolic pathways and GFRs and the influence of urinary pH on the extent of renal tubular reabsorption of unchanged drug. In any species, urinary pH depends mainly on diet. The usual pH of the urine of carnivores is acidic (pH 5.5 to 7.0), while that of herbivores is alkaline (pH 7.2 to 8.4). In humans, the urine is generally acidic but can vary over a wide pH range (4.5 to 8.2). Since a larger fraction of the dose is excreted unchanged in urine of humans than of animals, particularly herbivorous species, the influence of urinary pH on the overall rate of elimination (half-life) of drugs is greatest in humans. Comparison of the 24-h cumulative urinary excretion of trimethoprim (an organic base, pK_a 7.6), for example, showed that less than 5% of the dose was excreted unchanged in ruminant species (cattle and goats), 10% in horses, 20% in dogs, and 69% $\pm$ 17% in humans. The half-lives of trimethoprim in these species are 0.7 to 1.0 h (cattle and goats), 3.2 h (horse), 4.6 h (dog), and 11 $\pm$ 1.4 h (human).

Renal excretion is the principal process of elimination for drugs that are predominantly ionized at physiologic pH (7.35 to 7.45) and for compounds (polar drugs and drug metabolites) with limited solubility in lipid. The renal clearance of inulin (marker substance) or creatinine (endogenous or preferably exogenous) provides a useful measure of the GFR, which varies among species. On the basis of GFR, it can be predicted that the half-lives of drugs eliminated solely by glomerular filtration will be shorter in dogs and cats than in horses, assuming the apparent volumes of distribution are similar. The half-life of gentamicin, for example, is in the range 1.05 to 1.35 h in dogs and cats, compared with 2.5 to 3.25 h in horses. Gentamicin half-lives reflect the relative rather than actual rates of glomerular filtration in the different species and are unrelated to urinary pH reaction. Species variations in the half-lives of drugs that are mainly eliminated by renal excretion are less pronounced than for drugs that are extensively metabolized. Digoxin is a notable exception in that the half-life is 7.8 h in cattle, 23.1 h in the horse, 28.0 h in the dog, and 39 ± 13 h in humans.

Mammary excretion is quantitatively of greater consideration in veterinary medicine than in human medicine because the dairy industry has selected cattle and goats that are genetically predisposed to producing high volumes of milk relative to body weight. Dairy cattle can produce in excess of 12,000 kg of milk in a 305-day lactation, which translates into approximately 6% of a cow's body weight in milk each day. A 50-kg woman would have to produce 3 kg of milk, or slightly less than a gallon of milk, each day for an equivalent production by body weight. It follows that mammary excretion can contribute to a much greater extent to the overall elimination of a drug in dairy animals than in women.

The Henderson-Hasselbalch equation will often predict the milk:plasma concentration ratio of lipid-soluble drugs. Both nonpolar lipid-soluble compounds and polar molecules that possess sufficient lipid solubility passively diffuse through the predominantly lipoidal blood-milk barrier. Their rate of transfer is directly proportional to the concentration gradient across the barrier and the lipid solubility of the drug. The equilibrium concentration ratio is determined by the degree of ionization (determined by pK_a) of the drug in blood and milk, the charge on the ionized moiety, and the extent of binding to plasma proteins and macromolecules in the milk. It has been shown that only the lipid-soluble, nonionized moiety of a weak organic electrolyte that is free (unbound to proteins) in the plasma can diffuse into milk. Although the milk is usually more acidic than blood plasma, the stage of milk production and presence of mastitis influence the pH of milk and thus the drug concentrations attained in the milk. Mammary excretion of systemically administered weak organic bases that are lipid-soluble can represent a significant fraction of their excretion from the body.

Excretion of drugs in sweat, saliva, and tears collectively comprises a relatively minor contribution. Since the volume of saliva excreted by ruminant animals is large (e.g., 90 to 190 L/day in cattle), salivary excretion of drugs may be more important in the ruminant species. Elimination by these routes, like mammary excretion, takes place mainly by passive diffusion of the nonionized, lipid-soluble moiety through the epithelial cells of the glands and is pH-dependent. The concentration of some drugs in saliva may parallel that in plasma. This is more likely to be the situation in horses than in other domestic animal species, since the pH of saliva in the horse is similar to that of plasma. The pH of saliva secreted by ruminant animals is alkaline (pH 8.1 to 8.4) in reaction.

Half-life is determined by both the process of elimination and extent of distribution of a drug. Extensive ($>80\%$) binding to plasma proteins and enteropatic circulation prolong the half-life. Avid binding to tissues delays the ultimate elimination of a drug. The half-life of this terminal phase of elimination may not be a feature of dosage estimation but is critical in the prediction of withdrawal time. Moreover, tissue binding may contribute to the toxicity of a drug, particularly when multiple doses are administered.

Pulmonary excretion is the major process of elimination for drugs that are volatile at body temperature—e.g., the inhalational anesthetic agents. However, metabolism by enzymes in the liver and other tissues contributes to the elimination of some inhaled anesthetics, in particular halothane and methoxyflurane. The speed of recovery from inhalational anesthesia depends on the rate of removal of anesthetic from the brain after the inspired concentration has been decreased. The factors that govern the speed of recovery include the pulmonary blood flow (cardiac output) and the magnitude of ventilation as well as the solubility of the anesthetic in the tissues and the blood (blood:gas partition coefficient). The blood:gas partition coefficient at 37°C is a useful index of solubility and defines the relative affinity of an anesthetic for the blood compared to air. This coefficient is characteristic of each anesthetic (e.g., nitrous oxide, 0.47; isoflurane, 1.4; enflurane, 1.8; halothane, 2.5; methoxyflurane, 15) and determines whether pulmonary blood flow or pulmonary (alveolar) ventilation, or both of these factors, will substantially contribute to the speed of recovery from anesthesia. In the case of inhalational anesthetics with low solubility in blood (nitrous oxide), recovery from anesthesia is rapid and determined by pulmonary blood flow. The opposite applies to anesthetics with high solubility in blood (methoxyflurane) and speed of recovery is mainly determined by pulmonary ventilation. Both of these physiologic parameters contribute to the speed of recovery from anesthesia with inhalational agents of intermediate solubility in blood (isoflurane, enflurane, halothane), although pulmonary ventilation has a relatively greater influence on recovery from halothane.

The duration of exposure to an inhalational anesthetic can have a marked effect on the speed of recovery, especially in the case of anesthetics with high solubility in blood (methoxyflurane). Prolonged exposure to halothane, which is of intermediate solubility, can delay recovery from anesthesia.

An appreciation of the relationship between solubility in blood (blood:gas partition coefficient) and the physiologic parameters that determine the speed of induction and recovery is important in both the use and the development of inhalational anesthetic agents.

C. Interspecies Scaling of Pharmacokinetic Parameters

Species variations in the pharmacokinetic parameters associated with drug disposition generally follow a pattern that can be described mathematically. The allometric relationship between a pharmacokinetic parameter, like that of physiological variables (such as organ weight, hepatic blood flow, and GFR), and body weight is described by the heterogonic equation:

$$Y = aB^b$$

where Y is the pharmacokinetic parameter (or physiological variable), B is average body weight of the animal species, and a and b are the allometric coefficient and exponent, respectively. The exponent b denotes the proportionality between the pharmacokinetic parameter (Y) and body weight (B) of the animal species. Least-squares regression analysis is used to determine the allometric terms (a and b) and correlation coefficient. This method will assess the feasibility of interspecies scaling for a drug while double logarithmic plots will verify the linearity of the relationship between pharmacokinetic parameters and body weight of the animal species.

The predictive value of allometry as a technique for interspecies scaling of drug elimination depends on selection of the appropriate pharmacokinetic parameter. The elimination process for a drug largely determines which parameter to select. Even though half-life is a hybrid pharmacokinetic parameter and many drugs are eliminated by both renal excretion and hepatic metabolism, half-life correlates less well than systemic (body) clearance with body weight of animal species. Comparison is made between the use of half-life and systemic clearance in allometric scaling of theophylline elimination in nine mammalian species (Table 19). The allometric exponent describing the relationship between half-life of drugs and body weight of mammalian species is generally close to 0.25, which represents that for energy expenditure in mammalian species (Kleiber, 1975) and the turnover time of endogenous processes (Boxenbaum, 1982). The exponent relating clearance to body weight reflects the functional capacity of the organs of elimination and generally lies in the range 0.75 to 0.95 for mammalian species.

The allometric relationship between systemic (unless otherwise indicated) clearance of unrelated drugs and body weight of various mammalian species (including the human) shows good correlation (Table 20). For drugs that are eliminated entirely by a single organ, the use of refinements of systemic clearance would provide greater accuracy of predictions. For example, hepatic intrinsic clearance would be the appropriate parameter for allometric scaling of drugs that are eliminated by hepatic metabolism. Hepatic intrinsic clearance of

Table 19 Pharmacokinetic Parameters Describing Elimination of Theophylline in Various Species

Species	Body weight (kg)	$t_{1/2}$ (min)	Cl_B (ml/min·kg)
Rat	0.25	188	3.32
Guinea pig	0.55	206	2.02
Rabbit	3.6	295	2.58
Cat	3.9	468	0.68
Dog	15	342	1.67
Human	68	486	0.69
Pig	90	660	0.63
Cattle	200	384	1.52
Horse	375	890	0.67
Allometric term			
Coefficient		250	1.98
Exponent		0.169	0.829
Correlation coefficient		0.863	0.978
Level of significance		$P < .01$	$P < .001$

Source: modified from Gaspari and Bonati (1990).

unbound drug (Cl_{Uint}) may be calculated from the relationship (Wilkinson and Shand, 1975):

$$Cl_{Uint} = \frac{Q \cdot Cl_H}{f_u(Q - Cl_H)}$$

where Q is liver blood flow, Cl_H is hepatic clearance, and f_u is the fraction of drug in blood that is unbound to plasma proteins.

The hepatic intrinsic clearance of a drug is a measure of the maximal ability of the liver to irreversibly remove a drug from the systemic circulation in the absence of any flow limitations. It therefore reflects the sum of the rate-limiting processes associated with drug elimination, which are uptake into the hepatocytes, metabolic reactions, or biliary excretion. Based on knowledge of the process(es) of elimination of a drug and the judicious selection of animal species, the allometric technique of interspecies scaling could be usefully applied during the preclinical phase of drug development.

D. Changes in Drug Disposition

Disposition is the term used to describe the simultaneous effects of distribution and elimination, that is, the processes that occur subsequent to absorption of a

Table 20 Allometric Terms Showing Relationship Between Clearance (ml/min) of Drugs and Body Weight (kg) of Various Animal Species (including human)

Drug	No. of species	Allometric Coefficient	Exponent	Correlation coefficient (Significance)
HI-6[a]	7	9.80	0.76	0.986 (P < .001)
Gentamicin	8	3.07	0.84	0.986 (P < .001)
Ampicillin	6	6.36	0.925	0.971 (P < .01)
Ceftizoxime[b]	5	11.35	0.59	0.995 (P < .001)
Oxytetracycline	5[c]	8.25	0.77	0.988 (P < .01)
Digoxin	6[d]	9.18	0.86	0.9174 (P < .01)
Ketamine	7	30.51	0.97	0.943 (P < .01)
Theophylline	9	1.98	0.83	0.978 (P < .001)
Antipyrine[e]	10[f]	0.008	0.885	0.989 (P < .001)

[a] Data from Baggot (1994).
[b] Data from Mordenti (1985).
[c] Human not included.
[d] Cattle excluded.
[e] Hepatic intrinsic clearance of unbound drug (L/min) (Boxenbaum, 1980).
[f] Human excluded.

drug. Even though therapeutic agents are used predominantly in diseased animals, there are relatively few studies of the influence of disease states on drug disposition and dosage. The disposition kinetics of a drug is largely influenced by the capacity of the drug to penetrate cellular barriers and to accumulate (ion trapping mechanism) in certain body fluids. This is determined by the pK_a and lipid solubility of the drug and by the pH gradient between blood plasma and the body fluids where ion trapping occurs. Other factors affecting drug disposition include the extent of binding to plasma proteins (mainly albumin) and extravascular tissue constituents, activity of drug-metabolizing enzymes (determine rates of metabolic pathways), and efficiency of excretion (mainly renal) mechanisms. The relative influence of these factors will vary with the chemical nature and physicochemical properties of the drug. Some disease states (e.g., fever, dehydration, chronic liver disease, renal impairment), certain physiologic conditions (e.g., the neonatal period), prolonged fasting (48 h or longer), or pharmacokinetic-based drug interactions (plasma protein binding displacement, inhibition/induction of drug metabolic pathways, or competition for carrier-mediated excretion processes) may alter the disposition of drugs.

Both hypoproteinemia and uremia decrease plasma protein binding of most acidic drugs. In renal disease, decreased binding is associated with a decrease in the apparent affinity of drug for protein binding site(s), which is presumably caused by the accumulation of endogenous competitors, and may be partly attributed to hypoalbuminemia (Tozer, 1984). Uremia not only decreases plasma albumin binding of acidic drugs but also decreases the activity of some metabolic pathways, in particular, hydrolysis by plasma pseudocholinesterase. Plasma protein binding is also decreased in chronic liver disease, especially cirrhosis with associated hypoalbuminemia. The clinical significance of decreased binding to plasma proteins on the disposition and pharmacologic effect of a drug depends largely on the extent to which the drug is normally bound and its apparent volume of distribution. It is mainly for acidic drugs, which bind extensively ($>80\%$) to plasma albumin and have relatively small volumes of distribution (<300 ml/kg), that decreased protein binding alters disposition and the dose-effect relationship.

Attempts to correlate changes in disposition of drugs that undergo extensive hepatic metabolism with liver function tests have been generally unsuccessful. It has been shown that in human patients with chronic liver disease (serum albumin <3g/dl) the disposition of indocyanine green and antipyrine is altered (Branch et al., 1976). Systemic clearance of both marker substances was significantly decreased ($P < .001$), while volume of distribution was not significantly changed. Since the mechanisms of elimination of these substances by the liver are different, interpretation of the altered clearances can be made. Indocyanine green has a high hepatic extraction ratio ($E > .6$) and is eliminated by biliary excretion; it does not undergo enterohepatic circulation. Decreased clearance of indocyanine green could be due to decreased liver blood flow and/or reduced hepatobiliary transport of the marker substance. Antipyrine has a low hepatic extraction ratio ($E < .3$) and is eliminated by hepatic microsomal oxidation. Since clearance of antipyrine is independent of changes in liver blood flow, decreased clearance may be attributed to a reduction in mass of microsomal drug-metabolizing enzymes. In chronic liver disease, serum albumin concentration might serve as a prognostic indicator of hepatic drug-metabolizing enzyme activity.

Unlike the poorly quantifiable situation associated with liver disease, endogenous (or preferably exogenous) creatinine clearance can be used to estimate decreases in renal function (GFR). Calculation of altered renal clearance of a drug can be based on the fraction of normal renal function remaining in the patient ($Cl_{cr,patient}/Cl_{cr,normal}$), where Cl_{cr} is creatinine clearance, but requires knowledge of the fraction of dose usually excreted unchanged in the urine. The altered clearance of the drug can be used to adjust the dosing rate so that the usual steady-state plasma concentrations can be maintained.

Changes in volume of distribution may occur in disease states where membrane permeability is altered or when drug binding to plasma proteins and/or extravascular tissue sites is altered. In the presence of *E. coli* endotoxin-induced fever in dogs and etiocholanolone-stimulated fever in humans, serum concentrations of gentamicin were lower than in the afebrile state (Pennington et al., 1975). This could be attributed to increased extravascular distribution of gentamicin, since renal clearance of the antibiotic was not significantly changed. It is known that penicillin G distributes more widely in febrile than in afebrile animals (Fig. 12). The systemic clearance of orally administered pranoprofen, which is mainly eliminated by glucuronidation, was reduced during fever associated with upper respiratory tract infection in elderly subjects (Fujimura et al., 1989). No significant difference was observed in the time to maximum concentration (t_{max}), maximum plasma concentration (C_{max}), or apparent volume of distribution (V_d/F). There was a positive correlation ($r = .448$; $.05 < P < .10$) between the body temperature and apparent half-life of pranoprofen.

Even though infectious diseases have in common the presence of fever, the character of the altered disposition varies with the pathophysiology of the disease state. In a study of the disposition kinetics of the antiprotozoal drug imidocarb in diseased and control goats, it was observed that the diseases caused significant changes in both the volume of distribution at steady-state and systemic clearance of the drug, while the half-life did not significantly change (Abdullah and Baggot, 1986). Fever induced by *E. coli* endotoxin or infectious bovine rhinotracheitis virus (IBR) infection caused a similar pattern of changes in pharmacokinetic parameters, while the changes caused by *Trypanosoma evansi* infection were distinctly different (Table 21). Altered disposition of a drug caused by disease states can be due to changes in either or both of the basic parameters, volume of distribution and clearance; half-life, which is a derived pharmacokinetic parameter, will not necessarily reflect an anticipated change in drug elimination.

Table 21 Comparison of Pharmacokinetic Parameters (mean ± SD) Describing Disposition of Imidocarb (4 mg/kg, IV) in Control and Febrile Goats

Pharmacokinetic parameter	Control (n = 8)	*E. coli* endotoxin (n = 6)	IBR virus (n = 6)	*T. evansi* (n = 6)
$t_{1/2}$ (min)	251 ± 94	370 ± 391	208 ± 31	254 ± 94
Cl_B (ml/min · kg)	1.62 ± 0.50	0.76 ± 0.28	0.92 ± 0.09	4.10 ± 1.20
$V_{d(ss)}$ (ml/kg)	492 ± 82	222 ± 29	257 ± 41	1295 ± 333

When assessing changes in the disposition of drugs the relevant pharmaco-kinetic parameters to consider are volume of distribution at steady state and systemic clearance of the drug; half-life reflects the relationship between vol-ume of distribution (area method) and systemic clearance. In addition, the plasma concentration-time curves in healthy and diseased animals should be compared. The volume of the central compartment and the ratio of the first-order transfer rate constants describing transfer of the drug between compartments of the pharmacokinetic model can be informative. However, the ratio of k_{12}/k_{21} (two-compartment open model) reflects the tissue:plasma level ratio only at the time of peak tissue level. The changes in pharmacokinetic behavior that result from drug interactions are often difficult to predict, particularly in the presence of disease states. The concurrent use of quinidine and digoxin in horses results in drug interaction in which the potential toxicity of digoxin is increased due to altered disposition of the cardiac glycoside.

Quinidine has been shown to decrease the apparent volume of distribution of digoxin, due to displacement from tissue binding sites, and decrease its re-nal clearance (Leahey et al., 1981; Jogestrand et al., 1984; Parraga et al., 1995). The net effect of these changes in disposition is an increase in plasma digoxin concentrations, which manifest toxic effects > 2.5 ng/ml. The therapeutic range of plasma digoxin concentrations is 0.6 to 2.4 ng/ml, indicating that this drug has a narrow margin of safety.

Changes in pharmacodynamic activity that are related only indirectly to phar-macokinetic behavior may occur in some disease states. An increased sensitiv-ity to the pharmacologic effects of certain drugs may occur and could be the result of functional or morphologic modification of receptor sites or be due to decreased excretion of endogenous substances. The anesthesia-inducing dose of thiopental, for example, is substantially lower in uremic than in healthy animals. This could be partly attributed to decreased binding of thiopental to plasma proteins.

E. The Neonatal Period

The neonatal period is generally considered to be the time span from birth to 1 month of age, although it varies among domestic animal species (Baggot and Short, 1984). Because the adaptive changes in physiologic variables take place rapidly, absorption and disposition of drugs are most "unusual" during the first 24 h after birth. Increased bioavailability and altered disposition (wider distri-bution and slower elimination) affect the plasma concentration-time profiles of drugs and the concentrations attained at receptor sites. These circumstances account for the clinical observation that neonatal animals are often more "sen-sitive" to the pharmacologic effects of drugs, particularly drugs that act on the central nervous system.

Characteristic features of the neonatal period include increased absorption from the gastrointestinal tract, lower binding to plasma proteins, increased apparent volume of distribution of drugs that distribute exclusively in extracellular fluid or total body water, increased permeability of the "blood-brain" barrier, and decreased rate of elimination (both metabolism and excretion) of the majority of drugs.

Alterations in the extent of distribution may be related to the difference in the relative volumes of the body fluid compartments coupled with the lower binding to plasma proteins (due to relative hypoalbuminemia) in the neonatal animal. There are marked deficiencies in some of the prominent processes of elimination (hepatic microsomal metabolic pathways and renal excretion mechanisms) in neonatal animals, particularly during the first 5 days after birth. Because of slow elimination, many drugs have low systemic clearances and prolonged half-lives during the neonatal period. Some examples are trimethoprim, chloramphenicol, theophylline, phenytoin, and phenobarbital.

The systemic availability of orally administered drugs may be higher in neonatal than adult animals. For example, the systemic availability of amoxicillin administered orally as a 5% suspension of the trihydrate is 30% to 50% in 5- to 10-day-old foals compared with 5% to 15% in adult horses. Since the rumen takes 8 to 12 weeks to develop and become functional, the bioavailability profile of drugs administered orally to neonatal calves is similar to that obtained in monogastric species.

The influence of age on the disposition of antipyrine in calves (Table 22) reflects the pattern of change in the apparent volume of distribution of the marker

Table 22 Influence of Age on the Disposition of Antipyrine (50 mg/kg, IV) and Ceftiofur (2.2 mg/kg, IV) in Calves

Age in days (No. of animals)	$t_{1/2}$ (h)	$V_{d(area)}$ (ml/kg)	Cl_B (ml/min · kg)
Antipyrine			
1 (4)	26.9 ± 9.1	724 ± 59	0.33 ± 0.08
15 (6)	11.5 ± 4.0	715 ± 53	0.79 ± 0.23
42 (5)	7.3 ± 1.8	722 ± 59	1.20 ± 0.31
Cattle (7)	3.1 ± 0.6	649 ± 61	2.49 ± 0.31
Ceftiofur sodium[a]			
7 (12)	16.1 ± 1.54	345 ± 61.6	17.8 ± 3.25
30 (12)	17.2 ± 3.08	335 ± 91.9	16.7 ± 3.10
90 (12)	8.22 ± 2.84	284 ± 49.0	30.3 ± 14.5
180 (12)	5.95 ± 1.15	258 ± 71.5	39.8 ± 14.5
270 (12)	7.00 ± 2.26	300 ± 138	33.0 ± 5.52

[a]Brown et al. (1996).

substance and, to a greater extent, the rate of development of hepatic microsomal oxidation. In 6-week-old calves the systemic clearance of antipyrine was approximately 50% of that in adult cattle. This indicates that microsomal oxidative activity develops over a period of weeks, probably 8 to 12 weeks, in the bovine species.

Renal excretion mechanisms (glomerular filtration and proximal tubular secretion) are immature at birth. Glomerular filtration rate approaches adult values at 2 days in calves; 2 to 4 days in foals, lambs, kids, and piglets; and may take more than 14 days in puppies. Proximal tubular secretion matures within 2 weeks after birth in foals, ruminant species, and piglets, but may take up to 8 weeks in puppies. Although neonatal renal function is "immature" relative to that of the adult animal, it serves its physiologic demand.

The meager pharmacokinetic data available on drug bioavailability and disposition in neonatal animals preclude the making of other than general recommendations on dosage adjustment. The situation is complicated by the species variations in the rate at which drug elimination processes (metabolic pathways and excretion mechanisms) develop.

F. Implications of Stereoisomerism

Since many biologically active synthetic drugs contain chiral centers and a chiral environment exists within the body, the significance of stereoisomerism should be considered in formulating veterinary drug dosage forms. The vast majority of these drugs contain a single chiral center and thus exist as pairs of enantiomers, S(+) and R(-). Some examples of chiral drugs are ibuprofen, ketoprofen, naproxen, moxalactam, ketamine, pentobarbital, propranolol, and atenolol. It is usually the racemate (an equimolar mixture of the two enantiomers) that is used in formulating dosage forms. However, cognizance should be taken of the fact that enantiomers may differ in pharmacodynamic activity, which involves drug-receptor interaction, and in pharmacokinetic behavior. Species variations in the extent to which chiral inversion occurs and in the ratio of the enantiomers in the systemic circulation are complicating factors.

Enantiomeric discrimination in drug absorption, distribution, and elimination depends on the process under consideration (Caldwell et al., 1988). Absorption, distribution, diffusion from the systemic circulation into milk and saliva and from hepatic parenchymal cells into bile, glomerular filtration, and renal tubular reabsorption are passive processes that do not differentiate between enantiomers. Whether carrier-mediated renal tubular secretion and transport into bile are stereoselective does not appear to have been determined. The extent of protein binding, to plasma and tissue proteins, and the rate of metabolism can show a high degree of stereoselectivity. Metabolic pathways differ in their degree of stereoselectivity, but may substantially influence the systemic availability of the more active enantiomer of an orally administered racemic drug. Whether a racemate or an enantiomer should be used in formulating dosage forms of a drug would depend on the relative pharmacodynamic activity and toxicity of the single enantiomers, their pharmacokinetic profiles, and, importantly, the proportions formed in the target animal species. When both enantiomers show distinct and desirable effects or their effects are not stereoselective, the use of a racemate may be entirely justifiable (Caldwell, 1992). However, the pharmacokinetic profiles of the enantiomers should be determined using stereospecific assay methods (Foster and Jamali, 1987; Foster et al., 1988; Delatour et al., 1991; Pasutto, 1992; Carr et al., 1992; Levine et al., 1992; Landoni et al., 1995). The use of single enantiomers (e.g., R[-] propranolol, S[+] naproxen, S[+] ibuprofen) may increase selectivity of action, reduce total exposure to the drug, and simplify dose-response relationships. When a single enantiomer is used in formulating dosage forms, it must be optically pure. A generic drug must correspond to the innovator drug, whether a racemate or single enantiomer.

G. Tissue Residues

Tissue residue studies are most often conducted sequentially, first using radiolabeled drug to determine the metabolite profile and the appropriate target tissue and marker residue for residue monitoring, and second using the formulated product to determine the appropriate time after the last injection that concentrations in tissues decrease to below safe and therefore allowable levels with statistical confidence. Although other schemes are possible, this is the standard approach used by pharmaceutical companies.

   1.   Radiolabeled drug studies first require that the position of the radiolabel on the molecule be essentially not labile; otherwise monitoring of the radiolabel could not be equated with monitoring of parent drug and/or all associated metabolites. Therefore, the radiolabel is usually placed in an inert portion of the molecule, usually within a ring structure of the compound. Next, animals from the target species are dosed with radiolabeled drug to determine the tissue with the highest total radioactivity at the approximate anticipated withdrawal time, usually by combustion of the tissue and monitoring radioactivity. From that information the target tissue, which is the tissue with residues above the toxicologically determined safe concentration for the longest time, can be identified. Finally, the metabolite profile in the target tissue can be determined by means of separating the metabolites by chromatography and monitoring the radioactivity over time in the effluent. Identification of the metabolites is useful, although regulatory requirements dictate that only the majority of metabolites be identified. From that metabolite profile, the metabolite that can be used to predict the depletion of total residues in the target tissue is identified as the

marker residue, and analytical methods are developed to identify and quantify the marker residue in the target tissue from animals treated with nonradiolabeled product.

2. Tissue residue decline studies. Once an analytical method for the nonradiolabeled (i.e., "cold") marker residue in the target tissue has been developed, cold residue depletion studies in the target species can be conducted to calculate the appropriate withdrawal time. Animals are dosed with final formulation (i.e., the formulation intended for commercialization) at the anticipated label-prescribed dosage regimen, and four to six animals are slaughtered at predetermined times after administration of the last dose. For registration in the United States, typically six animals at each of four slaughter times are used, with at least one or two of those slaughter times such that concentrations of marker residue in the target tissue are below the anticipated tolerance for the marker residue. From those data, various statistical approaches are used to calculate a withdrawal time after administration of the last dose such that there is a minuscule chance that human beings will consume edible products containing residues above the toxicologically determined safe concentration. In the United States, the calculation of the withdrawal time uses the upper limit of the 95% confidence interval on the 99th percentile of the population. Other countries may utilize other statistically based approaches or may use a single point method such that all observed residues at a particular point are below the level determined to be allowable.

H. Injection Site Tolerance

Traditional methods of evaluating injection site tolerance focus on postmortem gross and microscopic evaluation of the injection sites at several intervals after injection. Gross evaluation is directed toward measuring the size of the irritated area (length, width, and thickness) and describing the lesion grossly. Histologic evaluation of injection sites almost always indicates inflammation, which can simply be the result of the trauma of injection rather than the formulation itself. Injection of sterile saline intramuscularly results in almost no injection site irritation apart from trauma associated with the needle tract, and histologically very slight inflammation is observed.

Newer methods for evaluating injection sites have focused on antemortem methods for evaluating the extent of irritation and rate of resolution at the injection site. Such methods, although not yet completely validated, include use of ultrasonography or circulating creatine phosphokinase in plasma to monitor injection site resolution (Toutain et al., 1995; Banting and Baggot, 1996). Both of these methods offer the promise of being quantitative, reliable indicators of the extent of injection site irritation and the rate and/or extent of resolution of the lesion.

VI. CONCLUSIONS

As can be appreciated, a plethora of factors can be involved in the choice of formulation for use in an animal species. Knowledge of the animal's anatomy and physiology, formulation properties, economics, and level of convenience, and inherent disposition of the compound and its pharmacokinetic profile in formulation, all may impact on the decision-making process. This practical framework of knowledge is vital in the decision making regarding specific formulation properties that will be discussed in the next few chapters. Together, this knowledge can lead to more effective treatment of animal diseases, thereby promoting the general health of the animal population and the wholesomeness of the human food supply.

REFERENCES

Abdullah AS, Baggot JD, (1986). Influence of induced disease states on the disposition kinetics of imidocarb in goats. J Vet Pharmacol Ther 9:192–197.

Adam HK, Kay B, Douglass EJ (1982). Blood disoprofol levels in anaesthetized patients: correlation of concentration after single or repeated doses with hypnotic activity. Anaesthesia 37:536–540.

Alexander F (1972). Certain aspects of the physiology and pharmacology of the horse's digestive tract. Equine Vet J 4:166–169.

Ali DN, Hennessy DR (1995). The effect of temporarily reduced feed intake on the efficacy of oxfendazole in sheep. Int J Parasitol 25:71–74.

Allen WM, Drake CF, Sansom BK, Taylor RJ. (1979). Trace element supplementation with soluble glasses. Ann Rech Vet 10:356–358.

Anderson N, Laby RH (1979). Activity against *Ostertagia ostertagi* of low doses of oxfendazole continuously released from intraruminal capsules in cattle. Aust Vet J 55:244–246.

Andrews ED, Isaacs CE, Findlay RJ (1958). Response of cobalt deficient lambs to cobaltic oxide pellets. NZ Vet J 6:140–146.

Angelucci L, Petrangeli B, Colletti P, Favilli S (1976). Bioavailability of flufenamic acid in hard and soft gelatin capsules. J Pharm Sci 65:455–456.

Ariens EJ (1984). Stereochemistry, a basis for sophisticated nonsense in pharmacokinetics and clinical pharmacology. Eur J Pharmacol 26:663–668.

Baggot JD (1977). Principles of Drug Disposition in Domestic Animals: The Basis of Veterinary Clinical Pharmacology. Philadelphia: Saunders. pp. 144–218.

Baggot, JD (1980a). Gastrointestinal absorption and bioavailability of drugs. In: Anderson NV, ed. Veterinary Gastroenterology. Philadelphia:Lea and Febiger, pp. 292–310.

Baggot JD (1980b). Distribution of antimicrobial agents in normal and diseased animals. J Am Vet Med Assoc 176:1085–1090.

Baggot JD (1983). Clinical utility and limitations of pharmacokinetics. In: Ruckebusch Y, Toutain P-L, Koritz GD, eds. Veterinary Pharmacology and Toxicology (Proceedings of the 2nd Symposium of EAVPT). Lancaster, U.K.: MTP Press, pp. 397–414.

Baggot JD (1988). Veterinary drug formulations for animal health care: an overview. J Controlled Release 8:5–13.

Baggot JD (1992). Bioavailability and bioequivalence of veterinary drug dosage forms, with particular reference to horses: an overview. J Vet Pharmacol Ther 15:160–173.

Baggot JD (1994). Application of interspecies scaling to the bispyridinium oxime HI-6. Am J Vet Res 55:689–691.

Baggot JD, McKellar QA (1994). The absorption, distribution and elimination of anthelmintic drugs: the role of pharmacokinetics. J Vet Pharmacol Ther 17:409–419.

Baggot JD, Short CR (1984). Drug disposition in the neonatal animal, with particular reference to the foal. Equine Vet J 16:364–367.

Baggot JD, Wilson WD, Hietala S (1988). Clinical pharmacokinetics of metronidazole in horses. J Vet Pharmacol Ther 11:417–420.

Ballard BE (1978). An overview of prolonged action drug dosage forms. In: Robinson JR, ed. Sustained and Controlled Release Drug Delivery Systems. New York: Marcel Dekker, pp. 1–69.

Banting AL, Baggot JD (1996). Comparison of the pharmacokinetics and local tolerance of three injectable oxytetracycline formulations in pigs. J Vet Pharmacol Ther 19:50–55.

Bates TR, Sequeira JA (1975). Bioavailability of micronized griseofulvin from corn oil-in-water emulsion, aqueous suspension, and commercial tablet dosage forms in humans. J Pharm Sci 64:793–797.

Benet LZ (1984). Pharmacokinetic parameters: which are necessary to define a drug substance? Eur J Respir Dis 65(suppl 134):45–61.

Benet LZ, Galeazzi RL (1979). Noncompartmental determination of the steady-state volume of distribution. J Pharm Sci 68:1071–1074.

Bevill RF (1972). Subcellular distribution and pharmacokinetics and dihydrostreptomycin in rats and swine. Doctoral thesis, University of Illinois.

Bevill RF, Dittert LW, Bourne DWA (1977). Disposition of sulfonamides in food-producing animals. IV. Pharmacokinetics of sulphamethazine in cattle following administration of an intravenous dose and three oral dosage forms. J Pharm Sci 66:619–623.

Blom L, Rasmussen F (1976). Tissue damage at the injection site after intramuscular injection of drugs in hens. Br Poult Sci 17:1–4.

Boettner WA, Aguiar AJ, Cardinal JR, et al. (1988). Morantel sustained-release trilaminate: a device for the controled ruminal delivery of morantel to cattle. J Controlled Release 8:321–335.

Bogan JA, Galbraith A, Baxter P, Ali NM, Marriner SE (1984). Effect of feeding on the fate of orally administered phenylbutazone, trimethoprim and sulphadiazine in the horse. Vet Rec 115:599–600.

Bogan JA, Marriner SE (1983). Pharmacokinetics of albendazole, fenbendazole and oxfendazole. In: Ruckebusch Y, Toutain P-L, Koritz GD, eds. Veterinary Pharmacology and Toxicology (Proceedings of the 2nd Symposium of EAVPT). Lancaster, U.K.: MTP Press, pp. 235–240.

Bogan J, Marriner SE, Galbraith EA (1982). Pharmacokinetics of levamisole in sheep. Res Vet Sci 32:124–126.

Bowman KF, Dix LP, Riond J-L, Riviere JE (1986). Prediction of pharmacokinetic profiles of ampicillin sodium, gentamicin sulfate, and combination ampicillin sodium-gentamicin sulfate in serum and synovia of healthy horses. Am J Vet Res 47:1590–1596.

Boxenbaum H (1980). Interspecies variation in liver weight, hepatic blood flow, and antipyrine intrinsic clearance: extrapolation of data to benzodiazepines and phenytoin. J Pharmacokinet Biopharm 8:165–176.

Boxenbaum H (1982). Interspecies scaling, allometry, physiological time, and the ground plan of pharmacokinetics. J Pharmacokinet Biopharm 10:201–227.

Boxenbaum H, Battle M (1995). Effective half-life in clinical pharmacology. J Clin Pharmacol 35:763–766.

Branch RA, James JA, Read AE (1976). The clearance of antipyrine and indocyanine green in normal subjects and in patients with chronic liver disease. Clin Pharmacol Ther 20:81–89.

Brander GC (1975). Adapting antibiotic formulations to conform with the requirements of mastitis control programmes. Proceedings of the International Dairy Federation Seminar on Mastitis Control. Reading University Bulletin Document 85:341–344.

Brooker PJ and Goose J (1975). Dermal application of levamisole to sheep and cattle. Vet Rec 96:249–250.

Brown SA, Coppoc GL, Riviere JE, Anderson VL (1986). Dose-dependent pharmacokinetics of gentamicin in sheep. Am J Vet Res 47:789–794.

Brown SA, Riviere JE (1991). Comparative pharmacokinetics of aminoglycoside antibiotics. J Vet Pharmacol Ther 14:1–35.

Brown SA, Riviere JE, Coppoc GL, Hinsman EJ, Carlton WW, Steckel RR (1985). Single intravenous and multiple intramuscular dose pharmacokinetics and tissue residue profile of gentamicin in sheep. Am J Vet Res 46:69–74.

Brown SA, Jaglan PS, Banting A (1991). Ceftiofur sodium: disposition, protein-binding, metabolism, and residue depletion profile in various species. Acta Vet Scand Suppl 87:97–99.

Brown SA, Chester ST, Robb EJ (1996). Effects of age on the pharmacokinetics of single dose ceftiofur sodium administered intramuscularly or intravenously to cattle. J Vet Pharmacol Ther 19:32–38.

Brown SA, Chester ST, Speedy AK, et al. (1997). Bioequivalence of ceftiofur sodium by the intramuscular and subcutaneous routes of administration. J Vet Pharmacol Ther.

Brown SA, Zaya MJ, Nappier JL, et al. (1990). Tissue concentrations of clindamycin after oral administration to cats. J Vet Pharmacol Ther 13:270–277.

Bywater RJ, Buswell JF, Wylie DF (1977). Oral absorption of different forms of ampicillin. Vet Rec 101:432.

Cabana BE, Willhite LE, Bierwagen ME (1969). Pharmacokinetic evaluation of the oral absorption of different ampicillin preparations in Beagle dogs. Antimicrob Agents Chemother 3:35–41.

Caldwell J (1992). The importance of stereochemistry in drug action and disposition. J Clin Pharmacol 32:925–929.

Caldwell J, Winter SM, Hutt AJ (1988). The pharmacological and toxicological significance of the stereochemistry of drug disposition. Xenobiotica 18:59–70.
Campbell CB (1979). Ophthalmic agents: ointments or drops? Vet Med Small Anim Clin
74:971–972.
Carr RA, Foster RT, Lewanczuk RZ, Hamilton PG (1992). Pharmacokinetics of sotalol
enantiomers in humans. J Clin Pharmacol 32:1105–1109.
Carver MP, Riviere JE (1989). Percutaneous absorption and excretion of xenobiotics after
topical and intravenous administration to pigs. Fund Appl Toxicol 13:714–722.
Chandrasekaran SK, Benson H, Urquhart J (1978). Methods to achieve controlled drug
delivery—the biomedical engineering approach. In: Robinson JR, ed. Sustained
and Controlled Release Drug Delivery Systems. New York: Marcel Dekker, pp.
557–593.
Chien YW (1982). Novel Drug Delivery Systems. New York: Marcel Dekker, pp. 149–
217.
Chien YW (1987). Development of transdermal controlled release drug delivery systems:
an overview. In: Kydonieus AF, Berner B, eds. Transdermal Delivery of Drugs,
Vol. I. Boca Raton, FL: CRC Perss, pp. 81–100.
Christie JD, Prentice MA, Upatham ES, Barnish G (1978). Laboratory and field trials
of a slow-release copper molluscicide in St. Lucia. Am J Trop Med Hyg 27:616–
622.
Cobb LM (1979). Adverse drug reactions. In: Yoxall AT, Hird JRF, eds. Pharmacological Basis of Small Animal Medicine. Oxford: Blackwell, pp. 29–39.
Cockshott ID, Briggs LP, Douglass EJ, White M (1987). Pharmacokinetics of propofol
in female patients: studies using single bolus injections. B J Anaesthesia 59:1103–
1110.
Colburn WA (1979). A pharmacokinetic model to differentiate preabsorptive, gut epithelial, and hepatic first-pass metabolism. J Pharmacokinet Biopharm 7:407–415.
Colburn WA, Gibaldi M (1978). Pharmacokinetic model of presystemic metabolism.
Drug Metab Dispos 6:193–196.
Coldman MF, Paulsen BJ, Higuchi T (1969). Enhancement of percutaneous absorption
by the use of volatile:nonvolatile systems as vehicles. J Pharm Sci 58:1098–1102.
Craigmill AL, Brown SA Wetzlich SE, Gustafson CR, Arndt TS (1997). Pharmacokinetics of ceftiofur and metabolites after single intravenous and intramuscular
administration and multiple intramuscular administrations of ceftiofur sodium to
sheep. J Vet Pharmacol Ther 20:139–144.
Curr C (1977). The effect of dermally applied levamisole against the parasitic nematodes
of cattle. Aust Vet J 53:425–428.
Dalvi RR, Nunn VA, Juskevich J (1987). Hepatic cytochrome P-450 dependent drug
metabolizing activity in rats, rabbits and several food-producing species. J Vet
Pharmacol Ther 10:164–168.
Davis LE, Neff CA, Baggot JD, Powers TE (1972). Pharmacokinetics of chloramphenicol
in domesticated animals. Am J Vet Res 33:2259–2266.
Davis LE, Westfall BA (1972). Species differences in biotransformation and excretion
of salicylate. Am J Vet Res 33:1253–1262.
Dayton PG, Cucinell SA, Weiss M, Perel JM (1967). Dose dependence of drug plasma
level decline in dogs. J Pharmacol Exp Ther 158:305–316.

DeBacker P, Braeckman R, Belpaire F, Debackere M (1980). Bioavailability and pharmacokinetics of phenylbutazone in the cow. J Vet Pharmacol Ther 3:29–33.

Delatour P (1987). Pharmacokinetics of albendazole administered by an intraruminal pulse release electronic device in cattle. Res Vet Sci 43:284–286.

Delatour P, Garnier E, Benoit E, Claude I (1991). Chiral behaviour of the metabolite albendazole sulphoxide in sheep, goats and cattle. Res Vet Sci 50:134–138.

DeWeck AL, Schneider CH (1972). Allergic reactions in man, horse and cattle due to the presence of carboxymethylcellulose in drug formulations. In: Proceedings of the European Society for the Study of Drug Toxicity, Vol. XIII. Amsterdam; Excerpta Medica, Toxicological Problems of Drug Communications, pp. 203–204.

Dewey DW, Lee HJ, Marston HR (1958). Nature (Lond) 181:1367–1371.

Dittert LW (1977). Pharmacokinetic prediction of tissue residues. J Toxicol Environ Health 2:735–756.

Dobson A (1967). Physiological peculiarities of the ruminant relevant to drug distribution. Fed Proc 26:994–1000.

Dowrick JS, Marsden JL (1975). U.S. Patent 3,912,806.

Eltom SE, Guard CL, Schwark WA (1993). The effect of age on phenylbutazone pharmacokinetics, metabolism and plasma protein binding in goats. J Vet Pharmacol Ther 16:141–151.

Fetisov VI (1977). Trudy Vsesoyuznogo Instituta Gel'mintologii imeni Akademika K.I. Skryabina: 23:135, through Vet Bull 49:803.

Firth EC, Nouws JFM, Driessens F, Schmaetz P, Peperkamp K, Klein WR (1986). Effect of the injection site on the pharmacokinetics of procaine penicillin G in horses. Am J Vet Res 47:2380–2384.

Forsyth BA, Gibbon AJ, Pryor DE (1983). Seasonal variation in anthelmintic response by cattle to dermally applied levamisole. Aust Vet J 60:141–146.

Foster RT, Jamali F (1987). High-performance liquid chromatographic assay of ketoprofen enantiomers in human plasma and urine. J Chromatogr 416:388–393.

Foster RT, Jamali F, Russell AS, Alballa SR (1988). Pharmacokinetics of ketoprofen enantiomers in healthy subjects following single and multiple doses. J Pharm Sci 77:70–73.

Fraunfelder FT, Hanna C (1977). Trends in topical ocular medication. Ocular Ther 10:85–97.

Fijimura A, Kajiyama H, Ebinara A (1989). The influence of fever on the pharmacokinetics of pranoprofen in elderly subjects. J Clin Pharmacol 29:500–503.

Fischell RE (1988). A programmable implantable medication system (PIMS) as a means for intracorporeal drug delivery. In: Type P, ed. Drug Delivery Devices, Fundamentals and Applications. New York: Marcel Dekker, pp. 261–284.

Fulper LD (1991). Evaluation of formulation parameters on the milkout profile of intramammary mastitis formulation in the bovine. Doctoral dissertation, University of Mississippi.

Gaenssler J-G, Wilkins CA, O'Donnovan WM (1978). The divided dosage/low dosage concept using fenbendazole. J S Af Vet Assoc 49:345–349.

Gibaldi M, Perrier D (1982). Pharmacokinetics. 2nd ed. New York: Marcel Dekker, pp. 145–198, 409–417, 433–444.

Gillespie, WR (1991). Noncompartmental versus compartmental modelling in clinical pharmacokinetics. Clin Pharmacokinet 20:253–262.

Gingerich DA, Baggot JD, Yeary RA (1975). Pharmacokinetics and dosage of aspirin in cattle. J Am Vet Med Assoc 167:945–948.

Givens DI, Cross PJ, Shaw WB, Knight PE (1979). Cobalt deficiency in growing lambs: a comparison of three forms of treatment. Vet Rec 104:508–509.

Govil SK (1988). Transdermal drug delivery devices. In: Type P, ed. Drug Delivery Devices, Fundamentals and Applications. New York: Marcel Dekker, pp. 385–419.

Gyurick RJ (1988). Rumen retention devices. In: Type P, ed. Drug Delivery Devices, Fundamentals and Applications. New York: Marcel Dekker, pp. 549–561.

Heerema JC, Friedenwald JS (1950). Retardation of wound healing in the corneal epithelium by lanolin. Am J Ophthalmol 33:1421–1427.

Hennessy DR, Ali DN, Tremain SA (1995). The partition and fate of soluble and digesta particulate associated oxfendazole and its metabolites in the gastrointestinal tract of sheep. Int J Parasitol 24:327–333.

Higuchi T (1960). Physical chemical analysis of percutaneous absorption process from creams and ointments. J Soc Cosmet Chem 11:85–97.

Hogben CAM, Tocco DJ, Brodie BB, Schanker LS (1959). On the mechanism of intestinal absorption of drugs. J Pharmacol Exp Ther 125:275–282.

Holler H (1970). Studies on the secretion of the cardiac gland zone in the pig stomach. Studies on the influencing of spontaneous secretion of the isolated carciac gland zone, fluid and electrolyte secretion in the isolated stomach containing different fluids. Z Vet (A) 17:857–873.

Hom FS, Miskel JJ (1970). Oral dosage form design and its influence on dissolution rates for a series of drugs. J Pharm Sci 59:827–830.

Hostetler VB, Bellard JQ (1970). Capsules. I. Hard capsules. In: Lachman L, Lieberman HA, Kanig JI, eds. The Theory and Practice of Industrial Pharmacy. Philadelphia: Lea and Febiger, pp. 346–359.

Houghton GW, Richens A (1974). Rate of elimination of tracer doses of phenytoin at different steady-state serum phenytoin concentrations in epileptic patients. Br J Clin Pharmacol 1:155–161.

Huber WG, Reddy VK (1978). Physiologic and anatomic features of monogastric and ruminant animals. In: Monkhouse DC, ed. Animal Health Products, Design and Evaluation. Washington, D.C.: American Pharmaceutical Association, Academy of Pharmaceutical Sciences, pp. 12–21.

Hungate RE (1966). The Rumen and Its Microbes. New York: Academic Press, p. 218.

International Commission on Radiological Protection (1975). Report of the Task Group on Reference Man. 1st ed. Oxford: Pergamon.

Jacobs DE, Fox MT, Gowling G, Foster J, Pitt SR, Gerrelli D (1987). Field evaluation of the oxfendazole pulse release bolus for the chemoprophylaxis of bovine parasitic gastroenteritis: a comparison with three other control strategies. J Vet Pharmacol Ther 10:30–36.

Jaglan PS, Roof RD, Yein FS, Arnold TS, Brown SA, Gilbertson TJ (1994). Concentrations of ceftiofur metabolites in the plasma and lungs of horses following intramuscular treatment. J Vet Pharmacol Ther 17:24–30.

Jogestrand T, Schenck-Gustafsson K, Nordlander R, Dahlqvist R (1984). Quinidine-induced changes in serum and skeletal muscle digoxin concentration: evidence of saturable binding of digoxin to skeletal muscle. Eur J Clin Pharmacol 27:571–575.

Johnson BF, Bye C, Jones G, Sabey GA (1976). A completely absorbed oral preparation of digoxin. Clin Pharmacol Ther 19:746–751.

Kern DL, Slyter LL, Leffel EC, Weaver JM, Oltjen RR (1974). Ponies vs. steers: microbial and chemical characteristics of intestinal ingesta. J Anim Sci 38:559–564.

Kleiber M (1975). Metabolic turnover rate: a physiological meaning of the metabolic rate per unit body weight. J Theor Biol 53:199–204.

Klotz U, Antonin K-H, Bieck PR (1976). Pharmacokinetics and plasma binding of diazepam in man, dog, rabbit, guinea pig and rat. J Pharmacol Exp Ther 199:67–73.

Korsrud GO, Boison JO, Papich MG, et al. (1993). Depletion of intramuscularly and subcutaneously injected procaine penicillin G from tissues and plasma of yearling beef steers. Can J Vet Res 57:223–230.

Kunin CM (1967). Clinical significance of protein binding of the penicillins. Ann NY Acad Sci 145:282–290.

Kydonieus AF (1987a). Fundamentals of transdermal drug delivery. In: Kydonieus AF, Berner B, eds. Transdermal Delivery of Drugs. Vol. I. Boca Raton, FL: CRC Press, pp. 3–16.

Kydonieus AF (1987b). Transdermal delivery from solid multilayered polymeric reservoir systems. In: Kydonieus AF, Berner B, eds. Transdermal Delivery of Drugs. Vol. I.. Boca Faton, FL: CRC Press, pp. 145–156.

Laby RH (1974). U.S. Patent 3,844,284.

Landoni MF, Cunningham FM, Lees P (1995). Pharmacokinetics and pharmacodynamics of ketoprofen in calves applying PK/PD modelling. J Vet Pharmacol Ther 18:315–324.

Landoni MF, Lees P (1995). Influence of formulation on the pharmacokinetics and bioavailability of racemic ketoprofen in horses. J Vet Pharmacol Ther 18:446–450.

Lanusse, CE, Ranjan S, Prichard RK (1990). Comparison of pharmacokinetic variables for two injectable formulations of netobimin administered to calves. Am J Vet Res 51:1459–1463.

Leahey EB, Bigger JT, Butler VP, et al. (1981). Quinidine-digoxin interaction: time course and pharmacokinetics. Am J Cardiol 48:1141–1146.

Lee VH-L, Robinson JR (1978). Methods to achieve sustained drug delivery—the physical approach: oral and parenteral dosage forms. In: Robinson JR, ed. Sustained and Controlled Release Drug Delivery Systems. New York: Marcel Dekker, pp. 123–209.

Lees P, Higgins AJ, Mawhinney IC, Reid DS (1986). Absorption of phenylbutazone from a paste formulation administered orally to the horse. Res Vet Sci 41:200–206.

Levine MAH, Walker SE, Paton TW (1992). The effect of food or sucralfate on the bioavailability of S(+) and R(–) enantiomers of ibuprofen. J Clin Pharmacol 32:1110–1114.

Levy G (1968). Kinetics and implications of dissolution rate limited gastrointestinal absorption of drugs. In: Ariens EJ, ed. Physicochemical Aspects of Drug Action.

Proceedings of the 3rd International Pharmacological Meeting, Vol. 7. Oxford: Pergamon Press, pp. 33–62.

Lindenbaum J (1977). Greater bioavailability of digoxin solution in capsules: studies in the postprandial state. Clin Pharmacol Ther 21:278–282.

Lo PKA, Fink DW, Williams JR, Blodinger J (1985). Pharmacokinetic studies of ivermectin: effects of formulation. Vet Res Commun 9:251–268.

Loscher W (1978). Serum protein binding and pharmacokinetics of valproate in man, dog, rat and mouse. J Pharmacol Exp Ther 204:255–261.

Mader DR, Conzelman GM Jr, Baggot JD (1985). Effects of ambient temperature on the half-life and dosage regimen of amikacin in the gopher snake. J Am Vet Med Assoc 187:1134–1136.

Magrane WG (1977). Ocular therapeutics. In: Canine Ophthalmology. Philadelphia: Lea and Febiger, pp. 31–56.

Maibach HI, Feldmann RJ (1967). The effect of DMSO on percutaneous penetraiton of hydrocortisone and testosterone in man. Ann NY Acad Sci 141:423–427.

Maitho TE, Lees P, Taylor JB (1986). Absorption and pharmacokinetics of phenylbutazone in Welsh Mountain ponies. J Vet Pharmacol Ther 9:26–39.

Mallis GI, Schmidt DH, Lindenbaum J (1975). Superior bioavailability of digoxin solution in capsules. Clin Pharmacol Ther 18:761–768.

Malone T, Haleblian JK, Paulsen BJ, Burdick KH (1974). Development and evaluation of ointment and cream vehicles for a new topical steroid, fluclorolone acetonide. Br J Dermatol 90:187–195.

Marshall AB, Palmer GH (1980). Injection sites and drug bioavailability. In: Trends in Veterinary Pharmacology and Toxicology. Proceedings of the 1st Congress of EAVPT. Amsterdam: Elsevier, pp. 54–60.

Marston HR (1962). U.S. Patent 3,056,724.

Masson MJ, Phillipson AT (1951). The absorption of acetate, propionate and butyrate from the rumen of sheep. J Physiol (Lond) 113:189–206.

Masson MJ, Phillipson AT (1952). The composition of the digesta leaving the abomasum of sheep. J Physiol (Lond) 116:98–111.

Matthews CA, Swett WW, McDowell RE (1975). External form and internal anatomy of Holsteins and Jerseys. J Dairy Sci 58:1453–1475.

McBeath DG, Preston NK, Thompson F (1979). Studies in sheep on the efficacy of fenbendazole administered via a feed-block carrier. Br Vet J 135:271–278.

Mercer HD, Baggot JD, Sams RA (1977a). Application of pharmacokinetic methods to the drug residue profile. J Toxicol Environ Health 2:787–801.

Mercer HD, Garg RC, Powers JD, Powers TE (1977b). Bioavailability and pharmacokinetics of several dosage forms of ampicillin in the cat. Am J Vet Res 38:1353–1359.

Metzler CM (1988). Statistical methods for deciding bioequivalence of formulations. In: A, Yacobi, E Halperin-Walega, eds. Oral Sustained Release Formulations: Design and Evaluation. New York: Pergamon Press, pp. 217–238.

Meyer MC, Guttman DE (1968). The binding of drugs by plasma proteins. J Pharm Sci 57:895–918.

Miller RW, Gordon CH (1972). J Econ Entomol 65:455–458.

Mordenti J (1985). Pharmacokinetic scale-up: accurate prediction of human pharmacokinetic profiles from animal data. J Pharm Sci 74:1097–1099.

Neff-Davis C, Davis LE, Powers TE (1975). Comparative body composition of the dog and goat. Am J Vet Res 36:309–311.

Nouws JFM (1984). Irritation, bioavailability, and residue aspects of ten oxytetracycline formulations administered intramuscularly to pigs. Vet Q 6:80–84.

Nouws JFM (1990). Injection sites and withdrawal times. Ann Rech Vet 21(suppl 1):145s–150s.

Nouws JFM, van Ginneken CAM, Hekman P, Ziv G (1982). Comparative plasma ampicillin levels and bioavailability of five parenteral ampicillin formulations in ruminant calves. Vet Q 4:62–71.

Nouws JFM, Smulders A, Rappalini M (1990). A comparative study on irritation and residue aspects of five oxytetracycline formulations administerd intramuscularly to calves, pig and sheep. Vet Q 12:129–138.

Ohshima Y, Yoshikawa H, Takada K, Muranishi S (1985). Enhancing effect of absorption promoters on percutaneous absorption of a model dye (6-carboxyfluorescein) as poorly absorbable drugs. I. Study on the absorption promoting effect of azone. J Pharmacobiodyn 8:900–905.

Oishi H, Ushio Y, Narobara K, Takebara M, Nakagawa T (1976). Effect of vehicles on percutaneous absorption. I. Characterization of oily vehicle by percutaneous absorption and transepidermal water loss test. Chem Pharm Bull (Tokyo) 24:1765–1773.

O'Moore LB, Smyth PJ (1958). The control of cobalt deficiency in sheep by means of a heavy pellet. Vet Rec 70:773–774.

Ostrenga J, Steinmetz C, Paulsen B (1971a). Significance of vehicle composition. I. Relationship between topical vehicle composition, skin permeability, and clinical efficacy. J Pharm Sci 60:1175–1179.

Ostrenga J, Steinmetz C, Paulsen B, Yett S (1971b). Significance of vehicle composition. II. Prediction of optimal vehicle composition. J Pharm Sci 60:1180–1183.

Parraga ME, Kittleson MD, Drake CM (1995). Quinidine administration increases steady-state serum digoxin concentrations in horses. Equine Vet J Suppl 19:114–119.

Pasutto FM (1992). Mirror images: the analysis of pharmaceutical enantiomers. J Clin Pharmacol 32:917–924.

Paulsen BJ, Young E, Coquilla V, Katz M (1968). Effect of topical vehicle composition on the in vitro release of fluocinolone acetonide and its acetate ester. J Pharm Sci 57:928–933.

Pavan-Langston D (1976). New drug delivery systems. In: Symposium in Ocular Therapy, Vol. 9. New York: John Wiley & Sons, pp.17–32.

Pennington JE, Dale CD, Reynolds HY, MacLowry JD (1975). Gentamicin sulfate pharmacokinetics: lower levels of gentamicin in blood during fever. J Infect Dis 132:270–275.

Phillipson AT, McAnally RA (1942). Studies on the fate of carbohydrates in the rumen of the sheep. J Exp Biol 19:199–214.

Piperno E, Ellis DJ, Getty SM, Brody TM (1968). Plasma and urine levels of phenylbutazone in the horse. J Am Vet Med Assoc 153:195–198.

Pitman IH, Rostas SJ (1981). Topical drug delivery to cattle and sheep. J Pharm Sci 70:1181–1194.

Ponting LM, Pitman IH (1979). A rapid in vitro method for evaluating pour-on dose forms of levamisole for sheep. Aust J Pharm Sci 8:15–19.

Pope DG (1975). Factors to be considerd in the formulation of pharmaceuticals for oral administration to ruminant animals. Aust J Pharm Sci NS4:65–78.

Pope DG (1978). Animal health specialized delivery systems. In: Monkhouse DC, ed. Animal Health Products, Design and Evaluation. Washington, DC: American Pharmaceutical Association, Academy of Pharmaceutical Science, pp. 78–114.

Pope DG (1980a). Accelerated stability testing for prediction of pharmaceutical product stability. I. Drug Cosmet Indust 127:54–62, 116.

Pope DG (1980b). Accelerated stability testing for prediction of pharmaceutical product stability. II. Drug Cosmet Indust 127:48–66, 110–116.

Pope DG, Tsuji K, Robertson JH, McGeeter MJ (1978). Pharm Technol 2:31.

Pope DG, Wilkinson PK, Egerton JR, Conroy J (1985). Oral controlled-release delivery of ivermectin in cattle via an osmotic pump. J Pharm Sci 74:1108–1110.

Prescott JF, Baggot JD (1988). Bovine mastitis. In: Antimicrobial Therapy in Veterinary Medicine. 1st ed. Boston: Blackwell, pp. 321–331.

Presson BL, Yazwinski TA, Greenway T, Pote L, Newby T (1984). Controlled field trial on the anthelmintic efectiveness of the morantel sustained-release bolus in grazing calves. Am J Vet Res 45:2628–2630.

Prichard RK, Gascon LH, Lanusse CE (1991). Selection of route of administration in the treatment of gastrointestinal roundworm infections. Acta Vet Scand 87(Suppl):65–74.

Prichard RK, Steel JW, Lacey E, Hennessy DR (1985). Pharmacokinetics of ivermectin in sheep following intravenous, intra-abomasal or intraruminal administration. J Vet Pharmacol Ther 8:88–94.

Punch PI, Costa ND, Edwards ME, Wlcox GE (1987). The release of insoluble antibiotics from collagen ocular inserts in vitro and their insertion into the conjunctival sac in cattle. J Vet Pharmacol Ther 10:37–42.

Punch PI, Slatter DH, Costa ND, and Edwards ME (1985). Investigation of gelatin as a possible biodegradable matrix for sustained delivery of gentamicin to the bovine eye. J Vet Pharmacol Ther 8:335–338.

Rasmussen F (1978). Tissue damage at the injection site after intramuscular injection of drugs. Vet Sci Commun 2:173–182.

Rasmussen F (1980). Tissue damage at the injection site after intramuscular injection of drugs in food-producing animals. In: Trends in Veterinary Pharmacology and Toxicology. Proceedings of the 1st Congress of EAVPT. Amsterdam: Elsevier, pp. 27–33.

Rasmussen F, Hogh P (1971). Irritating effect and concentrations at the injection site after intramuscular injection of antibiotic preparatoins in cows and pigs. Nord Vet 23:593–605.

Rasmussen F, Svendsen O (1976). Tissue damage and concentration at the injection site after intramuscular injection of chemotherapeutics and vehicles in pigs. Res Vet Sci 20:55–60.

Riegelman S, Collier P (1980). The application of statistical moment theory to the evaluation of in vivo dissolution time and absorption time. J Pharmacokinet Biopharm 8:509–534.

Robinson EP, Sams RA, Muir WW (1986). Barbiturate anesthesia in Greyhound and in mixed-breed dogs: comparative cardiopulmonary effects, anesthetic effects, and recovery rates. Am J Vet Res 47:2105–2112.

Robinson D, Williams RT (1958). Do cats form glucuronides? Biochem J 68:23P–24P.

Rohde TD, Buchwald H, Blackshear PJ (1988). Implantable infusion pumps. In: Type P, ed. Drug Delivery Devices. Fundamentals and Applications. New York: Marcel Dekker, pp. 235–260.

Rowland M, Benet LZ, Graham GG (1973). Clearance concepts in pharmacokinetics. J Pharmacokinet Biopharm 1:123–136.

Rolands D, Shepherd MT, Collins KR (1988). The oxfendazole pulse release bolus. J Vet Pharmacol Ther 11:405–408.

Salmon SA, Watts JL, Yancey RJ Jr, (1996). In vitro activity of ceftiofur and its primary metabolite, desfuroylceftiofur, against organisms of veterinary importance. J Vet Diagn Invest. In press.

Sams RA, Muir WW, Detra RL, Robinson EP (1985). Comparative pharmacokinetics and anesthetic effects of methohexital, pentobarbital, thiamylal, and thiopental in Greyhound dogs and non-Greyhound, mixed-breed dogs. Am J Vet Res 46:1677–1683.

Schadewinkel-Scherkl AM, Rasmussen F, Merck CC, Nielsen P, Frey HH (1993). Active transport of benzylpenicillin across the blood-milk barrier. Pharmacol Toxicol 73:14–19.

Schou J (1961). Absorption of drugs from subcutaneous connective tissue. Pharmacol Rev 13:441–464.

Schwarz C, Steinmetzer K, Caithaml K (1926). Arch Ges Physiol 213:595. Cited by Dukes HH (1955). In: The Physiology of Domestic Animals. 7th ed. Ithaca, NY: Comstock/Cornell University Press, p. 330.

Shaw JE, Powers M, Gale R (1987). Rate-controlled transdermal therapy utilizing polymeric membranes. In: Kydonieus AF, Berner B, eds. Transdermal Delivery of Drugs, Vol. I. Boca Raton, FL: CRC Press, pp. 101–116.

Sinkula AA (1978). Methods to achieve sustained drug delivery—the chemical approach. In: Robinson JR, ed. Sustained and Controlled Release Drug Delivery Systems, New York: Marcel Dekker, pp. 411–555.

Skerman KD, Sutherland AK, O'Halloran MW, Bourke JM, Munday BL (1959). The correction of cobalt or vitamin B_{12} deficiency in cattle by cobalt pellet therapy. Am J Vet Res 20:977–984.

Slatter DH, Edwards ME, Wilcox GE, Ezekiel D (1982). Ocular inserts for application of drugs to bovine eyes—effects of hydrophilic contact lenses. Aust Vet J 59:1–3.

Speiser P (1966). Galenic aspects of drug action. Pharm Acta Helv 41:321–342.

Spiegel AJ, Noseworthy MM (1963). Use of nonaqueous solvents in parenteral products. J Pharm Sci 52:917–927.

Spinelli JS, Enos LR (1978). Veterinary therapeutics and drug interactions. In: Drugs in Veterinary Practice. St. Louis: C.V. Mosby, pp. 12–28.

Stabenfeldt GH, Edqvist L-E (1984). Female reproductive processes. In: Swenson MJ, ed. Dukes' Physiology of Domestic Animals. 10th ed. Ithaca, NY: Comstock/Cornell University Press, pp. 798–832.

Stanley JP (1970). Capsules II. Soft gelatin capsules. In: Lachman L, Lieberman HA, Kanig JI, eds. The Therapy and Practice of Industrial Pharmacy. Philadelphia: Lea and Febiger, pp. 359–407.

Stella VJ, Mikkelson TJ, Pipkin JD (1980). Prodrugs: the control of drug delivery via bioreversible chemical modification. In: Juliano RJ, ed. Drug Delivery Systems. Characteristics and Biomedical Applications. New York: Oxford University Press, pp. 112–176.

Stoughton RB, Fritsch W (1964). Influence of dimethylsulfoxide (DMSO) on human percutaneous absorption. Arch Dermatol 90:512–517.

Strong JM, Dutcher JS, Lee W-K, Atkinson AJ Jr (1975). Absolute bioavailability in man of N-acetylprocainamide determined by a novel stable isotope method. Clin Pharmacol Ther 18:613–622.

Sund RB, Schou J (1964). The determination of absorption rates from rat muscles: an experimental approach to kinetic descriptions. Acta Pharmacol Toxicol 21:313–325.

Sundlof SF, Whitlock TW (1992). Chlorsulon pharmacokinetics in sheep and goats following oral and intravenous administration. J Vet Pharmacol Ther 15:282–291.

Taylor SM, Malton TR, Blanchflower J, Kennedy DG, Hewitt SA (1993). Effects of dietary variations on plasma concentrations of oral flukicides in sheep. J Vet Pharmacol Ther 16:48–54.

Taylor SM, Mallon T, Green WP (1988). Efficacy of a levamisole bolus in field infections of bovine nematodes. Ann Rech Vet 19:111–118.

Theodorides VJ, DiCuollo CJ, Guarini JR, Pagano JP (1968). Serum concentrations of chloramphenicol after intraruminal and intra-abomasol administration in sheep. Am J Vet Res 29:643–645.

Ther L, Winne D (1971). Drug absorption. Ann Rev Pharmacol 11:57–70.

Toutain PL, Lassourd V, Costes G, et al. (1995). A non-invasive and quantitative method for the study of tissue injury caused by intramuscular injection of drugs in horses. J Vet Pharmacol Ther 18:226–235.

Tozer TN (1984). Implications of altered plasma protein binding in disease states. In: Benet LZ, Massoud N, Gambertoglio JG, eds. Pharmacokinetic Basis for Drug Treatment. New York: Raven Press, pp. 173–193.

Tozer TN (1994). Bioequivalence data analysis: evaluating the metrics of rate and extent of drug absorption. J Vet Pharmacol Ther 17:105.

Tranquilli WJ, Paul AJ, Seward RL (1989). Ivermectin plasma concentrations in Collies sensitive to ivermectin-induced toxicosis. Am J Vet Res 50:769–770.

Trier JS (1967). Structure of the mucosa of the small intestine as it relates to intestinal function. Fed Proc 26:1391–1404.

Trueblood JH, Rossomondo RM, Carlton WH, Wilson LA (1975). Corneal contact times of ophthalmic vehicles: evaluation by microscintigraphy. Arch Ophthalmol 93:127–130.

Tyle P, Kari B (1988). Iontophoretic devices. In: Type P, ed. Drug Delivery Devices. Fundamentals and Applications. New York: Marcel Dekker, pp. 421–454.

Vesell ES, Lee CJ, Passananti GT, Shively CA (1973). Relationship between plasma antipyrine half-lives and hepatic microsomal drug metabolism in dogs. Pharmacology 10:317–328.

Wagner JG (1961). Biopharmaceutics: absorption aspects. J Pharm Sci 50:359–387.

Wagner JG, Gerard ES, Kaiser DG (1966). The effect of the dosage form on serum levels of indoxole. Clin Pharmacol Ther 7:610–619.

Watson ADJ, Rijnberk A, Moolenaar AJ (1987). Systemic availability of O,P$_1$-DDD in normal dogs, fasted and fed, and in dogs with hyperadrenocorticism. Res Vet Sci 43:160–165.

Webb AI, Weaver BMQ (1979). Body composition of the horse. Equine Vet J 11:39–47.

Wertz EM, Benson GJ, Thurmon JC, Tranquilli WJ, Davis LE, Koritz GD (1988). Pharmacokinetics of thiamylal in cats. Am J Vet Res 49:1079–1083.

Wester RC, Maibach HI (1987). Clinical considerations for transdermal delivery. In: Kydonieus AF, Berner B, eds. Transdermal Delivery of Drugs, Vol. I. Boca Raton, FL: CRC Press, pp. 71–78.

Westlake WJ (1988). Bioavailability and bioequivalence of pharmaceutical formulations. In: Peace KE, ed. Biopharmaceutical Statistics for Drug Development. New York: Marcel Dekker, pp. 329–352.

Wheaton JE, Al-Raheem SN, Godfredson JA, et al. (1988). Use of osmotic pumps for subcutaneous infusion of growth hormone-releasing factors in steers and wethers. J Anim Sci 66:2876–2885.

Wilkinson GR, Shand DG (1975). A physiological approach to hepatic drug clearance. Clin Pharmacol Ther 18:377–390.

Williams RT (1967). Comparative patterns of drug metabolism. Fed Proc 26:1029–1039.

Wilson CD, Westgarth DR, Kingwill RG, Griffin TK, Neave FK, Dodd FH (1972). The effect of infusion of sodium cloxacillin in all infected quarters of lactating cows in sixteen herds. Br Vet J 128:71–86.

Wohrl H (1977). Effectiveness of Citarin-L spot-on (levamisole) against endohelminths of swine. Inaugural dissertation, Freie University, Berlin; through Vet Bull 1978; 48:2357.

Yeary RA, Swanson W (1973). Aspirin dosages for the cat. J Am Vet Med Assoc 163:1177–1178.

Zinn RD, Anderson GR, Skaggs JW (1961). The public health significance of staphylococcic infection in cattle. J Am Vet Med Assoc 138:382–386.

Ziv G (1978). Pharmacokinetics of antimastitis products. In: Monkhouse DC, ed. Animal Health Products, Design and Evaluation. Washington, DC: American Pharmaceutical Assocciation, Academy of Pharmaceutical Science, pp. 32–66.

Ziv G (1980). Drug selection and use in mastitis: systemic vs local drug therapy. J Am Vet Med Assoc 176:1109–1115.

Ziv G, Saran-Rozenzvaig A, Risenberg R (1973). Retention of antibiotics in dry udder secretion after the infusion of several "dry cow" antibiotic products. Z Vet B 20:415–424.

Zoran DL, Riedesel DH, Dyer DC (1993). Pharmacokinetics of propofol in mixed-breed dogs and Greyhounds. Am J Vet Res 54:755–760.

2

Formulation of Veterinary Dosage Forms

PAUL R. KLINK, THOMAS H. FERGUSON
Elanco Animal Health, A Division of Eli Lilly and Company, Greenfield, Indiana

JUDY A. MAGRUDER
Alza Corporation, Palo Alto, California

I. INTRODUCTION

In many ways the methodologies and techniques used in the development of human drug dosage forms are applicable to the development of veterinary drug dosage forms. For example, fundamental pharmaceutical science techniques used to generate the preformulation data package, drug stability assessment and expiry dating, sterility assurance (if required), drug pharmacokinetics, and manufacturing processes are common to both human and veterinary dosage forms. In addition, veterinary drug products are subject to the same good manufacturing practice (GMP) regulations by the Food and Drug Administration (FDA) as human drug products. As one reads other chapters in this book, the similarities between the development of human drug dosage forms and veterinary dosage forms will become obvious.

However, there are significant issues that make the development of veterinary dosage forms unique. The multiplicity of animal species has led to veterinary dosage forms (in some cases using the same drug substance) specific to a single species, based on either different species-dependent drug pharmacokinetics (Chap. 1), accepted and specialized dose administration devices (Chap. 5), accepted management practices, or unique routes of administration such as intramammary infusion, gavage, intraperitoneal, or via the drinking water. The companion pet market aside, animals raised for the production of food (meat, milk, eggs) are managed intensively (e.g., 10,000 to 30,000 broilers per broiler house) and extensively (hands-off). These animals are rarely treated as individuals, since a disease state in one animal is usually transmitted readily to all other animals in the closely housed environment. The sheer numbers and the labor and time involved necessitate mass medication techniques, many times using mechanical equipment in differing environmental conditions (drug stability?). These animals enter the food chain. Therefore, tissue residues of the drug and major metabolites in edible tissues are required to be addressed. Since the excreta of all meat-producing animals ultimately deposit in the environment, the safety of environmental drug and metabolite residues also needs to be addressed. Thus, the acceptance of veterinary dosage forms by the animal health industry (food-producing animals) depends not only on the therapeutic efficacy but also on the cost of the product, the ease or speed of administration, dosing frequency, environmental and handler safety, tissue residues, drug side effects, and impact on slaughter practices.

Other chapters in this book discuss biological, toxicological, and clinical trial and regulatory issues critical to the development and approval of a new animal drug product. This chapter deals primarily with the formulation development of veterinary dosage forms for the animal health industry.

II. PREFORMULATION

Of all the product development activities that occur for a veterinary product, the one activity that is crucial, but underresourced and therefore receives the least amount of time and attention, is the preformulation data package for the drug and envisioned product formulation. All too often, the formulation scientist receives a limited quantity of the drug of interest, preliminary ideas for the desired formulation, and an admonishment to thrust a product formulation into field trials and for toxicological testing as soon as possible. Frequently, as the product development process progresses and as the bulk drug manufacturing process improves and is scaled up, drug physicochemical properties come to light perhaps necessitating a formulation or manufacturing change. Additional development time and resources are required to rectify a situation that might have been overcome with additional time and information in the preformulation stage of product development. Thus, it becomes imperative that technical communications between the manufacturing site (for both drug and dosage form) and product development occur early and persist through the bulk drug and product manufacturing process development.

A suggested list of preformulation data required for veterinary dosage forms is shown in Table 1. It is organized in order of priority as to the importance of each piece of data and with the concession that only limited quantities of drug may exist in the early product development stages. Not all preformulation data need be generated for all dosage forms, and an attempt has been made to sug-

Table 1 Suggested Preformulation Information for Veterinary Dosage Forms

Test	Method/function/ characterization	Dosage forms[a]
1. UV spectroscopy	Simple quantitative assay	O, P, T, CR
2. Solubility	Phase solubility/purity	
Aqueous	pH effects	O, P, T, CR
	in water—pH of saturated solution	O, P, T, CR
	in 0.9% NaCl	P
	in 0.1N HCl	O
	in 0.05M, pH 7.4 phosphate buffer	P, CR
pK_a	Solubility control (in aqueous and/or 66% DMF)	O, P, T, CR
	Salt formation	
Salts	Solubility, hygroscopicity, stability	O, P, T, CR
Solvents	Vehicle identification and formulation extraction	O, P, T, CR
Partition coefficient	Lipophilicity, structure-activity relationship	O, P, T, CR
Dissolution	Bioavailability	O, P, CR
3. Polymorphism/ crystallinity	DSC, TGA, x-ray powder pattern, microscopy	O, P, CR
4. Hydration state	KF water, water sorption/desorption isotherms, x-ray powder patterns	O, P, T, CR
5. Assay development	UV, HPLC, LC-MS, chiral,	O, P, T, CR
6. Stability	thermal, hydrolysis, pH,	O, P, T, CR
In solution	oxidation, photolysis and	
In solid state	metal ions	
7. Microscopy/ laser light scattering/ image analysis	Particle size, shape, morphology; milling needs	O, P, T, CR
8. Bulk density	Tablet and capsule formation	O
9. Flow properties	Tablet and capsule formation	O
10. Compression properties	Aid excipient choice	O, P, T, CR
11. Excipient compatibility	Screen by DSC, HPLC	O, P, T, CR

[a]O = oral (e.g., premixes, medicated blocks, drenches, soluble powders, liquid feed supplements, tablets, boluses); P = parenteral (e.g., injections, implants); T = transdermal (e.g., dips, pour-ons, spot-ons, patches); CR = controlled release (e.g., injections, implants, boluses, ear tags).
(*Source*: Refs. 1–3.)

gest and identify those data necessary for specific dosage forms—e.g., oral, parenteral, transdermal, and parenteral controlled-release dosage forms.

Usually, the first step in preformulation is to establish a simple analytical method so that all future measurements can be quantitative (1). This may be a simple UV spectrophotometric method or a HPLC method. Most drugs absorb light in the ultraviolet wavelengths (190 to 390 nm). The functional group chromophores with their characteristic absorption maxima are readily available in the literature (1). This initial preformulation step may be most efficiently accomplished with the collaboration of analytical chemists. Analytical chemists generate their own data to support regulatory assay development and validation for both the bulk drug and product dosage form. With the caveat that the confirmation of drug structure having been already established by the synthetic organic chemist, the analytical data include identity (NMR, IR, UV, TLC, DSC, ORD), purity (amounts of water, solvents, salts, heavy metals, organic impurities), quality (physical appearance, odor, color, pH of slurry or saturated solution, melting point), and assay development and characterization of degradation products (HPLC, LC-MS) (1). In the absence of a functioning preformulations group, the formulation chemist, the analytical chemist, and the manufacturing site all have a vested interest in the preformulation data package and should therefore collaborate closely to generate it.

The various physicochemical properties examined in preformulation have a number of important interrelationships that can be utilized for the efficient collection of preformulation data. These interrelationships impact on product development and manufacturing scale-up and affect the performance (efficacy) or therapeutic effectiveness of the drug or drug dosage form. The physicochemical properties and interdependencies that lead to possible differing effects on biological activity are summarized in Figure 1. The first physicochemical properties that probably should be measured are pK_a and solubility since these largely influence all additional preformulation and product development work. Solubility data help determine which initial formulations are used in initial efficacy or toxicology studies. The pK_a allows the use of pH to manipulate drug solubility in initial formulations and the selection of salts so that improved bioavailability, stability, and powder properties (e.g., bulk density, flowability, compressibility) can be achieved. All of the physicochemical properties ultimately affect either stability or solubility; two major issues of importance in the development of a drug dosage form. Adequate drug or dosage form stability (e.g., shelf life) assures that the biological activity or performance of the dosage form does not deleteriously change over the proposed shelf life of the dosage form. Solubility, under sink conditions, is directly proportional to dissolution rate, and thus has an important effect on bioavailability, hence activity or efficacy. Both drug and dosage form stability and drug solubility need to be controlled so that bioequivalence between lots is assured. When the drug physicochemical parameters are not well controlled or not understood, significant dosage form bioinequivalencies may result (3).

Imperative in preformulation is the evaluation of the stability of the drug under severe or stressful conditions. First, data from these stress tests indicate compound sensitivity to different chemical, manufacturing, and use environments

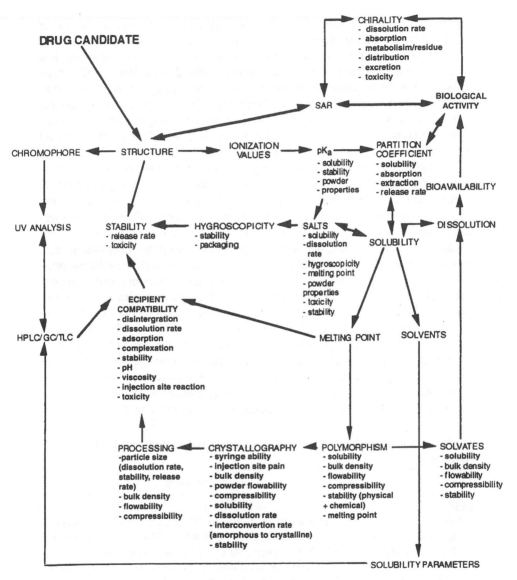

Figure 1 Interrelationship of drug physicochemical properties examined in pre-formulation and listed dependent effects that could create differing biological activities. (Adapted from Refs. 1,3,4.)

(e.g., pelleted feeds, range blocks, drenches, soluble powders, liquid feed supplements, dips, pour-ons) and therefore provide useful information to aid in the selection of proper storage conditions, formulation approaches, excipient choices, and the need for protective additives or packaging. Second, stress testing can be used to validate stability indicating analytical methods to assure their ability to resolve, detect, and quantify degradation products. Suggested stress testing conditions include thermal, acid-base, oxidation, light, physical stress and humidity conditions, summarized in Table 2. Additional stability tests depend upon the results of the preformulation stress testing and the dosage form utilized. The Center for Veterinary Medicine's *Drug Stability Guidelines* (5) details the stability requirements for various veterinary drugs and dosage forms. Other references regarding stability testing of drug and drug products should be consulted for additional information (1,6–13).

Of significant importance, yet poorly utilized because of economic or technological issues, in veterinary dosage forms is drug stereochemistry or chirality. Pharmaceutical production techniques are evolving to the point where economic stereospecific drug synthesis and chiral separations are possible on a commercial scale. In all living things there is chirality, a "handedness," even at the molecular level (4). Thus, one should expect that absorption, distribution, metabolism, and excretion of chiral compounds, together with their interactions with target tissues and receptors, would manifest themselves in a variety of stereospecific biological effects (14). Some well publicized examples document this point as shown in Table 3. Stereospecific biological differences include taste, odor, different magnitude of effects, different activities, and different toxicology or pharmacology. It is estimated that approximately one-half of the most commonly prescribed pharmaceutical drugs contain at least one chiral center and that 90% of these chiral drugs are marketed as racemates which exhibit enantioselectivity (4). The FDA's recent policy statement concerning stereoisomeric drugs outlines the FDA's expectations of sponsors in the development

Table 2 Suggested Stress Testing Conditions Used in Preformulation Stability Assessment

Test	Conditions
Solid	
Heat	Amber bottle @ 4, 20, 30, 37, 50, and 85°C, 1–30 days
Heat and humidity	Amber bottle @ 37°C/75% RH, 52°C/75% RH, 1–30 days
Moisture uptake	30, 45, 60, 75 and 90% RH @ room temperature
Light	Clear bottle or open dish @ RT, > 1,000 Lux, 7–30 days
Physical stress	Ball milling, air milling (polymorphic change?)
Aqueous	
Acid-base	pH 1, 3, 5, 7, 9 (and 11) @ RT and 37°C in different buffers 1–30 days, or 0.1N HCl, 0.1N NaOH, buffer pH 8 in amber bottles @ 40°C, 1–30 days
Oxidation	Slurry in amber bottle in 0.3% H_2O_2 @ RT for 24 h

(*Source*: Refs. 1,2.)

Table 3 Biological Differences Due to Enantioselectivity

Asparagine	(S) bitter taste	(R) sweet taste
Carvone	(S) caraway flavor/odor	(R) spearmint flavor/odor
Chloramphenicol	(R, R) antibacterial	(S, S) inactive
Propranolol	(S) 100× activity of (R)	
Ethambutol	(S, S) tuberculostatic	(R, R) causes blindness
Fluazifop butyl	(S) inactive	(R) herbicide
Paclobutrazol	(2R, 3R) fungicide	(2S, 3S) plant growth reg.
Warfarin	(S) 5-6× activity of (R)	
Clozylacon	(αS, 3R) fungicidal activity	
Labetalol	(R, R) beta blocker	(S, R) α-antagonist
Mianserin	(R) toxic	(S) nontoxic
Hexobarbital	(S) reduced oral clearance in elderly	
Etodolac	(S) anti-inflammatory activity	(R) reduced activity
Verapamil	(S) cardiovascular effects	(R) cancer therapy
Hydoxychloroquine	(R) > (S) absorption stereospecific	
Flurbiprofen	(S) analgesic/antiinflammatory activity	
Atenolol	(S) negative chronotropic/inotropic responses	
Thalidomide	(R) sedative	(S) tetratogen
Allethrin	(1S *trans*, R) 188× activity of (1R *trans*, S) (1RS *cis/trans*, RS) 2× oral toxicity of (1R *cis*, RS)	
Fenvalerate	(S, S) >350× activity of (R, RS) (R, RS) >100× oral toxicity of (S, S)	
Norepinephrine	(R) 100× activity of (S)	
Isoproterenol	(R) 1000× activity of (S) @ β_1 adrenoceptor (R) 600–800× activity of (S) @ β_2 adrenoceptor (R) 300× activity of (S) @ β_3 adrenoceptor	
α-Methyldopamine	(S) 9× activity of (R) @ α_2 adrenoceptor	
Ketorolac	(S) Active	(R) essentially inactive

(*Sources*: Refs. 4,16,17,19,20.)

of pharmaceutical chiral drugs (15). Very simply, the FDA's position is that development of a single enantiomer should be considered when the enantiomers are pharmacologically active but differ significantly in potency, specificity, or maximum effect. When one enantiomer is essentially inert, or when the effects of the enantiomers are not stereospecific, the use of the racemate may be justifiable and may be developed (4). Intuitively, it makes sense to learn from the biology and to mimic it by understanding the stereochemistry of potential veterinary drug products early in the development phase (e.g., preformulation), leading to optimized efficacy and potentially minimizing tissue residues. The references (4,14–19) should be consulted for additional information on chiral separations and the importance of chirality on biological activity.

Many excipients (e.g., cyclodextrins, cellulose derivatives, phospholipids, ascorbic acid, anhydrous dextrose, mannitol, sorbitol, lactic acid, tartaric acid, malic acid) are chiral and can interact preferentially with drug enantiomers resulting in stereospecific absorption and release from the formulation (4). The dissolution rate of the individual enantiomers from a formulation containing a racemate and a chiral excipient may differ (4). Stereospecific metabolism may result. Isomerism may also be introduced into otherwise nonisomeric drugs by metabolism (e.g., oxidation of nortriptyline) potentially affecting systemic clearance (21). To date, little attention has been given to the stereochemical implications of bioequivalence, metabolism, tissue residues, and, in the case of topically applied veterinary dosage forms (e.g., pyrethroids), environmental safety and toxicology in nontarget living organisms.

The preformulation requirements for protein and protein dosage forms are covered in Chapter 3. However, an emerging technology that is gaining increasing interest in the animal health and veterinary industries is that of oligonucleotide or gene therapy (22,23). The preformulation requirements for oligonucleotides, plasmids, or genes are ill-defined, but fundamentally the requirements should be similar to those of recombinantly derived proteins. Issues such as stereochemistry, solution structure, interaction with proteins, cell membrane permeability, binding selectivity with the target, resistance against nuclease digestion, stability (inherent, and in the presence of delivery vehicle and during manufacture), purity, endotoxin level, residual solvents, and sterility should be evaluated (24).

Preformulation data, the determination of the fundamental physical and chemical properties of the candidate drug molecule, are essential to the development of quality veterinary drug products. These data are particularly valuable when more than one dosage form is developed as a product (e.g., ivermectin—pour-on, injection, drench, premix, tablets; tetracyclines—injection, soluble powder, premix; tylosin—injection, soluble powder, premix; monensin—premix, controlled-release bolus, range block, briquettes) and/or utilized in more than one animal species or used in combination with other drugs. There are excellent reviews on preformulation and the impact on product development that should be consulted for additional information (25–30).

III. DOSAGE FORM SELECTION CRITERIA

Veterinary dosage forms are utilized in a complex industry where animal management practices, ease of use, safety, and economic considerations may have more impact on the acceptance of a product than scientific and technical uniqueness. The $12.9 billion (in 1994) industry is fragmented by animal species and by the type of drugs or products used (Table 4). Sales in the three major food production species—cattle, swine and poultry—account for approximately 74% of the total market. Most of the products utilized in these three species are administered orally, either in premixes and drinking water additives, or as licks, powders, solutions, tablets, capsules, or boluses. This is documented in Table 5, where approximately 52% of the 1987 U.S. animal health sales were oral dosage forms. Second in animal health sales were injectables and implants com-

Table 4 World Animal Health Market (1994) by Species and by Product Segment (percent of specie market)

Cattle: $4.114 billion
    Antibacterials—19.6%
    Biologicals—15.1%
    Medicinal feed additives—9.1%
    Nutritionals—21.6%
    Parasiticides—18.4%
    Performance enhancers—7.3%
    Other pharmaceuticals—8.9%
Swine: $3.048 billion
    Antibacterials—21.3%
    Biologicals—10.2%
    Medicinal feed additives—22.7%
    Nutritionals—32.5%
    Parasiticides—6.7%
    Other pharmaceuticals—6.6%
Pets/other: $2.583 billion
Poultry: $2.335 billion
    Antibacterial—4.9%
    Biologicals—19.5%
    Medicinal feed additives—33.6%
    Nutritionals—37.7%
    Other pharmaceuticals—4.3%
Sheep: $820 million

(*Source*: Ref. 31.)

Table 5 U.S. Animal Health Market by Dosage Form
Based on 1987 Sales

Dosage form	Percent of sales
Feed premixes	35
Injectables	33
Oral tables, capsules, boluses	9
Oral liquids/powders	8
Topicals	5
Implants	4
Paste/gels	3
Intramammary	2
Other	1

(*Source*: Ref. 32.)

bined for approximately 37% of the sales. Clearly, if a drug has oral bioavailability and oral therapeutic efficacy, the industry prefers this route of administration. This is because it is the most economical, most convenient, and
the easiest method to administer substances to large numbers of animals (every
animal eats and drinks water every day unless in a diseased or stressed condition).

Having established (at least historically) that the preferred route of administration in the animal health industry is the oral route, it is important to recognize that there are many other routes of administration that have been utilized
and there are a variety of dosage forms. Potential routes of administration and
common dosage forms (including for companion animals) are listed in Table
6. Several factors influence the decision on which administration route to pur-
sue. These will be discussed more fully later in this chapter and elsewhere in
this book. As technological advances continue to occur for the treatment of
diseases in the pharmaceutical industry, the future offers the potential for other
innovative administration routes (e.g., to the fetus) and cures for diseases in
animals (e.g., nondrug: using oligonucleotides, genes, affecting the genetic tem-
plate). This application of new technologies to the animal health industry will
become important as the world's population continues to increase, the demand
for high-quality and inexpensive food increases, while the land area available
for the production of food continues to decrease. These factors lead to more
intensive, efficient, and hence more extensive management of food-producing
animals.

Often the administration route and the dosage form for a particular drug are
chosen on the basis of the customary means of administering the medication to
the animal with little regard to a rational process for making that decision. Table
7 shows several questions that should be answered in the decision process for

Table 6 Routes of Administration and Typical Dosage Forms for Veterinary Products

---

Oral
  By mouth (premixes, solutions, suspensions, pastes)
  By stomach tube (solutions, suspensions, pastes)
  Intraruminally (injections, tablets, capsules, boluses, pastes)
Parenteral
  Intravenous (solutions, suspensions)
  Intraperitoneal (solutions, suspensions, implants)
  Intramuscular (solutions, suspensions, implants)
  Subcutaneous (solutions, suspensions, implants)
  Intra-arterial (solutions, suspensions)
  Epidural (solutions)
  Intrapulmonary (solutions, suspensions)
  Intracardiac (solutions)
  Intra-articular (solutions)
  Intramedullary (solutions)
  Intradermal (solutions, suspensions)
Inhalation (gas, pulmonary with particulates, vaporized)
Topical
  Skin or hair coat (ear tags, spot-ons, pour-ons, dips, collars, dust bags)
  Mucous membranes
    Sublingual (tablets, depots)
    Rectal (suppositories)
    Nasal (gas, aeorsols)
    Urethral
    Uteral
    Vaginal (suppositories)
    Bladder
    Intramammary (solutions, suspensions)
    Ophthalmic (powders, solutions, suspensions, ocular inserts, ointments)
  Transdermal
    Passive (pour-ons, spot-ons, dips, ointments, powders, linaments)
    Iontophoresis
    Sonophoresis

---

determining the best route of administration for a veterinary dosage form. Intimately woven into the decision process for the best administration route and dosage form is the information derived from the preformulation studies, particularly those drug properties (e.g., solubility, partition coefficients) that affect the pharmacokinetics (e.g., absorption, distribution, metabolism, excretion) of the drug in the body (see Fig. 1). An appreciation and knowledge of physiological differences between animal species and physiological changes within

Table 7 Factors Involved in Deciding
Route of Administration

Concentration of drug needed?
Where in the body is the drug needed?
How fast is the action needed?
For how long is the action needed?
Any problems with this route?
Safety of the treatment?
Cost of the treatment?

---

an animal species due to disease states, stress, sex, age, and the environment need also to be integrated into the decision process for determining the appropriate administration route and dosage form.

The blood plasma is the predominate body fluid into which the drug is absorbed and distributed to the tissues. Usually by passive diffusion, drugs penetrate biological membranes (drug solubility, partition coefficient, and degree of ionization are important properties) and are absorbed into the systemic circulation to be distributed throughout the tissues and the highly perfused organs, the liver and the kidneys, where metabolism and excretion occur. Because these organs are highly perfused, they continually receive a major fraction of the amount of drug in the plasma. In addition, veterinary dosage forms administered via the oral and intraperitoneal routes release their medicaments into the portal venous blood, in which they are conveyed directly to the liver, where metabolism can occur (the so-called first-pass effect) before entering the systemic circulation.

While in the plasma, drugs (usually acidic drugs) that bind to plasma proteins (usually albumin) decrease the drugs' accessibility to sites of action. Protein binding influences metabolism (because the drug is protected from rapid catabolism), distribution, and elimination of many drugs (33).

Thus, drug binding to plasma proteins usually decreases the maximum intensity of the therapeutic effect, but prolongs the duration of drug action (33). In addition, there have been at least six binding sites identified on albumin, two of which are major drug binding sites (33,34). Binding of drugs to albumin may be stereospecific and competitive with other drugs (33,34). Drugs may also be bound or sequestered (because of lipophilicity) in tissues (such as fat). Assuming the binding interaction with plasma proteins and the sequestering in tissues are reversible, both these phenomena may serve as a reservoir to replenish free drug that is lost, maintaining the therapeutic concentration, and actually may be of some benefit in sustaining the biological action from a single administration. Couple the above factors with metabolism and excretion, only a small fraction of the drug dose (variable from drug to drug) administered ever reaches the site of action. The factors that influence the concentration of the drug or medicament in the blood plasma, and hence the amount of free drug available at the site of action, are represented in Figure 2.

More specifically, Figure 3 outlines major administration routes for drugs and medicaments and the interaction between the various distribution and storage areas within the body and the major elimination or excretion routes from the blood plasma. Many parenteral routes of administration deliver their active agents directly into the blood plasma, unless these routes of administration involve a specific organ (e.g., heart, lungs, mammary gland, skin). The concentration of the drug in the blood plasma with parenteral administration routes depends on the ability of the drug to cross endothelial membranes and the capillary blood flow at the site of administration. Only the intravenous and intra-arterial routes of administration ensure that the full drug dose enters the systemic circulation. These two routes thus provide the most rapid effects but also have the shortest duration of response and potentially the greatest danger of adverse effects. Absorption of the administered drug must precede the distribution of the drug by any other route of administration. Drug absorption may vary widely within animal species among different injection sites on the animal and among parenteral routes of administration, such as intramuscular or subcutaneous administration.

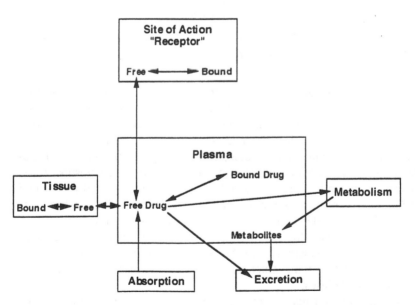

Figure 2 Physiological factors that influence the amount of drug available to the site of action.

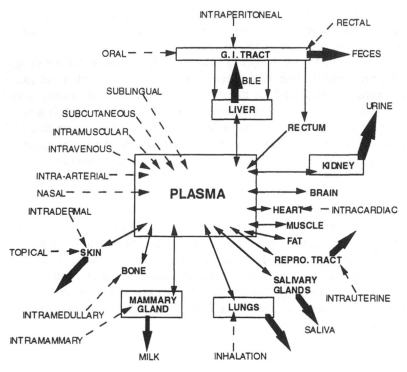

Figure 3 Drug routes of administration (−−→), drug distribution (←→) and major routes of drug elimination or excretion (➔).

Species differences, genetic makeup, environmental conditions, and other factors that influence drug absorption, distribution, metabolism, and excretion, and thus drug action, are shown in Figure 4. Clearly, differences across species in gastrointestinal function either in dietary habit or anatomical arrangement are evident. Certain species (horse, cattle, sheep, goat) acquire their nutritional needs via a bacterial fermentation process in the gastrointestinal tract. Differences in anatomical arrangement further fragment the herbivores into ruminants (cattle, sheep, goat) and monogastric (horse). Monogastric non-herbivore species include the pig, dog, and cat. Chickens and turkeys have gastrointestinal tracts and different functions in a class of their own! Oral drug administration is most widely used in the animal health industry; formulations range from feed and drinking-water additives to licks, pastes, drenches, tablets, capsules, and boluses.

Gastric emptying is a major factor governing the drug absorption rate for orally administered formulations. In monogastric species, where the small intestine is the principal site of absorption, the transit time through the stomach

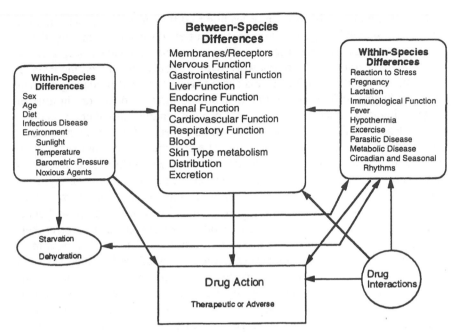

Figure 4 Between-species and within-species factors that affect the performance of drug formulations and thus drug actions within the animal.

often dictates the extent of drug degradation due to either hydrolysis or enzymatic degradation. In addition, drug solubility, drug pK_a, drug lipophilicity, and intestinal blood flow can all alter the rate of absorption of drugs, and thus drug action (35). In ruminants, the reticulorumen is the principal distinguishing feature of digestive physiology. This voluminous organ (100 to 225 L in cattle and 6 to 20 L in sheep and goats) creates an environment where the drug is immediately diluted and exposed to varying pHs, content viscosities, and bacterial enzymes. Rumen pHs (pH 4 to 7) and content viscosities vary according to diet, e.g., alfalfa hay versus concentrate, which in turn influence the turnover rate of the ruminal contents, drug residence time in the rumen, and ultimately affects the rate of drug absorption from the abomasum. The time for absorption may also be significantly reduced by esophageal groove closure allowing the drug to bypass the rumen and go directly into the abomasum. This occurs frequently with oral administration of anthelmintic drenches to ruminants. Between-species differences in gastrointestinal anatomy and function—e.g., volume and content composition, content viscosities, pHs, intestinal blood flow—clearly have the potential to affect both the rate and the extent of drug absorption from orally administered drug products.

Other between-species differences leading to different species-dependent drug actions are shown in Figure 4. Receptor affinities, distribution, and population densities may vary among species, leading to species dependent drug actions. This is demonstrated by the recent work with β agonists for increasing production efficiency and improving carcass characteristics of food-producing animals (36–39). Species differences in drug metabolism/biotransformation (in liver or in gastrointestinal tract mucosa), plasma protein binding, distribution, and excretion lead to different drug half-lives among species (35). In many cases a metabolite may have therapeutic activity. The anatomy and physiology of skin, the amount of hair, and the climate affect the topical performance and transdermal absorption of drugs among the different species. Cattle, sheep, pig, dog, and cat do not have the ability to sweat profusely—in contrast to the horse, which has highly developed and effective sweat glands (35). Cattle, sheep, and goats exude large quantities of lipoid material from sebaceous glands to protect their skin (35). Pigs have an extensive layer of keratin (35). These differences must be considered when designing topically applied formulations for either systemic or topical indications. The between-species differences in anatomical and physiological function leading to variations in drug systemic availability, distribution, metabolism, and excretion make it a prerequisite that potential therapeutic agents and formulations be studied in the target species early in product development.

Aside from physicochemical factors of the drug that may contribute to within-species variations of drug action (see Fig. 1), there are physiological, disease, and environmental factors that contribute to different within-species drug actions. These include age, sex, diet (e.g., protein levels in feed), pregnancy, dehydration, starvation, lactation (e.g., stage of lactation and response to bovine somatotropin), interaction with other drugs in vivo, various disease states, levels of stress (e.g., shipping fever), and various environmental conditions such as heat and barometric pressure. Most cattle feedlots change their feed rations and ionophore levels in the feed in anticipation of deep low-pressure systems to minimize stress on the animals and maintain production efficiency. A more complete discussion of the between-species and within-species differences that contribute to variations in drug response or therapeutic action can be found in the literature (35) and in Chapters 1 and 7 of this book.

In summary, marketing, packaging, and economic considerations aside, the selection of a veterinary drug dosage form condenses down to four product characteristics: the amount of drug to be dosed, the route of administration, and the frequency and duration of administration (Fig. 5). These product characteristics are governed by the physicochemical properties of the drug and its interaction with formulation excipients (see Fig. 1) so as to achieve appropriate pharmacokinetics of the drug, e.g., absorption, distribution, metabolism, storage and excretion, to answer the questions "Where in the body is the drug needed?," "How much?," and "For how long?" (see Table 7). "How much?"

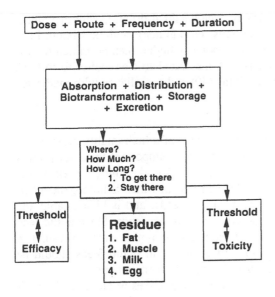

```
┌─────────────────────────────────────────────────┐
│  Dose  +  Route  +  Frequency  +  Duration        │
└─────────────────────────────────────────────────┘
              │        │        │
              ▼        ▼        ▼
┌─────────────────────────────────────────────────┐
│      Absorption  +  Distribution  +               │
│      Biotransformation  +  Storage                │
│                +  Excretion                       │
└─────────────────────────────────────────────────┘
```

Where?
How Much?
How Long?
　　　1. To get there
　　　2. Stay there

Threshold	Residue	Threshold
↑	1. Fat	↑
↓	2. Muscle	↓
Efficacy	3. Milk	Toxicity
	4. Egg	

Figure 5 Summary schematic and criteria for the selection of a veterinary dosage form.

is usually answered by the measurement of plasma drug concentrations, specific tissue levels of the drug if required, and volume of distribution of the drug. "How long?" is usually answered by the measurement of peak blood levels and the half-life of the drug in the plasma during the distribution and elimination phases after administration. Once the pharmacokinetics of a drug dosage form are understood and presumably controllable via the physicochemical properties of the drug or formulation, the dosage form can be optimized for therapeutic action balanced by minimization of tissue residues and toxicological effects.

IV.   ORAL DOSAGE FORMS

A.   Type A Medicated Articles (Premixes)

1.   Definition

Prior to the advent of the second generation regulations for medicated feed products as published on March 3, 1986, in the Federal Register (51 FR 7392), type A medicated Articles were commonly referred to as premixes both in the United States and worldwide. For purposes of the discussion in this section the terms "type A medicated articles" and "premixes" will be considered synonymous. A type A medicated article is intended solely for use in the manufacture of another type A medicated article or a type B or type C medicated feed. It should never be fed directly to animals without diluting to the approved use level in type C medicated feed. It consists of a new animal drug(s), with or without carrier (e.g., calcium carbonate, rice hulls, corn gluten), with or without inactive ingredients. The drug(s) is (are) at a potency higher than permitted in type B feed levels. The manufacture of a type A medicated article requires form FDA 356 approval under 21 CFR § 514.105(a) (40) and compliance with medicated premix current GMP regulations.

2. Dry Formulations

Listed in Table 8 are some of the characteristics of a high-quality premix formulation. Both physical and chemical attributes are important for a dry premix to handle and perform properly. The ultimate goal in designing a premix formulation is to render the active ingredient in a form which is homogeneous itself, mixes homogeneously into type B and C feeds, and is bioavailable and chemically stable both as premix and when mixed into feeds. Since liquid feed supplements and other liquid forms of feeds are becoming more prevalent in the animal feed industry, performance of premixes in liquid feeds should be included in the evaluation of candidate dry premix formulations.

Dry premixes range in composition from simple mixes of an active ingredient with a suitable diluent-to-complex matrices designed to impart desirable physical properties and/or chemical stability. The active ingredient may be in a purified form or may be incorporated into a premix formulation as a dried

Table 8   Characteristics of a High-Quality Premix Formulation

Bioavailable
Active constituent is physically and chemically stable
Active constituent is stable when incorporated into either dry or liquid feeds
Active constituent is stable during feed processing such as pelleting, extrusion
    and expanding
Appropriate concentration of active constituent for intended use
Homogeneous
Free-flowing
Dust-free
Elegant
Mixes homogeneously into feeds
Does not segregate from feeds during transport or conveying
Not electrostatic
Not hygroscopic
Economical manufacturing process
Safe to animals and feed mill operators
Efficacious
Does not carry over into subsequent batches of feeds
Environmentally friendly

fermentation broth. Many drugs and feed additives are produced by a fermentation process. Specific microorganisms under controlled conditions of pH, temperature, oxygen, and nutrients are used to produce pharmacologically active compounds. Frequently the concentration of the compound in the fermentation broth is low and chemical isolation of the pure material cannot be economically justified. Prior to making a decision to not isolate the active compound from the fermentation broth, it is essential that no mutagenic effects are observed when the dried fermentation broth is subjected to standard genetic toxological testing. When the active compound is not isolated, the mycelia and other components of the fermentation broth serve as part of the diluent for the final premix formulation. In addition to being a more economical method of manufacture, use of fermentation broth in the final premix formulation offers several other advantages:

1. The drug is evenly dispersed throughout the mycelia residue
2. Nutrients are retained in the premix formulation
3. Disposal of the fermentation residue is eliminated
4. The use of organic solvents commonly used in extraction procedures is eliminated

For the dried fermentation broth to function well as a carrier, it is important for the dried residue to be low in oil content and not hygroscopic; otherwise the premix will agglomerate and become lumpy during storage. To minimize the residual oil and lipids in the dried residue, the fermentation process must be controlled such that most of the fats and lipids added as nutrients are consumed by the time harvesting commences.

There are a number of factors to consider in the design of a premix formulation to ensure successful mixing into feeds. First, the density, particle size, and geometry of the premix particles should match as closely as possible those of the feed in which it will ultimately be mixed. Most animal feeds, except some aquaculture feeds and those containing high levels of roughage such as some cattle rations, range in density from 15 to 50 lb per cubic foot. The particle size of poultry rations is usually smaller than for swine rations, which is usually smaller than the particle size in cattle rations. The shape of the premix particles will many times determine how well it mixes into feed and how well it stays mixed during transport or transfer of feeds through augurs and pneumatic conveying systems. For example, drug-containing particles that are smooth and spherical in shape, particularly larger-size particles, will tend to roll to the outside or segregate from feed containing smaller-size particles.

Dust associated with drug premixes has become a major issue from not only a nuisance aspect but also from feedmill operator safety and economic loss. Laing (41) has shown that as little as a 0.8% loss of dust during feed manufacture represents an economic loss assuming that granulation increases the premix cost by $0.006 per gram. A variety of techniques are available to measure dustiness (42). However, one of the more commonly accepted international methods for evaluating dustiness in premixes is the Stauber-Heubach method (43). A schematic of the Stauber-Heubach device is shown in Figure 6. Dust is contained in the rotating drum of the device, which mechanically agitates the particles. Aerosolized particles are then entrained in an air stream flowing through the drum and after passing through a settling chamber the remaining dust is collected on a filter.

A schematic of the Carlson laboratory dust disperser (44), an alternative dustiness testing device, is shown in Figure 7. In the Carlson device, dust is dispensed from a hopper, falls by gravity through a tube, and is entrained in an air stream for sampling. When there is concern for human inhalation exposure the Carlson device is the preferred method since the results correlate well with the inhalable dustiness index (dust levels associated with particles ≤ 15 μm).

Two common methods are employed to achieve dust reduction in the manufacture of premix formulations. Perhaps the simplest method, which requires the least capital equipment, is the addition of an antidusting oil to the formulation. Hydrocarbon oils, such as light mineral oil or vegetable oils (e.g., soybean oil), have been used successfully. However, with the proposed ban on hydrocarbons in feedstuffs in Europe, alternative oils should be considered for those markets. Several factors should be considered in the selection of the appropriate antidusting oil and the volume quantity to be used in the formulation. A spreading oil is preferred over a nonspreading oil for most applications. A spreading oil will cover a greater surface area of dust particles at a lower concentration and will not tend to "ball" or agglomerate as much as a nonspreading oil. Care must be exercised in using nonspreading oils such as soybean oil be-

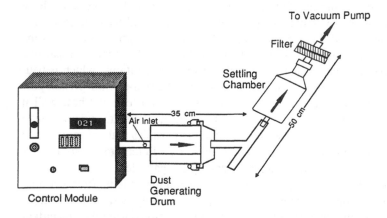

Figure 6   Schematic of the Stauber-Heubach device. (From Ref. 44.)

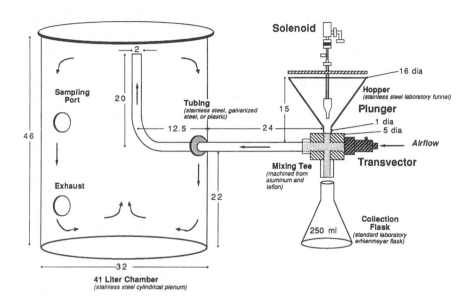

Figure 7   Schematic of the Laboratory Dust Dispenser (dimensions in cm). (From Ref. 44.)

cause of the tendency for the oil to smear and build up on mixer parts which eventually could result in mixing problems or cross-contamination from batch to batch. If the only purpose of the antidusting oil is to control dust during manufacture, then a volatile hydrocarbon oil is very effective. However, within several weeks to a few months the effectiveness of the oil in controlling dust when the premix is poured from the bag or container by the end user will be significantly diminished. If dust control is desired throughout the shelf life of the product, then a less volatile oil, such as light mineral oil or soybean oil, should be employed. When these oils are employed, the concentration of the oil in the formulation is more critical. Too low a concentration will be ineffective or will not distribute homogeneously throughout the batch. Too high a concentration will result in "caking" or lumpiness in the premix, particularly after storage for a period of time at the bottom of stacked materials.

The concentration of the antidusting oil is dependent on the level of dust control desired, the extent of agglomeration that can be tolerated, the extent of bleeding of the oil through the premix bag or container, the type of oil selected, cost, availability, etc. Concentrations in the range of 1% to 3% will frequently be effective for most premix formulations. Formulations should be evaluated not only at the time of manufacture but also throughout the shelf life of the products. It is not uncommon for a formulation to perform quite satisfactorily at the time of manufacture and later, during storage at various temperatures, to find that the oil may cause the ink print on the bag to "bleed" or smear and/ or the product becomes lumpy and does not flow well when used in microingredient machines or mixed into feed.

Granulation is another technique that can be employed to reduce premix dustiness. Granulation has come to be recognized as an important step in the production of quality feed additives and premixes. Granulated premixes offer several advantages as compared to a powdered form of the same drug. Powders, whether dried fermentation broths or highly purified pharmaceuticals, do not usually flow well and may bridge in microingredient machines commonly used in automated feedmills and feedlots. Granulation increases flowability, reduces dust, and lowers the risk of carryover contamination from one batch of feed to the next. Some premixes in powder form tend to compact and adhere to storage bins. Granular premixes are less affected by humidity and high temperatures and many times are more stable, if formulated properly, because there is less surface area of the drug exposed to the environment.

A variety of high-volume, low-cost manufacturing methods can be used to granulate both purified compounds and dried fermentation broths containing active compounds. Granulation usually involves either compaction, employing binding agents and/or absorbing material unto a presized particle or granule. Examples of each of these granulation techniques will be discussed briefly below.

Granulation by compaction can be achieved by using devices such as the pellet mill, extruders, and roller compactors. When using these devices, it is usually necessary to perform a size reduction step in which larger particles are reduced in size and a sieve sizing operation in which the dust and smaller particles are recycled through the compaction device. The pellet mill is particularly effective as a granulation technique when ingredients in the formulation contain natural binders such as starches. Heat and steam are used in the pelleting process to activate the natural binders or added binders in the manufacture of durable, high-quality granules. For a more thorough discussion of the pelleting process, the reader is referred to the American Feed Manufacturers Association's *Pellet Mill Operators Manual* (45).

Extruders perform a similar function to that of a pellet mill in the manufacture of granular premixes except that equipment and processing conditions can be varied quite substantially to produce granules of various densities, sizes, shapes and texture. Frequently, a doughlike material is forced through a die and then dried to an acceptable moisture level to ensure stability during storage. Depending on the design, extruders can be used to granulate materials that are considerably higher in moisture than can be accommodated by a pellet mill. Cooking extruders can be used to manufacture premixes which either float or sink for the aquaculture industry. To prepare a floating formulation, the drug

can be applied to a floating matrix such as popped popcorn or can be manufactured with a cooker extruder which has a barrel and screw configuration
which will permit the addition of steam and water at a rate of up to 8% of the
dry feed rate. Expansion of the extrudate after it passes through the die should
result in a bulk density of 320 to 400 g/L. Floatability of a dry expanded product
can be improved by drying at elevated temperatures to a low residual moisture
content of less than 8%. In the production of sinking aquatic premixes, the extruder is modified to yield a product with a bulk density of 450 to 550 g/L by
using less steam in a cooking extruder or by using a forming extruder. To
achieve uniform granules, extruded material may be subsequently introduced
into a marumerizer. The marumerizer is capable of effectively breaking long
stringy extrudate into smaller pieces which are then rolled into elliptical-shaped
granules by the spinning action of a rotating disk. Figures 8 and 9 show a product that has been spheronized by marumerization for 30 sec and several minutes, respectively. Figure 10 shows a cross-sectional view of a marumerizer.
Figure 11 shows how a continuous extrusion and batch marumerization process
fits together in a production mode. Significant improvements have been made
in extrusion equipment in recent years such as dome dies. *Pharmaceutical Pelletization Technology* (46, 47) contains a good overview of the extrusion processes and its application to premix manufacture. Premixes such as Tylan (tylosin phosphate, Elanco) and Apralan (apramycin sulfate, Elanco) are
manufactured by the extrusion/marumerization process.

Another compaction device that finds utility in the manufacture of premixes
is the roller compactor such as a Chilsonator. Roller compactors utilize two

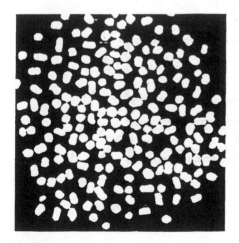

Figure 8 Product spheronized by marumerization for 30 sec. (Courtesy of LCI Corporation, Charlotte, NC.)

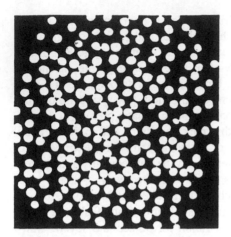

Figure 9 Product spheronized by marumerization for several minutes. (Courtesy of LCI Corporation, Charlotte, NC.)

rollers that revolve toward each other, as illustrated in Figure 12. Material to be compacted is fed between the rollers maintained at a constant pressure by a hydraulic ram. The rollers may be smooth or serrated, and may contain pockets or various other designs including liquid cooled rollers to dissipate heat. The feeding mechanism may consist of both a horizontal and a vertical screw. The

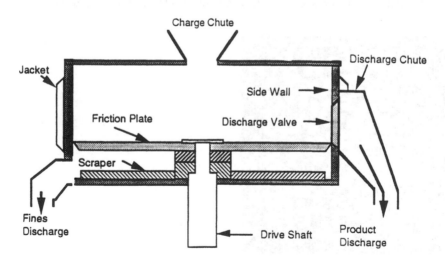

Figure 10 Cross-sectional view of a marumerizer-type spheronizer. (Courtesy of LCI Corporation, Charlotte, NC.)

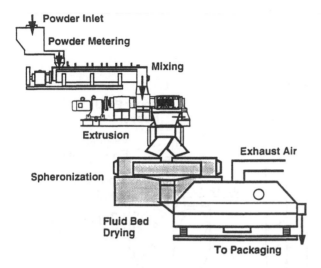

Figure 11 Continuous extrusion and batch marumerization process. (Courtesy of LCI Corporation, Charlotte, NC.)

horizontal screw maintains a uniform flow of material from the hopper to the vertical screw. The vertical screw delivers the powder to the compaction rolls and serves to deaerate the powder and maintain a constant flow onto the compaction rolls. Granulation is achieved by starting with an appropriate formulation which may or may not contain added binders and then regulating the compaction force. The compaction force is regulated by the force of the hydraulic ram, the rotational speed of the rolls, and the feed rate. Once the material is compacted, it must be cracked to size and the fines recycled for further compaction. Figure 13 shows a complete roller compactor granulation system including a Chilsonator, granulator, screener, and recycle system. Examples of formulations that can be granulated by roller compaction are given in U.S. Patent 4,447,421 (48). Not all materials granulate well by this method.

Since one of the major purposes of the granulation process is to reduce or eliminate fines and dustiness, it may be necessary to add a binder to the formulation to enhance cohesion among the particles within a premix granule. Premix ingredients vary widely in cohesive properties. Furthermore, the form in which ingredients are added to the formulation may substantially alter the cohesive nature of the formulation. For example, drugs and/or binders added as dry ingredients may provide different binding properties from addition of the same materials as liquids. It is not always predictable as to which form or which binder will yield the best-quality granules. Listed in Table 9 are examples of binders that have been used in the manufacture of premixes and animal feeds.

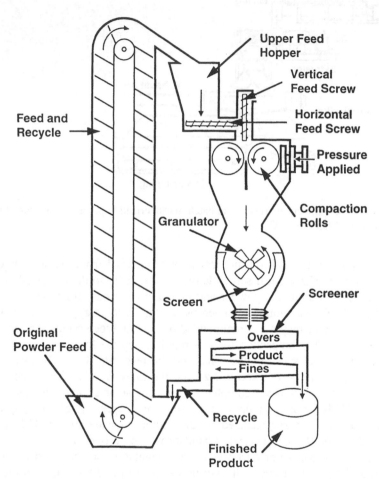

Figure 12 Schematic of a dry compaction operation. Dry material is compacted under pressure between two rolls, then granulated and sieved to obtain the desired particle size. (Courtesy of Fitzpatrick Company, Elmhurst, IL.)

Figure 13   A small Chilsonator roller compactor in a production unit. (Courtesy of Fitzpatrick Company, Elmhurst, IL.)

Care must be taken in the selection of a binder so as not to reduce the bioavailability of the drug(s) either due to the hardness of the granules or due to binding of the drug. Binders such as calcium magnesium montmorillonite, bentonite (see Table 10), and magnesium mica have been shown to compromise efficacy and/or interfere with the analysis of certain drugs (49–51). The concentration of the binder in the formulation should be at the lowest level that provides sufficient cohesiveness to prevent attrition of granules during mechanical processing and handling.

Table 9 Typical Binders Used in Premixes
and Animal Feeds

Pregelatinized starches
Wheat gluten
Vegetable gums
Sodium alginate
Carrageenan
Modified milo
Collagen
Calcium magnesium montmorillonite
Urea formaldehyde resin calcium sulfate
Sodium bentonite
Distiller dried grain sugars
Hydrogenated tallow glycerides
Gelatin and gelatin byproducts
Magnesium mica
Lignin sulfonate
Dolime
Attapulgite
Maltodextrins

In addition to compaction methods and the use of binders, a third granulation method is to apply a liquid or powder form of a drug to presized particles. A good cost-effective absorbent material to use as a carrier for liquid drugs is presized ground corncobs. The woody ring of corncobs is a very hard, dense,

Table 10 Drugs with
Reduced Efficacy and/or
Assay Interference in the
Presence of Bentonite

Amprolium
Buquinolate
Carbadox
Decoquinate
Morantel tartrate
Nequinate
Oleandomycin
Pyrantel tartrate
Robenidine hydrochloride
Thiabendazole
Tilmicosin
Tylosin

absorbent material which when ground and sized through a series of aspirators and screens results in a uniform, dust-free, flowable carrier. Corncobs are environmentally friendly in that they are biodegradable, chemically free, inert, and derived from a renewable resource. The primary sources of cobs are ear corn and hybrid seed corn farming. Presized ground corncobs are commercially available worldwide from companies such as Andersons (Maumee, Ohio) in the U.S. and Eurema in France. For feed additive premixes, usually the most desirable particle size fraction is 250 to 500 μm unless the addition of the drug substantially increases the size of the corncob particles. Then it may be desirable to use a smaller particle size fraction.

Ground corncobs are capable of absorbing up to 40% by weight of water and still flow sufficiently well to allow processing and drying. When liquids containing drugs are applied to corncobs, the drug is absorbed into the pores of cob, which effectively reduces the surface area of the drug exposed to the environment. For this reason it is not uncommon for corncob formulations to be more stable than powder formulations of the same drug, particularly for antibiotics with limited shelf life. When sizing process equipment, it is important to size the wet mixer at least 1.5 times the volume required for the dry cobs because of the expansion that occurs during the addition of liquids to the dry cobs. Although cobs may appear dry and handle well, it is essential that the moisture content be reduced sufficiently such that water activity (discussed later) is below that which will support mold or mildew growth. However, care must be exercised to avoid overdrying because of the inherent static nature of overdried cobs. An alternative absorbent material that has also been used successfully to manufacture dry premixes from liquid drugs is verxite. Verxite is a food-grade-quality vermiculite that has been expanded. Expanded vermiculite is inherently very dusty. However, the addition of liquid drugs, particularly those liquids that are somewhat viscous or sticky, will frequently result in a satisfactorily low dust formulation.

Drugs in the dry state can also be applied to presized ground corncobs or other suitable, less absorbing presized materials such as limestone, clays, peanut hulls, sugar beads, urea prills, cereal grains, and byproducts of the cereal grain industry. For a drug in the dry state to adhere to presized granular particles, it is usually necessary to incorporate a sticking agent, wax, or coating material into the manufacturing process. A very effective method of manufacturing an elegant, free-flowing, granular premix formulation from a dry drug substance is to add a wax such as the polyethylene glycols, hydrogenated tallow, triglycerides, fatty acids, fatty alcohols, or stearates to the formulation. The mixture is then heated to above the melting point of the wax to allow the wax to thoroughly coat the presized granules. Using a mixer equipped with a chopper such as the Littleford Day mixer shown in Figure 14, the granulation is vigorously mixed as it cools. The drug becomes incorporated into the melted

Figure 14 Littleford Day Mixer equipped with a heating and cooling jacket, plow-shaped mixing elements, and high-speed choppers (not shown). (Courtesy of Littleford Day, Inc., Florence, KY.)

wax which in turns coats the presized carrier particles. The resulting premix formulation is a homogeneous granular product. By varying the processing conditions and selecting the proper wax or combination of waxes, unmanageable powders can be converted into very elegant premix formulation containing minimal dust. Powders, which inherently have a static charge, can frequently be converted into granules with very little static charge using this method.

A variation of the granulation method using wax and presized carrier particles is a patented process developed by Washington University Technology Associates (52). The drug powder and wax are applied to a rotating disk. The disk is heated to melt the wax. As the powder moves across the disk it becomes incorporated into the wax, which forms droplets as the material moves off of the rotating disk by centrifugal force. The droplets fall through a cooling tower and are collected as granules. Granule size can be regulated by varying the processing conditions such as shape of the rotating disk, temperature, rotational speed, choice of wax, feed rate, etc. The rotating disk method offers several advantages in that it is a continuous process, has a very short processing time (< 1 min), can accommodate high-viscosity materials, can achieve high drug concentrations in the granules, and is a high-volume, inexpensive manufacturing method. Because of the short processing time, it is a suitable method for labile materials.

Premix manufacturing cost may be reduced by formulating high-potency granules which are then blended with a suitable diluent to produce the final premix formulation. The quantity of material that must be granulated is thus reduced. Frequently, binders and other granulation excipients are more expensive than diluents which can be added after granulation. In addition, because of the lower volume of material that must be granulated, capital equipment costs are lower as are processing time and labor costs. Because of variations sometimes observed in the potency of a granulated drug, another added advantage of using the high-potency granule method is that the final product potency can be accurately achieved by adjusting the quantity of diluent added to the formulation.

a. Diluents. Depending on cost and availability, a wide array of diluents have been used in the manufacture of premixes (see Table 11). Diluents may be included in a premix formulation as means of standardizing potency, increasing the bulk of material that is mixed into a medicated feed ration, reducing granulation costs, reducing shipping costs of the active component to international markets, adding flexibility for local country manufacture, enhancing mixibility in final feeds, and complying with regulations. As was discussed previously, potency may vary from lot to lot of a fermentation product, granulation, or other form of a drug substance. A sufficient quantity of diluent should be included in a premix formulation such that variations in the quantity of diluent needed to standardize potency does not substantially alter the physical appearance, handling properties, or chemical stability of the formulation.

Determination of the quantity of diluent needed in the final formulation may also be influenced by the need to provide sufficient mass or bulk of material to permit adequate mixing into supplements and final feeds. As drugs become more and more potent, smaller and smaller quantities of the active drug are required per ton of feed. As a general rule, it is desirable to design a formulation such that at the lowest use level at least 4 oz. or 100 g is added to each ton of feed. It is preferable that at least 0.5 to 1 lb (250 to 500 g) of premix be added to a ton of feed to enhance mixibility of the drug in feed. Certain countries such as

Table 11 Premix Diluents

Ground rice hulls
Soybean mill run
Solvent extracted soybean feed
Sized corn germ meal
Distillers grains with solubles
Corn gluten feed
Ground grains
Limestone
Sodium sulfate
Toasted soy flour
Wheat red dog
Kaolin
Peanut hulls
Ground corn cobs
Wheat middlings
Vermiculite (Verxite)
Calcium montmorillonite and other clays

France have local regulations that a minimum of 5 kg of a premix formulation must be added to a ton of feed. Thus an adequate quantity of a diluent must be added to the formulation to reduce the drug concentration to an acceptable level for use in the feed industry and/or comply with local regulations. Sometimes it makes good economic sense to ship highly concentrated forms of a drug to international markets and complete the manufacture of the premix by adding locally available diluents. Not only are shipping costs reduced but locally available diluents may frequently be less expensive.

Cost, availability, consistency, water activity, and chemical stability of the premix formulations are all important in the selection of a suitable diluent. However, one of the most important criteria in any premix formulation is homogeneity and absence of segregation during transport and handling. Round, dense drug granules and light, fluffy diluents do not mix well and may segregate during transport or handling. The density, size, and shape of the diluent should match as closely as possible that of the drug substance whether granular or powder.

b. Water Activity.   To prevent the growth of microorganisms such as bacteria, yeasts, and molds in a premix formulation, it is essential that the water activity (Aw) of the formulation be below the limit where growth is possible. Typical growth limits of various microorganisms are listed in Table 12 (53). Traditionally, moisture content expressed as percent water or percent "loss on drying" has been used as an indicator of the quantity of water that could be

Table 12 Typical Growth Limits of Various Microorganisms as a Function of Water Activity

Microorganism	Water activity (Aw)
Most bacteria	0.91–0.95
Staphylococcus aureus	0.86 salt
Staphylococcus aureus	0.89 glycerol
Most yeasts	0.88
Most mildew	0.80
Halophile bacteria (high salt tolerance)	0.75
Xerophile mildew (low water tolerance)	0.65
Osmophile yeasts (high sugar tolerance)	0.61
No growth of any organism	<0.60

tolerated in a premix formulation. However, moisture content is not always a good indicator of "free water," which is responsible for mold growth. Aw expresses the active part of moisture content or free water as opposed to the total moisture content, which includes bound water.

Water activity is defined as the ratio of vapor pressure of water in the product to the vapor pressure of pure water at the same temperature. It is numerically equal to 1/100th of relative humidity generated by the product in a closed system. Instruments for measuring Aw are commercially available through companies such as Rotronic Instrument Corp. (160 E. Main St., Huntington, NY 11743) and Decagon Devices, Inc. (P.O. Box 835, Pullman, WA 99163). Figure 15 shows a dual-station Rotronic water activity system. The sample chambers are jacketed to provide temperature control.

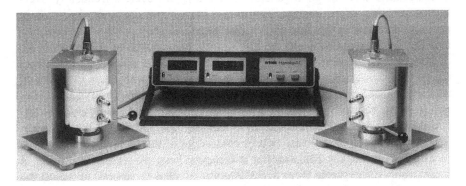

Figure 15 Rotronic Model D2102 Water Activity System. (Courtesy of Rotronic Instrument Corp., Huntington, NY.)

The U.S. Food and Drug Administration (54) has adopted the concept of Aw for establishing moisture limits beyond which certain types of food are considered susceptible to microbial growth. Although moisture specifications for most premix formulations are based on total moisture or total volatiles, the concept of Aw provides a rational basis for determining the maximum allowable moisture level in a formulation.

   c. Mixing of the Drug Premix.   At the time of manufacture it is essential that the drug be mixed homogeneously throughout the lot. If the entire premix is granulated, segregation during transport or handling is not an issue; nor is homogeneity with the premix itself if the materials were mixed satisfactorily prior to granulation. However, when a diluent or other additive is included in a formulation containing a potent drug, proper mixing becomes essential. Mixing trials should be performed in the production size mixer to ensure adequate characterization of the formulation and mixer. Although there are many different mixer designs currently used to manufacture premixes, not all mixers perform equally satisfactorily (55,56). Typically, a minimum of 10 samples are obtained throughout a premix lot and assayed for the drug. The greater the number of samples per lot, the more reliable the evaluation. The precision of the analytical method can influence the results. Replicate assays on each sample can reduce the variability due to the assay method. Mixing times and the minimum and maximum batch sizes should be evaluated.

   The variability of drug distribution among the samples is determined by calculating the standard deviation (SD) using common statistical procedures. The coefficient of variation (CV or COV) or the relative standard deviation (RSD) is the ratio of the SD to the average or mean of the samples multiplied by 100. The COV for premix formulations should be less than 5% and must be within 5% of theory to make the manufacturing process and assay method acceptable. These criteria serve as good general guidelines. However, a frequently overlooked problem in the homogeneity evaluation of a mixer is the deviation in potency observed in the first and/or last bag(s) of premix from a production lot. The first several bags and the last several bags should be subjected to additional scrutiny in terms of performing additional drug assays and examining the physical appearance and texture of the premix. If there is any evidence of additional fines, high or low potency, or other peculiarities observed in material from the beginning or end of a bagging operation, then additional investigation is warranted. If it is determined that the first and/or last bag(s) is unacceptable due to potency variation or other reason, then a procedure needs to be established whereby that material is removed from the lot and discarded. It is usually not acceptable from a GMP point of view to rework or blend the unacceptable material into future lots. At the very least the procedure would have to be approved by the U.S. Center for Veterinary Medicine or other regu-

latory body governing the manufacturing operation. Additional information regarding mixer selection and blending trials may be found under the section on
type B feeds.

3. Liquid Concentrates

A limited number of type A medicated articles are available as liquid premixes
or liquid concentrate formulations. Generally, liquid type A medicated articles
are designed for mixing only into liquid feed supplements whereas dry type A
medicated articles can frequently be mixed into either liquid or dry supplements.
However, a variety of vitamins, minerals, and amino acids are available as liquids and may be applied to dry feeds through a dribble bar in the mixer or by
spraying onto the feed while mixing.

  a. Solutions. Liquid concentrates are usually either solution or suspension
formulations. The simplest and most economical liquid formulation is an aqueous solution. When considering a liquid formulation, stability and solubility are
the primary factors to determine whether a liquid formulation is feasible and
whether a solution or a suspension formulation is preferable. When a drug substance exhibits limited solubility, it may be possible to solubilize the drug by
one or more techniques. It is not uncommon for drugs to be more soluble when
two or more solvents are mixed together than when solubility is tested independently in each solvent. When selecting solvents and formulation additives
for liquid concentrates, it is important to select only those ingredients that are
GRAS (Generally Recognized As Safe); otherwise it will be necessary to conduct a full battery of toxological tests to demonstrate the safety of the solvent.
A list of GRAS items for food additives permitted in feed and drinking water
of animals may be found under 21 CFR § 573 and 21 CFR § 582. Yalkowsky
(57) has demonstrated that the solubility of certain materials may be increased
as much as a thousandfold or more by this technique.

  Surface-active agents (surfactants) may be used to solubilize drugs through
micellar solubilization. Materials such as the cyclodextrins, polyvinylpyrroli-
done, and urea may form a complex with a drug, resulting in enhanced solubility.

  Another technique to enhance solubility is to convert the insoluble acid or
base form of the drug to a salt form or convert an insoluble salt form to different salt, ester or pro-drug. As an initial screen, the simplest technique is to
determine the solubility of the drug over a wide pH range. If the drug is an
acid, add a mineral base such as sodium hydroxide to a slurry of the drug in
water. The insoluble drug slurry may readily dissolve, as a base is added with
stirring. If the drug is a base, then a mineral acid such as sulfuric, hydrochloric, phosphoric, etc., would be added to the drug slurry to form a soluble salt.
The reader is referred to Yakowsky's treatise *Techniques of Solubilization of*

Drugs (57) for a more exhaustive discussion on the topic of enhancing drug solubility.

Even though it is possible to solubilize otherwise insoluble drugs using one or more of the techniques described above, it is also incumbent upon the formulator to make certain that the solution performs satisfactorily when mixed into liquid feed supplements or sprayed onto dry feeds. For example, if a cosolvent or solvent other than water is used in the liquid concentrate formulation, the drug may precipitate out upon addition to a liquid feed supplement. This may be acceptable if the precipitated drug is bioavailable, disperses uniformly throughout the supplement, and does not settle out during storage and/or during use. However, some drugs form large crystals or may precipitate and adhere to agitator shafts, tank walls, etc. If a precipitate occurs upon addition to liquid feed supplements, it may be advantageous to formulate the product as a suspension, in which the drug particle size is well controlled.

b. Suspensions. A suspension type of formulation may be a viable alternative when lack of drug solubility precludes the manufacture of a solution of any kind or when the drug degrades in solution. Liquid concentrates formulated as suspensions contain finely divided drug particles distributed somewhat uniformly throughout a vehicle in which the drug exhibits a minimum degree of solubility. Factors that must be considered in the successful formulation of a suspension formulation include particle size of the insoluble ingredients, viscosity, dispersibility into liquid feed supplements, solubility of the drug(s) in the vehicle system, chemical stability, bioavailability, permanency of the formulation, and resuspensibility of ingredients if settling occurs.

At any given viscosity, particle size of the insoluble ingredients is probably the most important parameter to consider in developing a quality suspension with an acceptable sedimentation rate. The effect of reducing particle size on the sedimentation rate can be calculated using Stokes' Law:

$$v = \frac{d^2(\rho_1 - \rho_2)g}{18\eta}$$

where v = rate of settling (cm/sec), d = diameter of the insoluble particles (cm), ρ_1 = density of the particle (g/cm^3), ρ_2 = density of the vehicle system (g/cm^3), g = acceleration due to gravity (980.7 cm/sec^2), and η = viscosity of the vehicle system in poise (g/cm/sec).

From the equation it is apparent that the rate of settling of a suspended particle is greater for larger particles than it is for smaller particles, all other factors remaining constant. The preparation of a physically stable homogeneous suspension will usually require a solid with a particle size in the range of 5 to 10 μm. To achieve a particle in the 5- to 10-μm range it is usually necessary to micronize the drug by air-milling (sometimes referred to as jet-milling, or fluid energy grinding) or by a suitable wet-milling procedure. Both small laboratory models and large production models of air mills are commercially available. Inside an air mill the particles are accelerated to high velocities and collide with one another, resulting in fragmentation and thus particle size reduction.

The formulator must be aware that even though a small particle size of a relatively insoluble drug is used in a suspension, it is not uncommon to experience particle size growth over time, particularly when temperatures cycle up and down. Particle size growth may be due to two unrelated factors. First, particles may agglomerate into larger crystals or into masses which rapidly settle or form a cake which is difficult to resuspend. Methods to overcome this phenomenon will be discussed later. Secondly, a very slight solubility of the drug in the vehicle system may set up an equilibrium between drug in suspension and drug in solution. As some of the drug passes into and out of solution, crystal growth may occur in which relatively large particles may start to form in the suspension, thus destroying the physical characteristics of the suspension. Such a phenomenon has been known to occur at drug solubilities as low as 1% or less. In addition, if the drug is not stable in solution, slight solubilities may allow an otherwise stable formulation to degrade because, as the solubilized drug degrades, additional drug will dissolve to reestablish the equilibrium. Ultimately the level of degraded drug may exceed acceptable limits.

There are several ways to decrease the solubility of drugs. In some cases it may be possible to synthesize other derivatives or salts that are less soluble. When solubility is pH-dependent, a physically stable suspension may be simply a matter of adjusting the pH to the range in which the drug is least soluble. The solubility of a material that is slightly soluble in water or vehicle system may be decreased even further by the addition of a common ion, inorganic salts such as sodium chloride, or substances such as sucrose or sorbitol. However, it should be noted that added salts to a formulation sometimes increase the difficulty of obtaining a satisfactory suspension.

A properly formulated liquid concentrate suspension should not settle during long periods of storage; i.e., during the shelf life of the product or if settling does occur, the settled material should readily redisperse upon gently agitating or shaking. By adding a viscosity building agent and/or formulating a thixotropic suspension, it is possible to eliminate settling entirely throughout the shelf life of the product. A common viscosity building agent in liquid concentrates and also in liquid feed supplements is discussed later is xanthan gum. The concentration of xanthan gum should not exceed 1% or 2%; otherwise the material will be difficult to pour particularly under cold conditions and may not mix satisfactorily into liquid and dry supplements or feeds.

Attapulgite clay may be used to formulate thixotropic suspensions. A thixotropic suspension flows freely when moderately agitated but sets up like a gel or becomes very viscous when allowed to remain quiescent. The formulation should be designed such that the gelling or increased viscosity occurs within a few minutes to 1 h after agitation. The increased viscosity maintains the drug positionally stable in the suspension and prevents agglomeration and caking.

In addition to employing viscosity building agents and thixotropic agents to reduce settling and caking of a suspension, certain other ingredients may be used to create a flocculated suspension which resists settling and caking. A flocculated suspension is a suspension in which there is loose aggregation of the particles and in which there exists a lattice structure which resists settling. The most common ingredients in flocculated suspensions are the clays. However, the use of nonionic polyether surfactants and/or ionic surface active agents can induce the flocculation of particles. Adjusting the pH and adding electrolytes are other methods of inducing flocculation. It should be noted that small changes in surfactant concentrations, pH, electrolytes, etc., may produce dramatic changes in flocculation. Thus, simply testing high and low concentrations of surfactants, for example, may not be sufficient to identify suitable flocculating agents. Several intermediate concentrations should be tested as should combinations of surfactants.

Visual observation of both freshly prepared formulations and formulations aged at various temperatures ranging from freezing to 40°C is usually a satisfactory method for evaluating suspensions for sedimentation and caking. A differential manometer (58) may also be used to compare the pressure of a suspension near the top and bottom of a container. For additional information regarding suspension formulations, the reader is referred to the sixth edition of *Pharmaceutical Dosage Forms and Drug Delivery Systems* by Ansel et al. (59).

B. Type B Medicated Feeds

1. Definition

The second-generation regulations for medicated feed products established specific criteria which must be met in order for a medicated feed to be classified as type B. It is intermediate in concentration between a type A medicated article described in the previous section and a type C medicated feed, which is a complete feed to be fed directly to animals. The type B formulation contains a substantial quantity of nutrients including vitamins and/or minerals and/or other nutritional ingredients in an amount not less than 25% of the weight of the formulation. The maximum concentration of a drug in a type B feed is 200 times the highest continuous use (14 days or more) level for category I drugs and 100 times the highest continuous use level for category II drugs or, if the drug is not approved for continuous use, it means the highest level used for disease prevention or control. If the drug is approved for multiple species at different use levels, the highest approved level of use would govern under this definition. The manufacture of a type B medicated feed from a category II, type A medicated article requires an application approved under 21 CFR § 514.105(b) (40). The primary purpose of type B feed is to provide a drug formulation to the animal feed industry which is not considered a human risk drug source. A human risk drug source such as a type A medicated article requires a FDA form 1900 for it use in the U.S. Feedmills that choose not to use human risk drug sources are subject to a relaxed set of current GMP regulations and are exempt from mandatory FDA inspections, establishment registration, and drug assay requirements. The main disadvantage of type B feeds as a source of animal drugs is that the cost per gram of drug activity is usually greater when supplied as a type B feed as opposed to a type A medicated article.

2. Composition

The composition of type B medicated feeds varies from the least expensive diluents to conventional concentrates designed to provide protein, vitamins, minerals, amino acids, or other nutritive ingredients. In the case where the goal is simply to dilute a type A medicated article to a level in which the user is not required to have an approved FDA form 1900 on file, diluents such as limestone, ground rice hulls, soybean hulls, peanut hulls, and other diluents listed in Table 4 are commonly selected. However, when selecting inexpensive diluents the formulator must make certain to include 25% nutritional ingredients. Rice hulls, soybean hulls, and the like do not qualify as nutritional ingredients, but limestone does qualify because it is a calcium source. When using limestone in a type B formulation, care must be exercised not to exceed a quantity that will cause a calcium/phosphorus imbalance in the final feed ration. Distillers' grains with solubles or other nutritive ingredients should be substituted in situations in which high limestone levels are likely to cause imbalances in rations.

Type B medicated feeds may be a simple mix of a drug with suitable diluents, in which case the main concerns are homogeneity, segregation during transport and stability. However, all the factors discussed under premixes or type A medicated articles should be considered. Thus, dustiness, flow properties, water activity, ingredient compatibility, etc., are all important to ensure an acceptable product that will perform satisfactorily and not deteriorate during the expected shelf life. Many type B feeds are pelleted, in which case the composition may have to be varied for good pellet quality since rice hulls, limestone, etc., do not pellet well.

3. Mixing Dry Feeds

Satisfactory mixing plays a vital role in the manufacture of quality premixes and medicated feeds. As shown in Figure 16, "coefficient of variation for various types of mixers," not all mixers are created equal in terms of their ability and time required to mix premixes and feeds homogeneously. Obviously, shorter mixing times allow for more batches to be manufactured in the same time pe-

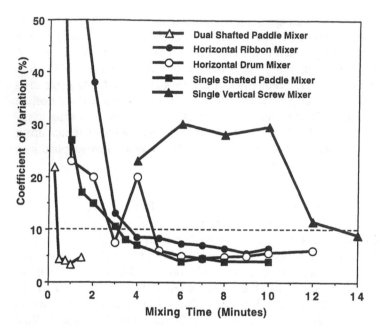

Figure 16 Coefficient of variation for various types of mixers. (Adapted from Refs. 56,60.)

riod. However, it should be equally obvious that mixing times should be based on data that demonstrate that a particular mixer is capable of producing a homogeneous mix of the premixes and feeds being manufactured. Periodic validations should be performed to verify that mixing times are adequate and the mixer is performing satisfactorily. Worn mixers frequently do not mix as efficiently as new mixers, resulting in the need to lengthen the mixing time and/or performing maintenance on the mixer. For example, as shown in Figure 17, replacing a worn ribbon of a horizontal ribbon mixer substantially reduces the coefficient of variation. Some mixers have shims which can be removed periodically to reduce the tolerance between a worn ribbon and the bottom of the mixer, thus improving mixing efficiency.

When conducting mixing trials, it is important to establish that the mixer is functioning properly; the number of revolutions per minute is adequate; worn ribbons, paddles and the like are replaced; and the mixer is not overfilled. Frequently, mixer capacities are thought of in terms of tons rather than cubic feet or cubic meters. When mixing materials of lower density, the number of pounds per batch may have to be reduced so as not to exceed the mixing capacity of the mixer. For example, when mixing in a horizontal ribbon blender, it is es-

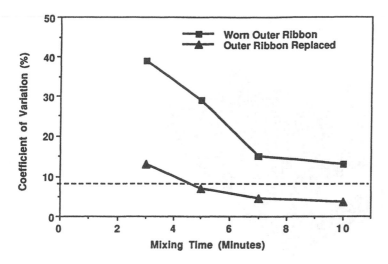

Figure 17 Effect of ribbon wear on mixing time in a 2-ton horizontal ribbon mixer. (Adapted from Ref. 56.)

sential that the outside ribbon extend above the surface of the material being mixed. Otherwise, inadequate mixing will occur.

Mixers are available in many sizes, shapes, designs, and configurations. When selecting a mixer to use in the manufacture of premixes and/or medicated feeds, cost, mixing efficiency, and capacity are all important issues. The dual-shafted paddle mixer illustrated in Figures 18 and 19 is perhaps one of the most efficient mixers in terms of producing homogeneous mixes in the shortest amount of time. Because of their mixing efficiency and short cycle time, it may frequently be possible to purchase a smaller-size mixer and still manufacture the required tonnage. The dual shafted paddle mixers have been shown (60) to be extremely tolerant to both underloading and overloading in terms of their ability to produce homogeneous mixes in less than 1 min of actual mixing time. The mixer can also be equipped with a flow distortion bar (FDB) which rotates to distort the fast flowing top material layer into a near horizontal flow pattern (see Fig. 19). This sheeting action exposes maximum material surface area for liquid addition. Dual-shafted paddle mixers are now available worldwide. See Table 13 for a sample listing of vendors.

4. Liquid Feed Supplements

Liquid feed supplements (LFS) have become an increasingly important means of supplementing the nutrition of cattle in the U.S. The American Feed Manufacturers Association recognizes the importance of LFS in the cattle industry by annually conducting a liquid feed symposium. In certain areas 50% or more

Figure 18 Dual-shafted paddle mixer. (Courtesy of Halvor Forberg AS, Norway.)

of the feed lot cattle are fed LFS. As a result, LFS have become a convenient method of delivering drugs to cattle provided the drug is stable in the LFS. The composition of LFS varies widely depending on locally available byproducts and relative cost of other ingredients. Consequently, great differences exist in nutritional content, specific gravity, dissolved solids content, pH, viscosity, and thixotropy. Frequently, byproducts such as distillers' solubles, fish solubles, corn steep liquor, fermentation liquors, lignin sulfonate, surfactants, emulsifiers, and suspending agents such attapulgite clay are incorporated into LFS. Some of these ingredients present special challenges in formulating LFS with satisfactory drug stability. Both the physical and chemical properties of the LFS and the drug must be considered carefully to ensure that the medicated product is chemically stable and also positionally stable if the product is not agitated prior to use.

Two basic types of LFS are marketed to the cattle industry—conventional formulation and a thixotropic formulation; i.e., the formulation becomes quite fluid when agitated but becomes viscous or forms a soft gel within a few minutes to an hour after agitation is discontinued. LFS containing a drug(s) must be agitated immediately prior to use unless it has been demonstrated that the drug is positionally stable throughout the batch and does not settle to the bottom or float to the surface during the expected storage period. If the drug is soluble in the continuous phase of the LFS, then it would be expected to be positionally stable. However, insoluble drugs and drugs formulated with insoluble carriers, such as mycelial products, may become stratified during stor-

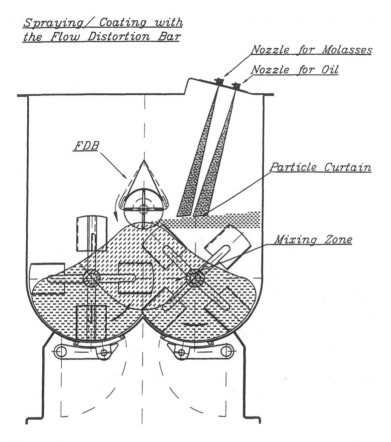

Spraying/ Coating with the Flow Distortion Bar

Nozzle for Molasses

Nozzle for Oil

FDB

Particle Curtain

Mixing Zone

Figure 19 Schematic of the mixing action of a dual-shafted paddle mixer equipped with a flow distortion bar and nozzles for the addition of molasses, oil, or other liquid ingredients. (Courtesy of Halvor Forberg AS, Norway.)

age. Stokes' Law, discussed under Liquid Concentrates suspension formulations, also governs the rate of settling that occurs in LFS containing insoluble drugs. Thus, the smaller the diameter of the drug particles and the more viscous the formulation, the slower the rate of settling.

Xanthan gum, certain carbohydrates, pectin, and other viscosity-building agents may be used to increase the viscosity of conventional LFS formulations such that the settling is not significant during the expected shelf life of the product. Agents such as attapulgite clay can be used to manufacture thixotropic formulations which have good flow properties when being pumped or sheared but will have sufficient viscosity when not being agitated to suspend limestone, minerals, drugs, etc., which would otherwise settle. To establish that a drug is

Table 13 Vendors of Dual Shafted Paddle Mixers

Name	Address	Country
Chemical Plant & Engineering	P.O. Box 160, Broadmeadows, Vic 3047	Australia
William Tatham Ltd.	Belfield Works, Rochdale OL16 5AU	England
N. Linos Ltd.	35, D. Moutsopoulou St GR-185 40 Piraeus	Greece
Incofe	Anillo Periférico 17-36 Zona 11	Guatemala
Toshmiwal Brothers Priv. Ltd.	267, Kipauk Garden Road, Madras-600 010	India
Cimma Ing. Morandotti Spa.	I-20145 Milano, Mi 046	Italy
KEMCO	1-8-1, Shinjuku, Shinjuku-ku Tokyo 160	Japan
Nippon Pneumatic Mfg.	Nabari Factory, Nabari-City, Mie-Pref.	Japan
Wijnveen Ede b.v.	P.O. Box 212, NL-6710 BE Ede	Netherlands
Star Union Engineering Pvt. Ltd.	63 Joo Koon Circle, Singapore 2262	Singapore
Gramec Pty. Ltd.	P.O. Box 890380, Johannesburg	South Africa
Sei Gee Corporation	C.P.O. Box 400, Seoul	South Korea
E. Bachiller B.S.A.	Parets del Valles, Barcelona	Spain
Idah Machinery	2F-1, 77, Keelung Rd., Sec. 2, Taipei	Taiwan, R.O.C.
Paul O. Abbe Inc.	Little Falls, NJ	U.S.
American Forberg Systems	Gurnee, IL 60031	U.S.
Dynamic Air Inc.	St. Paul, MN 55110	U.S.
Hough International Inc.	Alberville, Al 35950	U.S.
Industrial Screw Conveyors Inc.	P.O. Box 2019, Burleson, TX 78028	U.S.
Sackett Equipment	Baltimore, MD 21224	U.S.

positionally stable, the U.S. Center for Veterinary Medicine (CVM) requires that the active ingredient(s) remain dispersed throughout (no significant stratification or layering) the LFS after 8 weeks storage without agitation at 30°C and under refrigeration. This is accomplished by sampling the top, middle, and bottom of a sample container and assaying the drug. Positional stability in the same product must also be verified by a field study in which there is no agitation. Any requirements as to minimum viscosity, pH, or other restriction for safe and efficacious use must appear on the product label. For example, for a liquid feed supplement containing lasalocid to be positionally stable, the viscosity must be not less than 300 centipoises per second (61).

The CVM also requires data demonstrating that the drug ingredient(s) does (do) not chemically degrade. Three pilot batches are manufactured under production conditions and stored at 30°C and under refrigeration. Samples should be assayed in duplicate at 0, 2, 4, 6, and 8 weeks. A short freeze study for a period of at least 5 days should also be conducted to indicate the effect of freezing, if any, and the subsequent thawing. A one-lot field stability study is also required by the CVM. If stability is pH-dependent or dependent on other variables, such as moisture or liquid feed supplement ingredients, additional lots will need to be tested to establish acceptable ranges.

When two or more drugs are incorporated into the same LFS, it is necessary to test the stability of all of the drugs utilizing a high/low experimental design. One drug is incorporated into the formulation at the lowest use level while the other drugs are incorporated at the highest use level. All drugs are assayed initially (zero time point). However, only the drug present at the lowest level is assayed at subsequent time points. Additional lots are manufactured in which one of the other drugs is incorporated at the lowest use level followed by the same assay scheme mentioned previously. Typically, cattle are fed from 0.5 to 2.5 lb per head per day. Thus when medicating a LFS it is essential that the proper concentration of the drug(s) is (are) incorporated based on the intended feeding rate.

C. Medicated Blocks

The U.S. Center for Veterinary Medicine (CVM) defines medicated blocks as compressed feed material (which contain medication) that is shaped into a cubic form for animal free-choice use. Medicated blocks are considered as type C medicated products. A type C feed can be fed "as is," without further dilution. However, because medicated blocks are considered a free-choice feed, there are specific regulatory requirements to obtain approval to market the product—i.e., that the block is efficacious when compared with the non-medicated control and that the block is consumed within a stated range, usually expressed as pounds or kilograms per head per day. Thus, when formulating a medicated block it is essential to design into the block some feature or ingredient(s) which will limit the consumption when fed on a free-choice basis.

There are a number of techniques and patented processes used to manufacture blocks and to restrict the consumption of medicated blocks. Blocks can be poured, manufactured by compression, and/or produced by a chemical reaction to gel or harden the block. Consumption is restricted by formulating with a high level of salt, by degree of hardness and by ingredient composition. The earliest poured blocks were manufactured by cooking molasses to evaporate the water, much like candy manufacturing, and then poured into a mold (62). Poured blocks manufactured by the "candy" method are usually not satisfactory for the delivery of drugs because of lack of restricted consumption and because the blocks tend to soften in hot weather and erode when exposed to rain and adverse weather conditions. Poured blocks manufactured by chemical reaction with calcium and/or magnesium oxide are much more resistant to adverse weather. Blocks manufactured by compression are usually limited in size (≤ 50 lb), whereas poured blocks can range in size up to 500 lb or larger.

The binding properties of blocks manufactured by either the compression or pouring process must be sufficient to prevent the block from crumbling or breaking apart during rough treatment by cattle. In the manufacture of blocks by compression, it may be necessary to add binders such as lignin sulfonate to the formulation. Dry molasses frequently produce a better block than liquid molasses. Fine ingredients usually bind better than coarse ingredients. In addition, the proper amounts of steam, heat, and liquid are required to activate the binder. To produce a block of consistent hardness and size, the hydraulic ram pressure or extruder pressure must be controlled along with the temperature. Inadequate mixing of liquid and steam prior to entering the molding chamber will result in length variations and other quality problems. Variations in length are not only indicative of problems with quality, but also may cause difficulty in packaging, labeling, and palletizing. A 24- to 48-h curing period is required to achieve adequate hardness prior to significant handling. The hardness and thus the daily consumption rates, of poured blocks manufactured by the reactions of calcium oxide, phosphoric acid, and molasses can be modified substantially by the incorporation of varying amounts of magnesium oxide (63).

In the manufacture of poured blocks by the chemical reaction method, phosphoric acid is added to a suspension of calcium oxide and molasses. The resulting reaction is exothermic and requires a cooling jacket on the mix tank. Dry ingredients such as drugs, vitamins, minerals, urea and magnesium, required for hardness, are added and mixed into the formulation. The resulting mixture is then poured into molds which can range from a cardboard box to a large plastic tub which can be used to medicate a large group of animals. The exothermic reaction may continue for up to several days after packaging. Adequate air circulation in a warehouse or controlled temperature storage is thus required during the curing process. The blocks will slowly harden and may require a week or longer to reach final hardness depending on their size and formulation. Depending on the type of formulation, it may be necessary to seal the product sufficiently to prevent absorption of moisture due to the hygroscopic nature of molasses. If the surface starts to soften prior to presenting the product to animals, drug overdose may result because of overconsumption of the hydrated layer.

As with any medicated feed, uniformity from block to block is essential as is homogeneity within a block, particularly when the block is used as a method to daily deliver a precise quantity of drug to each animal. Since ingredients in block formulations frequently do not flow or mix well or are viscous, sufficient samples should be obtained during production start-up to demonstrate homogeneity within the batch. In addition, both surface and core samples should be taken from several of the blocks to determine uniformity on the surface and throughout the block. Core samples can be obtained by using a 1-in. wood drill bit and drilling at various locations. The drilling technique not only obtains a sample, but it also breaks up the material for drug assays.

The CVM requires that the stability of medicated blocks be demonstrated under both warehouse conditions and range conditions. Under simulated warehouse conditions, samples should be stored at 25°C and at 37 or 40°C for 6 months. Surface and core samples should be assayed initially and every month or so to determine stability. Blocks placed outdoors on the range are to be exposed to the same weather conditions (temperature, cloud cover, rain, snow, etc.) as would be expected during normal consumption of the product. As a general rule, the blocks should be in the field for at least 14 days but not more than 90 days. To obtain an accurate stability profile, it is important that moisture uptake of the product be understood. Thus in addition to assaying for the drug, it may be necessary to determine the moisture content of the sample and to calculate drug potency on a dry-weight basis.

D. Drinking Water Medications

1. Methods of Treating Drinking Water

Both manual and automatic methods are used to medicate the drinking water of animals. Prior to providing medicated drinking water to animals, it is important that other sources of water be removed and that animals do not have access to lakes, streams, or rainwater run-off. Otherwise, the animals either may not get the proper dose or, in many cases due to the off flavor produced by a drug in water, will not drink the medicated water if another water source is available. Manual methods for adding drug to drinking water vary from placing a single dose of drug in a trough of water for an animal to medicating a header tank designed to provide water for a large number of animals. Typically the instructions on containers of drinking-water medication indicate to dissolve or dilute the entire contents in a specified quantity of water. Errors commonly occur when the user attempts to weigh or measure a portion of the contents of a single-use product. For example, the drug concentration in soluble powders may vary from lot to lot. The manufacturer adjusts the fill dose to compensate for lot-to-lot variations and to guarantee that each bottle contains the precise quantity of active ingredient indicated on the label. Thus 1 g of soluble powder from lot A may not contain the same quantity of active ingredient as 1 g from lot B and most likely will contain less than 1 g of active compound.

Many products designed to medicate drinking water contain instructions to prepare a stock solution which is further diluted in the final drinking water either manually or using devices called proportioners, as illustrated in Figure 20. A proportioner is a device which pumps a precise quantity of stock solution into a chamber where it is mixed with the flow before passing through the outlet to the animal drinking station. Because of its volumetric design, the injector or porportioner device maintains a constant ratio of concentrate to water flow irrespective of water pressure and water flow. Most proportioners are designed so that the ratio of drug concentrate to water can be varied depending on the drug to be used and the intended drug concentration in the final drinking water. For example, the ratio of drug stock solution to water can be varied from 1:10 to 1:500 with the Dosatron International proportioner shown in Figure 20.

Figure 20 illustrates a typical installation of a proportioner. Water flow enters the plumbing at the top left of the figure and goes directly to the animals when the top valve is open and the other two valves are closed. When the drinking water is being medicated, the top valve is closed and the other two valves are opened, causing the water to flow through the proportioner which, in turn, pumps drug solution into the mixing chamber via a tube connected to the bottom of the proportioner. The device to the left of the proportioner is an in-line filter usually 80 μm or finer designed to remove particulates which may interfere with the operation of the proportioner and/or water fountains and drinking nipples. To determine whether the filter is plugging or restricting water flow, a pressure gauge may be installed before and after the filter. A pressure drop across the filter during water flow may indicate that the filter should to be replaced.

Proportioners are commonly used in poultry houses and are being installed in almost all new and/or renovated swine confinement facilities as well as in some cattle and sheep facilities. Multizone medication can be designed so that one proportioner can medicate selected pens rather than the entire facility. The reason for the increasing popularity of proportioners is that medication can be immediately introduced to the animals by simply turning a few valves, whereas

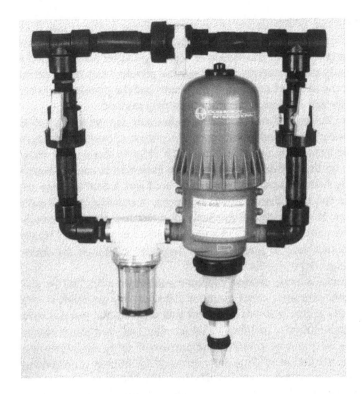

Figure 20 Drinking water proportioner used to medicate water by pulling a precise amount of drug solution into the bottom of the unit, accurately blending it with the water flow coming from the left before passing through the outlet on the right. The plumbing is designed so that the proportioner can be bypassed by opening the top valve on the horizontal pipe and closing the two valves on the vertical pipes. (Courtesy of Agri-Pro Enterprises, Iowa Falls, IA.)

when medication is administered via the feed, it is not only necessary to mix specially medicated feed, but the existing feed supply must be removed or consumed by the animal prior to commencing medication. In addition, cross-contamination and violative drug residues or carryover is a real concern with medicated feed.

2. Soluble Powders

Soluble powders are so named because of the requirement that the drug and any excipients included in the formulation be completely dissolved in water ranging in temperature from 5°C to 40°C. When formulating a soluble powder, both the solubility and the rate of dissolution are important. Some powders that have a high water solubility may be difficult to dissolve because they either do not "wet" sufficiently to permit rapid dissolution or form a gel layer which impedes dissolution. Frequently the dissolution rate of soluble powders can be modified by simply changing the order of addition of the water and the powder. In other cases it may be necessary to change the manufacturing method.

A variety of formulation and manufacturing techniques can be employed to produce soluble powder formulations with good dissolution properties. A small amount of surfactant or wetting agent will cause the drug to disperse in water rather than floating on the surface. An effervescent formulation can be manufactured by including both an acid such as tartaric acid and a base such as sodium bicarbonate in the same formulation. However, successful effervescent formulations must be free of water, manufactured under low-humidity conditions, and packaged in containers impervious to moisture. The addition of highly water-soluble ingredients such as granular sugar may also accelerate the dissolution rate of the drug.

Changes in the manufacturing process can have a dramatic effect on the dissolution rate of soluble powders. Lyophilizing or freeze-drying produces a very porous material which allows water to penetrate and dissolve the powder very quickly. Spray agglomeration is another technique that has been successfully employed to create a very porous powder. The process of spray agglomeration consists of spraying a solution of drug onto spray-dried drug in a controlled manner. A properly controlled process will cause the spherical spray-dried particle to agglomerate in such a manner as to leave relatively large void spaces within the agglomerated particles. Spray agglomeration is used when spray-dried material either dissolves too slowly or is too dusty to be an acceptable product. Soluble powders manufactured by a precipitation process or via conventional drying techniques are frequently more dense and dissolve more slowly than freeze-dried or spray-dried materials. However, by using suitable granulation techniques, some of the problems associated with the dissolution of powdered materials may be overcome. Dustiness of the formulation may also be reduced by granulation.

The spectrum of packaging for soluble powders ranges from foil pouches to plastic bottles, and from unit-dose containers to larger multiple-dose buckets. Regardless of the size or number of doses in a container, it is essential that the container protect the product both prior to opening and during use. Screw-cap wide-mouth bottles and buckets with snap lids are popular for products dispensed with a volumetric scoop or measuring cup; pouches are popular when the entire contents are to be mixed at one time.

During the development of the formulation, both the physical and chemical stability of the product should be evaluated. The effect of aging in the market container at various temperatures and humidity conditions must be assessed to make sure that the product is chemically stable and also to make sure that caking or changes in solubility characteristics do not occur. Soluble powders should remain free-flowing throughout the shelf life of the product and should not harden or separate during shipping. A dissolution test may be useful to determine whether there is any change in the dissolution rate throughout the shelf life.

In addition to establishing dry powder stability, studies should be conducted to determine both physical and chemical stability of the drug when diluted as stock solutions and when diluted to the final drinking-water concentrations. These studies should be performed over a range of temperatures (5°C to 40°C) including the effect of freezing and thawing as well as in hard water and water containing dissolved minerals such as iron, sulfur, calcium, magnesium, etc. If the drug precipitates or causes the dissolved minerals to precipitate, it will be necessary to reformulate. Reformulation may be as simple as adding a buffer so that the pH of the final drinking water is such that the minerals remain dissolved or it may mean that a different salt form of the drug is required so that insoluble inorganic salts such as calcium phosphate do not precipitate. Tests should be conducted over the entire range of anticipated concentrations that might be used for stock solutions and the final drinking-water concentration.

3. Solution Concentrates

Solution concentrates are frequently preferred to soluble powders when the drug is sufficiently stable and sufficiently soluble to permit a high-potency concentrate to be manufactured. Dust and handling problems associated with soluble powders are eliminated, as is the dissolution step when medicating drinking water. Because of the greater propensity for instability in solution relative to dry powders, it may be necessary to include formulation aids such as cosolvents, buffers, chelating agents, antioxidants, surfactants, preservatives, and/or stabilizers. As with soluble powders, both physical and chemical stability tests must be conducted on the marketed product as well as on the stock solution and final drinking water concentrations. In addition to testing for potency and degradation products, color, pH, odor, clarity, formation of precipitate, container integrity, and preservative effectiveness should be monitored. A sufficient level of preservatives will be required to demonstrate that the product meets or exceeds the requirements of the U.S. Pharmacopeia, British Pharmacopeia, or other governing pharmacopieae throughout the shelf life of the product. The stability testing program should also include freeze/thaw evaluation of both the formulation and the bottle or package.

Water is the preferred solvent for solution concentrates. When a drug substance exhibits limited solubility in water, it may be possible to solubilize the drug using one or more of the techniques described under the topic of solutions in an earlier section of this chapter dealing with type A medicated articles formulated as liquid concentrates. Perhaps one of the most effective means of increasing solubility is by the use of cosolvents. However, it is absolutely essential to establish that the drug will remain in solution when diluted with water at any level. Only those solvents which are Generally Recognized As Safe (GRAS) as food additives should be considered as candidates unless extensive toxicological testing is performed to demonstrate the safety of the solvent for the intended use. A list of GRAS items for food additives may be found under 21 CFR § 573 and 21 CFR § 582. Liquids such as glycerin, propylene glycol, polyethylene glycol 400, phosphoric acid, sorbitol, and ethyl alcohol are GRAS and may be considered for use in solution concentrate formulations.

Table 14 lists preservatives including antioxidants which are GRAS when used in accordance with good manufacturing or feeding practice. Restrictions, limitations, and conditions of use, if any, are also included in the table. For the preservative to be acceptable, the preservative must first of all be soluble in the solvent system used in the solution concentrate formulation. Secondly, the preservative must remain dissolved and not "oil out" when diluted with water at any level. Finally, the preservative must be compatible with the drug and other ingredients of the formulation.

Coloring agents may be used in solution concentrates. Two possible reasons for including a coloring agent in the formulation might be to identify the final drinking water as being medicated and to uniquely identify the product. However, the formulator is strongly encouraged to determine the regulatory status of the dyes or other coloring agents under consideration. Dyes have come under considerable scrutiny, and many formerly approved dyes have been de-listed from approved. There is considerable variation from country to country regarding approved dyes. There are very few water-soluble dyes which are acceptable worldwide. Thus, unless there is a real need for a dye in a solution concentrate, its inclusion is not recommended.

E. Products for Use in Lick Tanks

Lick tanks are of two basic designs. The most common design consists of a closed tank as shown in Figure 21 in which a wheel revolves through the contents of the tank. The top of the wheel is exposed so that cattle can lick the wheel and thus obtain the liquid product from the tank. A patented lick tank (64) designed to deliver a measured quantity according to an established schedule within which both quantity dispensed and times when dispensed are user definable is illustrated in Figure 22. Liquid feed is allowed to flow from a bulk storage tank by lowering a transfer pipe a predetermined distance into the contents of the tank. The liquid then flows by gravity to a feeding area. A microcomputer is programmed to dispense a specific quantity of liquid feed at set time intervals.

Table 14 Chemical Preservatives Generally Recognized as Safe (GRAS)

Preservative	Restrictions, limitations, and conditions of use
Ascorbic Acid	GMP[a]
Benzoic acid	NMT[b] 0.1%
Erythorbic acid	GMP
Propionic acid	GMP
Sorbic acid	GMP
Thiodipropionic acid	NMT 0.02% of fat or oil content
Ascorbyl palmitate	GMP
Butylated hydroxyanisole	NMT 0.02% of fat or oil content
Butylated hydroxytoluene	NMT 0.02% of fat or oil content
Calcium ascorbate	GMP
Calcium propionate	GMP
Calcium sorbate	GMP
Dilauryl thiodipropionate	NMT 0.02% of fat or oil content
Methylparaben	NMT 0.1%
Potassium bisulfite	GMP
Potassium metabisulfite	GMP
Potassium sorbate	GMP
Propyl gallate	NMT 0.02% of fat or oil content
Propylparaben	NMT 0.1%
Sodium ascorbate	GMP
Sodium benzoate	NMT 0.1%
Sodium bisulfite	GMP
Sodium metabisulfite	GMP
Sodium propionate	GMP
Sodium sorbate	GMP
Sodium sulfite	GMP
Stannous chloride	NMT 0.0015% as tin
Tocopherols	GMP

[a]Good manufacturing or feeding practices, meaning that the amount is reasonably required to accomplish its intended technical effect in food.
[b]Not more than.

Lick tanks have found utility in delivering nonprotein nitrogen, vitamins, minerals, and other nutrients as well as drugs to pasture and dairy cattle. As with medicated blocks, products sold for use in lick tanks are considered by the CVM to be free-choice feeds. Therefore, to obtain approval to market a drug product for use in a lick tank, it must be demonstrated that the product is efficacious when compared with a nonmedicated control and that the product is consumed within a stated range, usually expressed as pounds or kilograms per head per day. In addition, data must be provided indicating that the product is

Figure 21 Lick tank. The cattle lick the wheels which revolve through the contents of the tank to obtain the liquid from inside the tank. (Courtesy of Cargill's Molasses Liquid Products Division, Minneapolis, MN.)

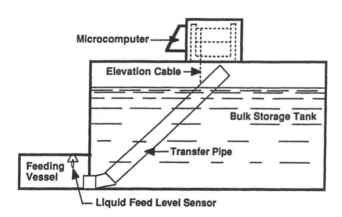

Figure 22 Portable programmable liquid-feed-dispensing device. The pipe inside the holding tank is lowered into the liquid a predetermined distance under the control of a microcomputer to dispense a specific quantity of liquid feed to the feeding area on the left. (From Ref. 64.)

homogeneous and that the active ingredient does not settle or separate during the time it is in the lick tank—i.e., that the drug concentration remains constant in the product starting with a full tank and ending with the contents of the tank having been consumed by the animals.

Because of the requirement that the product must be consumed within a specific range, various techniques have been used to restrict daily consumption. The patented lick tank described previously is the most sophisticated and most accurate delivery method. The least sophisticated method is the use of an adjustable slide which partially covers the exposed portion of the lick tank wheel. Reducing the exposed area of the wheel causes the cattle to work harder and longer to obtain the liquid from the wheel, thus slowing and regulating consumption. Consumption can also be regulated by varying the ingredients and/ or pH of the formulation. A high level of salt is typically used to reduce consumption. Certain sources of condensed molasses solubles are less palatable than others. Lowering the pH to 4 or below will also substantially reduce intake. These methods of controlling consumption are effective only if there is a consistent source of roughage and other nutrients available to the cattle.

For a formulation to be used successfully in a lick tank, all of the ingredients must remain dissolved over a range of temperatures, or, if the formulation is thixotropic, the ingredients should not settle or separate while in the lick tank. Typical ingredients in a lick tank formulation include corn steep liquor, molasses, condensed whey, condensed molasses solubles, brewers' condensed solubles, condensed distillers solubles, urea, salt, phosphoric acid, sulfur acid, dispersible vitamins, and soluble minerals. The only formulations acceptable for delivering insoluble drugs are a thixotropic formulation or a formulation with sufficient viscosity to prevent settling or movement of the drug particles. To minimize the settling of drug particles, the particle size should be reduced to a very fine powder. Surfactants and/or suspending aids may have to be employed to prevent agglomeration and subsequent settling. As with other liquid feed supplements, stability must be demonstrated throughout the expected shelf life or storage time under conditions simulating those to be used in field storage and field use. The laboratory portion of the stability study is the same as outlined previously for liquid feed supplements.

F. Semisolid Preparations

Veterinary semisolid preparations include mastitis infusion products, pastes, pumpable gels, creams, and ointments. Creams and ointments will not be discussed here since the topic is adequately discussed by Idson and Lazarus (65) in *The Theory and Practice of Industrial Pharmacy*. Numerous examples and formulae for various types of cream and ointment bases are included in their discussion.

1. Mastitis Infusion Products

The spectrum of commercially available intramammary mastitis infusion products ranges from aqueous gels to oleaginous mixtures and from low-viscosity formulations (<50 centipoise) to viscous pastes (>2500 centipoise). The drug may be dissolved in the vehicle system or may exist as a suspension. Insoluble forms of the drug, such as a benzathine salt, may be preferred to extend the persistence of the drug in dry cow therapy as compared to a more soluble form, such as a sodium salt, which may have application in lactating cow therapy. Milk from lactating cows cannot be used when the drug exceeds established safe limits. Thus lactating cow formulations should be designed to rapidly release the drug in a form which will dissolve in the milk and thus be milked out within 96 h of administration. The product label must indicate the period of time during which the milk has to be discarded (milk discard time) subsequent to treating a mastitic cow with an intramammary infusion product.

The CVM recognizes that proper product manufacturing is crucial to the milk-out properties of the finished product and has published a guideline (66) to aid the drug industry and the formulator in gathering the type of data which will demonstrate that an anti-infective bovine mastitis product: is safe for the cow; is effective; and fulfills human food safety, manufacturing, and environmental requirements. Because of the potential for differing efficacy, absorption, and elimination rates of the final product due to physicochemical characteristics such as dissolution coefficient, viscosity, particle size, shape, crystalline structure, and distribution, the guideline indicates that proper manufacturing process and control must be demonstrated. One way to demonstrate proper process control is to establish meaningful process specifications which, if not met, will result in a product that does not perform the same as the product possessing the critical physicochemical characteristics. Another method of demonstrating process control is to conduct bioavailability studies which may take the form of a milk-out study. However, as a control method for manufacturing, it is not practical to perform a milk-out study on each lot. Thus it behooves the formulator to develop in vitro test methods which can be used to ascertain changes in the manufacturing method. The guideline further states that the sponsor is encouraged to develop a complete set of physicochemical specifications on the final product and make a case that the product would not have bioavailability properties different from the original product. The arguments in some cases would likely require biological data.

Mastitis infusion products should be sterile. Since many antibiotics are heat-labile, terminal sterilization by autoclaving or dry heat is usually not acceptable. Terminal sterilization by irradiation sterilization may be a viable alternative. Otherwise, it will be necessary to sterilize the vehicle system and drug separately and combine them under aseptic conditions. The effectiveness of the sterilization process must be demonstrated by standard microbiological procedures, particularly in the case of oleaginous vehicle system. Details regarding the various sterilization techniques are presented in Section V, Sterile Dosage Forms. The effect of the sterilization process on the physicochemical properties of the formulation must also be determined.

In cases in which it is not possible to sterilize a mastitis infusion product without degrading the drug, the CVM requires that the product be free of potential mammary pathogens or endotoxins. Each lot must be tested for bacterial contamination. Samples that are positive for any microorganism should be subjected to bacterial or fungal identification to confirm that it is not one of the pathogens listed in Table 15 or any other microorganism with the potential to induce infection.

Although a few aqueous gels are currently marketed, most mastitis infusion formulations traditionally contain oleaginous vehicle systems. The main reason for selecting an oil-based formulation is for enhanced antibiotic stability. Other reasons include longer payout, inert vehicle system, and GRAS status of vegetable and mineral oils. Corn, sesame, peanut, and olive oils as well as liquid paraffin (heavy mineral oil) and caprylic/capric tryglycerides have been used successfully in commercial formulations. Since vegetable oils are susceptible to oxidation, it may be necessary to employ an antioxidant in the oil or formulation to prevent rancidity. Sesame oil is more resistant to oxidation than the other oils because of the presence of sesmol, a precursor of vitamin E, which acts as an antioxidant. Except for a few oil-soluble drugs such as erythromycin base, most oil formulations consist of a finely milled drug suspended in an oil carrier with the aid of thickening or suspending agents such as hydrogenated castor oil, aluminum mono- or distearate, colloidal silica, and waxes. In the case

Table 15 Pathogens to Be Tested For in Mastitis Infusion Products

---

Staphylococcus aureus
Streptococci spp.
Escherichia coli
Corynebacterium pyogenes
Yeast
Salmonella spp.
Mycoplasma spp.
Pseudomonas aeruginosa
Nocardia

---

of aqueous vehicle systems, ethylcellulose, sodium carboxymethylcellulose, and carboxyvinyl polymers—e.g., Carbopol 934 (B.F. Goodrich)—may be used as thickening agents.

Mastitis infusion products are usually packaged in single-dose tubes or cannula syringes such that the entire contents of the syringe is infused into a single teat or quarter. Research has suggested that partial insertion of a standard-length cannula into the teat canal results in a lower infection rate than full insertion (67). As a result, cannula syringe manufacturers have developed syringes with combination cannulae which permit either full insertion or partial insertion. Several different syringe designs with combination cannulae are shown in Figure 23.

2. Pastes and Pumpable Gels

Pastes and pumpable gels have become a convenient form for administering antidiarrheal agents, anthelminitics, and other drugs to small and newborn animals. Dosing a dog, cat, piglet, etc., with a liquid or tablet can be difficult, whereas a paste contained in a small syringe or tube or a pumpable gel contained in a pump bottle with a tube can be ejected into the mouth of a dog, cat, or piglet with relative ease. Once in the mouth the paste does not readily drip out (as compared to a liquid) or be ejected (as in the case of tablets).

Pastes can also be used to mass-medicate horses and cattle using a multiple-dose syringe or paste dispenser. Multiple-dose syringes are usually equipped with a stop mechanism permitting the desired dose to be preset for each animal before administration. Some manufacturers have developed their own dis-

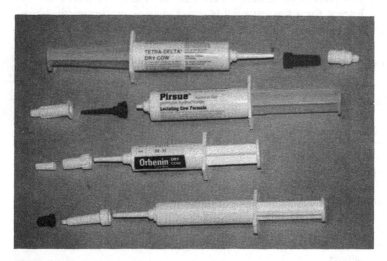

Figure 23 Cannula syringes designed to permit either full insertion or partial insertion of intramammary infusion products.

pensing devices which can be loaded with a prefilled tube or cartridge that will dose many animals without reloading. For horses or cattle, the syringe or dispenser should be equipped with a long tip which can be inserted into the interdental space, depositing the paste on the tongue at the rear of the mouth. A paste of the proper consistency and adhesive properties will not be readily dislodged from the tongue, and the animal will swallow it.

As the term implies, pastes should have some adhesive properties so that the material will adhere to a surface. Water is the preferred vehicle for a paste. An aqueous solution or suspension of a drug can be thickened to make a paste or pumpable gel. A variety of thickening agents are commercially available ranging from naturally occurring materials such as acacia, alginic acid, bentonite, cellulose, and tragacanth to synthetic or chemically modified materials such as xanthan gum, hydroxypropylmethyl cellulose, carbomer, colloidal magnesium aluminum silicate, and sodium carboxymethylcellulose. The particle size of a drug used in a paste can be as large as 100 µm. However, the particle size range must also be controlled to minimize batch-to-batch variation in viscosity. The degree of cohesiveness, plasticity, and ejectability can sometimes be modified by using a combination of thickening agents. Frequently, manufacturers of the thickening agents can provide recipes and guidance as to appropriate concentrations to obtain the desired effect.

Syneresis or separation of a liquid from the paste may occur over time. The addition of absorbing materials such as microcrystalline cellulose, kaolin, colloidal silica, starch, magnesium oxide, or dicalcium phosphate is useful to overcome this problem. When formulating a viscous formulation such as a mastitis product, paste, or pumpable gel, adequate shear is required for good mixing and complete hydration of the thickening agent. Otherwise, viscosity may increase with time to the point that the formulation is unacceptable. However, excessive shear may actually cause a loss of viscosity which may or may not be reversible with time.

Cosolvents such as alcohol, polyethylene glycol, propylene glycol, or glycerin may be used to increase the solubility of the drug. In addition, glycerin and the glycols will make the vehicle more viscous, decrease the amount of thickening agent required, and prevent the paste from drying out rapidly and forming a crust or hard mass at the tip of the dispenser. Carboxyvinyl polymers can be used to thicken polyhydroxy liquids such as glycerin, propylene glycol, and polyethylene glycol. Some interesting formulations can be prepared with the carboxyvinyl polymers and amine-neutralizing drugs or other agents. Pastes can also be formulated with vegetable or mineral oils and thickened with hydrogenated castor oil, aluminum mono- or distearate, colloidal silica, hydroxypropyl cellulose, petrolatum, waxes, xanthan gum, or other suitable thickening agent. Preservatives should be added as necessary to prevent microbial growth in the formulations.

G. Capsules and Tablets

Capsules and tablets are the most common method of administering human drugs primarily because of convenience and accuracy of the dosage form. However, administration of capsules and tablets to animals is difficult at best, and these dosage forms are often unsuitable for dosing animals on a weight basis. In addition, the amount of drug given to a large animal such as a calf or cow may be considerable. For example, the recommended dose of sulfamethazine is 16 g/100 lb body weigh. Thus, a 600-lb cow would receive 96 g of drug. The usual round tablet made to contain even a third of the active ingredient would be an unwieldy object cumbersome to administer. Thus the geometry is modified to an oblong, cylindrical-shaped tablet commonly called a bolus. To aid in the administration of a bolus, a balling gun is frequently used. Balling guns are designed to hold the bolus until discharged either in the back of the throat or in the esophagus.

There are numerous good reference sources dealing with tabletting technology (68–71). However, because of the size of the bolus, there are special problems associated with formulating and compressing the material. The high ratio of drug to inert material in a bolus may hinder the formulator's ability to use sufficient quantity of excipients to overcome inherent drug characteristics which adversely affect compression, hardness, disintegration, dissolution, etc. Consequently, it is imperative that the appropriate excipients be chosen to address the specific problem. Rudnic and Kottke (71) have developed tables of three Hiestand Tableting Indices which provide information as to how various excipients and drugs contribute to the physical characteristic and strength of a tablet or bolus. The strain index (SI) is a measure of the internal entropy, or strain, associated with a given material when compacted. The bonding index (BI) is a measure of the material's ability to form bonds and undergo plastic transformation to produce a suitable tablet. The brittle factor index (BFI) is a measure of the brittleness of the material and its compact. By performing the appropriate measurements on both the drug and excipients, it is possible to select those excipients most likely to compensate for problems associated with the drug.

Capping, or lamination, of a bolus is a phenomenon that occurs when air is trapped during the compression step and/or there is insufficient cohesiveness of the formulation. There literally can be small explosions seconds after the bolus leaves the table of the compression machine as the air escapes and lifts off the top of the bolus. This is particularly true of single-compression bolus machines. More modern machines use precompression rolls and multiple compression strokes to reduce capping and lamination due to entrained air. Other methods of reducing entrained or occluded air include changing the granule size, adjusting the moisture content, and adding hygroscopic excipients such as sorbitol, solid polyethylene glycol, or urea. Kaneniwa et al. (1988) reported the effect of particle size on the compaction characteristics of two sulfonamide drugs, one exhibiting brittle fracture and the other being compressed chiefly by plastic deformation (72).

As discussed in many of the standard textbooks (68–71) dealing with the subject of tableting, there are many different types of excipients which impact the performance of a formulation both in the tablet press and subsequently upon administration to an animal or human. Typical uses of excipients include fillers, binders (dry and wet), disintegrants, dissolution retardants, lubricants, glidants, antiadherents, wetting agents, antioxidants, preservatives, coloring agents, and flavoring agents. Although a wealth of information exists in the literature on tableting technology, it should be noted that because of the extreme depth and width of a die cavity in bolus production, formulations may have to be modified to enhance powder flow and at the same time enhance adhesion of the particles.

Frequently, during scale-up it is necessary to adjust formulation to obtain the proper adhesion in the final bolus. For example, the moisture level in the granulation step may have to be increased to yield the same bolus strength when going from pilot plant operation to production. It is important that the release profile of tablets and boluses be determined as part of formulation development and scale-up operations. Obviously, a bolus must have sufficient binding properties to prevent crumbling and flaking. However, upon administration the bolus must disintegrate and release the drug within a reasonable time. In vitro disintegration and dissolution tests should be developed which can be used as a quality control tool to detect significant variations between lots.

Although fewer capsules are used in the animal health industry than tablets and bolus products, tamping pistons (73) on capsule filling machines pack greater quantities of material into capsules than was previously possible. Thus there may be renewed interest in capsules for animal health applications. Capsules have the advantages that they are smooth, slippery, and easily swallowed, and that they provide a barrier for drugs with a bitter taste. Modern automatic capsule filling machines are capable of filling dry powders but can also fill beads or pellets, tablets, liquids, and pastes into capsules.

V. STERILE DOSAGE FORMS

On October 11, 1991, the FDA officially proposed new regulations for aseptic processing and terminal sterilization in the preparation of sterile pharmaceuticals for human and veterinary use (74). The agency's policy requires manufacturers to use terminal sterilization techniques (e.g., autoclave processes, gamma radiation, electron beam radiation, ethylene oxide) unless such a process adversely affects a drug product. When sterile products cannot undergo terminal sterilization and be manufactured by aseptic processing (e.g., filtration and filled into a presterilized package), manufacturers are required to explain in writing why terminal sterilization cannot be used. Scientific evidence with data showing the unacceptable degradation of the product or container with terminal sterilization needs to be presented to the agency (75). In cases where drug sensitivity to autoclave processing (e.g., proteins and complex biological products) has been established in the literature, the literature may be submitted to justify aseptic processing (75). It is clear that the FDA prefers terminal sterilization to aseptic processing; however, as engineering and control improvements in modern aseptic facilities have occurred, the FDA has become more receptive to aseptic processes (76).

In 1995, the Center for Veterinary Medicine issued a policy letter revising their Policy and Procedures Manual Guide 1240.4122 describing the sterility and pyrogen requirements for veterinary injectable drug products (77).

> It is the policy of the Center for Veterinary Medicine that (1) all injectable drug products (including intra-mammary products) be sterile except euthanasia products and ear implants for bovine and ovine species, and that (2) pyrogen levels in sterile veterinary drugs should not exceed established limits. Approved products (including intra-mammary) which fail to meet the requirements for sterility and pyrogens must be labeled as being "manufactured by a non-sterilizing process." The labeling requirement is effective at the time of the next label printing after the date of this Notice (77).
>
> Sterile veterinary products should be manufactured using validated processes that assure sterility and pyrogen levels that are within established levels (77).
>
> The Center believes that products produced in accordance with the controls associated with sterile processes will enhance quality. Nonetheless, CVM does recognize that there may be instances where our expectations regarding the manufacture of a sterile injectable drug product may not be feasible. In these instances, the effects of microbial contamination on the efficacy and safety of the product must be addressed as well as the issues associated with the inability of manufacturing a sterile drug product. The information necessary to support the inability to manufacture a sterile injectable drug product and the information to support the safety and efficacy of a non-sterile injectable product may be extensive. A sponsor proposing to take this alternative route should discuss these issues with the Center early in the drug product development process. CVM will work with sponsors in exploring options to collect acceptable information and data to address our manufacturing, safety and efficacy concerns. If this information is developed, a sponsor may apply for an exemption from the sterility requirements. The Center will make a decision based on the documentation provided to support the exemption. When an exemption from the sterility requirements is granted by CVM, we will determine other appropriate controls and specifications to assure the quality of the non-sterile drug product. For example, these controls and specifications could include bioburden

testing and limits, and pathogen testing, etc. A sponsor proposing an exemption to the sterility requirement should be aware that alternative standards may be as difficult to meet as the sterility requirements. Any injectable product manufactured by a non-sterilizing process that receives an exemption from the sterility requirements will also be expected to be labeled as being "manufactured by a non-sterilizing process" (77).

These policy statements summarize the agency's expectations of manufacturers of animal health products and veterinary pharmaceuticals in regard to sterile manufacturing processes and sterility requirements for injectables. Unless the drug formulation has an inherent antimicrobial effect, there is no mechanism to protect the drug or formulation from accidental introduction of microorganisms during manufacture or upon administration to the animal (e.g., feedlot situations). Usually, then, preservatives are added to inhibit the growth of microorganisms that may be introduced inadvertently during or subsequent to the manufacturing process. Typically, products exempt from terminal sterilization include freeze-dried products, sterile powders, and nonaqueous solutions or suspensions on the basis that they contain no appreciable water content and thus limited opportunity for microbial growth. It is clear that the issue of injectable sterilization method and manufacturing controls needs to be addressed early in the product development process.

A. Methods of Sterilization

In general, sterilization methods can be classified into two categories: physical methods, and chemical methods (78). Physical thermal methods include dry and moist (autoclave) heat treatments. Physical nonthermal methods include ultraviolet radiation, ionizing radiation, either gamma or electron beam, and filtration. Chemical methods include the use of gases, such as ethylene oxide, ozone, hydrogen peroxide vapors, formaldehyde, chlorine, etc., and liquids, such as phenols, iodine, formaldehyde solutions, alcohols, hydrogen peroxide, etc. All the above methods, with the exception of filtration, have been used as terminal sterilization methods with varying degrees of success. The classical method of achieving terminal sterilization has been autoclave processes (79), although ionizing radiation, where applicable, is gaining acceptance. References (78,80–89) should be consulted for additional information on potential affects of sterilization methods (e.g., gamma irradiation) on drugs, pharmaceutical excipients, biomedical polymers, and controlled-release systems.

Terminal sterilization is defined as a final processing step used to deliver a sterility assurance level of at least 10^{-6} after the product is placed in its final container-closure system (79). The product, container, and closure usually have a low bioburden at the time of assembly, but are not sterile prior to terminal sterilization. Aseptic processing is defined as a process in which the product is sterilized separately and is filled into presterilized packaging components in a controlled environment (79). Drug solutions can be sterilized by filtration (<0.2-µm filter) and filled into packaging components that have been sterilized by steam, dry heat, ethylene oxide, or radiation. Aseptic processing is generally used when the product or packaging components cannot withstand the conditions necessary for terminal sterilization, resulting in drug degradation or weakening, embrittlement, or color changes in packaging materials. It is the large input of energy necessary to effect sterilization and these resultant changes that occur in materials that limit terminal sterilization's usefulness with many products—e.g., proteins, biologicals. Aseptic processing utilizes more manipulations (usually by people) that places the sterility of the product in jeopardy. Advantages of aseptic processing include less effect on drug stability, lower particulate levels, greater packaging flexibility, less effect on preservative systems, less effect on physical properties, less effect on container-closure system integrity and extractables, less effect on chemical properties, and longer product shelf life (79). Advantages of terminal sterilization include lower processing costs and, in general, a higher sterility assurance level (the basis for the FDA's policy) (79). As engineering and control systems have improved and aseptic processing has evolved, human contact with the drug product has been reduced and aseptic processing has become a more controlled and robust process resulting in sterility assurance levels above 10^{-3}, approaching levels of 10^{-4} to 10^{-5}. With the institution of isolation or barrier technologies, sterility assurance levels of the order of 10^{-6} may be achieved, similar to those achieved by terminal sterilization.

In addition to the strategies of aseptic processing coupled with a less severe terminal sterilization (e.g., heat treatment) on the final sealed container or coupled with the inclusion of preservatives, new technologies and improved operations have been used to improve the sterility assurance of aseptic processes. These include barrier systems (LaCalhene type or fixed), high-speed continuous lines, nonsterile (dark side) maintenance of equipment, form-fill-seal or blow-fill-seal operations, the elimination or minimization of aseptic connections, filtration at point of filling, minimizing dwell time of open containers, automation of environmental monitoring, robotic handling of sterilized vial closures, check weighing, filling, and tray handling (76,79). Most of these new technologies have the goals of improving sterility assurance of aseptic processes by reducing the degree of human contact (a major source of contamination) with the product during processing and improving the economics, quality, and flexibility compared to conventional aseptic processing.

B. Injectables

Of the approximately 145 injectable products listed in the 21 CFR § 522 (90), approximately 80 are listed as sterile (most presumably by aseptic processing).

Of the remaining 65, at least 10 include preservatives, such as benzyl alcohol, phenol or parabens, to effect sterility. The three sustained-release bovine somatotropin products for improved lactation performance in dairy cows (Posilac, Monsanto, Optiflex, Elanco [outside U.S.], and Boostin, LG Chemicals [outside U.S.]) are aseptically processed to achieve sterility. Other commercial injectable sustained- (controlled-) release products in the pharmaceutical world include Parlodel (pLA microspheres of bromocriptine), probably sterilized by gamma radiation, and Lupron (pGLA microspheres of a LHRH acetate analog), probably sterilized by aseptic processing. It is evident from these commercial products that aseptic processing is currently the most viable means of achieving sterility with easily degradable drugs, such as peptides and proteins, even in light of the published literature suggesting low doses of gamma radiation to achieve sterility in products with low bioburdens (78,81,91).

C. Implants

The widely used, efficacious, and economical implants for cattle (Synovex family of implants, Ralgro, Revlor, and Compudose) are all nonsterile. Compudose implants have a coating of the antibiotic, oxytetracycline HCl, to inhibit local infections upon implantation into the ears. The CVM has argued for a sterility requirement for ear implants containing somatotropins (e.g., recombinant porcine somatotropin) on the basis that the implant is being placed in an environment ideal for microbial growth, providing a nutrient source for growth of bacteria. Aseptic processing of implants containing proteins will be difficult due to the number of manufacturing steps required and the thus increased probability of contamination during manufacturing. Terminal sterilization processes will most likely lead to protein degradation or formulation changes, thus affecting the performance of the implant.

VI. DERMAL/TRANSDERMAL DOSAGE FORMS

A. Dips

One of the oldest methods of treating animals, especially sheep and cattle, for parasites on the surface of the skin is by way of dips. By totally immersing an animal, all areas are covered by the insecticides. Because of human health risks, certain dips containing benzene hexachloride, lindane, and arsenic have been banned in many countries. Following the ban of these products, organophosphates, diazinon, propetamphos, rotenone, amitraz, and compounds from other chemical groups were introduced. However, these products are not without risk. Because of the large number of adverse reactions reported in the U.K. following use of organophosphates in dips, the sale of these products is restricted to individuals holding a certificate of competence.

Flumethrin was introduced in the U.K. by Bayer in 1986 as that country's first nonorganophosphate dip effective against ticks, although it does not control blowfly larvae. Since then Grampian Pharmaceuticals has introduced three synthetic pyrethroid dip (Crovect, Provinec and Robust) containing high *cis*-cypermethrin which are indicated for the control of blowfly *Psoroptes* (scabies), ticks, and lice.

Formulations intended for use in dips vary from aqueous solutions to emulsifiable concentrates to wettable powders. Stability in the formulation as well as in the final dip solution is required. Photostability is also a requirement. There are no specific rules in formulating dip preparations except that the active ingredient must remain in a form that does not "oil out," or separate from water in the dip tank. For example, since many insecticides and parasiticides are water-insoluble, they must be formulated as either an emulsifiable concentrate, a wettable powder, or a flowable. An example of an ectoparasiticide available in all three forms is coumaphos (an organophosphate), which is indicated for the control of flies, lice, ticks, and mites.

Because of the quantities of drug required for a dip and the increasing difficulty of disposing of the dip, many producers have discontinued using dips and now apply the same or similar formulation as a spray or pour-on. Sprays have advantages in being portable and requiring smaller amounts of chemical, though coverage may not be as complete. To compensate for incomplete coverage and incomplete soaking of the skin, the concentration of the drug in the spray solution is frequently increased. For example, dips containing crufomate are usually formulated at a concentration of 0.25%; the spray contains 0.375%. The amount delivered can also be altered depending on the parasite. For example, a higher concentration or higher application rate is used for mites and ticks than for hornflies.

B. Pour-Ons

The realization that drugs applied topically enter the bloodstream has stimulated the search for topical drug delivery systems to control external as well as internal parasites. Systemic insecticides such as the organophosphates can be used to control external parasites if the latter are bloodsuckers such as the sucking louse *Linognathus vituli*, or if the drug is secreted onto the skin via sweat or sebaceous glands (92). Pour-ons are used to deliver ectoparasiticides for controlling a wide range of parasites including blowflies, hornflies, warble flies, lice, mites, and ticks by applying the product to just one side of the midline down the animal's back from the shoulder area to the tail head and then back up to the shoulders, to just the other side of the midline. If face flies are a problem, a few milliliters can be applied to the face starting between the animal's eyes and ending near the muzzle.

The concentration of the active ingredient in a pour-on formulation is usually considerably higher than in a dip or spray. There are two general types of pour-on formulations. The first type is designed to spread the active ingredient over as much skin as possible. The goal of the second type is to deliver parasiticides to the systemic circulation via the topical route. In the first type, solvents, surfactants, and spreading agents are incorporated into the formulation to cause the active ingredient to migrate from the point of application to distal regions of the animal. In the development of the formulation a small quantity of the material is poured on, or "spotted on." Hair samples and/or skin samples are then obtained from various locations on the animal and assayed for the active component. The efficacy of these types of products are highly dependent on the vehicle system. Solvents used in these formulations range from xylene to dipropylene glycol methyl ether to various petroleum distillates. Synergists such as piperonyl butoxide and diethyltoluamide may be added to formulations containing naturally occurring pyrethrins to enhance control of lice and *Culicoides* (biting midges) on horses, and fleas on dogs and cats. Dimethyl phthalate and N-N-diethyl-m-toluamide (DEET) are often used in fly and mosquito repellent products.

The goal of the second general type of pour-on formulation is to have a topical dosage form which results in well-controlled systemic absorption. Substantial evidence in the literature indicates that the barriers to absorption of topically applied drugs on animals differ from barriers found in human skin (92). An understanding of the physicochemical properties of the barriers in the target animal is thus important in designing a topical dosage form that will enhance absorption of the active ingredient. For example, in humans the stratum corneum is the rate-limiting barrier to absorption; drugs are more likely to traverse sheep and cattle skin via hair follicles, sweat and sebaceous glands, and ducts. Pitman and Rostas (92) provide a good overview of the many factors that influence topical drug absorption in sheep and cattle.

Because cattle, sheep, and certain other animals secrete an oily sebum, it is usually necessary for a formulation to contain an organic solvent, a surfactant, and/or an absorption-enhancing agent unless the drug itself is soluble or miscible in the sebum. The sebum becomes emulsified with sweat and fills the follicle pores and coats the animal's skin and the lower parts of the hair and wool. The chemical composition of the emulsion varies from animal to animal and with such factors as the time of year and climate, which may explain why trial pour-on formulations of levamisole and phosmet showed reduced penetration through skin harvested in the winter as compared to the summer (92).

The ultimate test of efficacy is whether the formulation kills the parasite of interest. However, as a formulation screening tool, it is frequently possible to use blood levels and other parameters, such as reduction in cholinesterase activity in the case of organophosphates, to determine the rate and extent of drug absorption. Since there is very little information in the literature regarding the effect of various solvents and vehicle systems on absorption rate through animal skin, it is important to develop methods that employ the target animal or that correlate well with results in the target animal. Employing the target animal is important because substantial differences in absorption have been observed even when using penetration enhancing agents such as dimethylsulfoxide (DMSO).

A common mistake made when determining whether a chemical would potentially be absorbed through the skin is to formulate it in DMSO. DMSO carries many drugs across human skin but it actually slightly depresses the penetration rate of levamisole through cattle and sheep skins compared to aqueous solvents (92). Although mineral oil and soybean oil are poor solvents for drugs, they are sometimes included in formulations to resist rain wash-off. In October 1995, a long-acting, water-based permethrin formulation (Durasect-Pfizer) with a unique water-repellent was approved in the U.S.

The water repellent is a polymer that coats and binds the active ingredient permethrin to hair follicles on treated cattle and prevents water, sweat, sebum, etc., from diluting the insecticide. Consequently, the topical solution remains active for longer than other pour-on formulations because the permethrin is not removed as quickly from the hair coat. Other solvents used in pour-on formulations include isopropanol alcohol, 2-2-butoxyethoxyethanol, amyl alcohol, and aromatic solvents.

C. Spot-Ons

Spot-ons are used at lower dose rates in terms of actual volume of material than pour-ons. However the concentration of the active ingredient in the formulation is usually higher than in pour-ons. Consequently, users of spot-on products are exposed to increased risk of absorbing toxic materials from handling concentrated drug products that have been specifically formulated to promote absorption. Spot-ons are applied in one or two places on the back of the animal at a location that cannot be licked off by the animal's tongue. The quantity of formulation used per animal varies depending on the weight of the animal. Doses may range from a few milliliters to 25 or 30 ml. The obvious advantage of these products is that they are quick and easy to apply, and there is no requirement to cover the entire animal or the entire midline of an animal. Automatic applicator guns are available that measure appropriate doses accurately. Spotton (fenthion-Miles Inc.) is an example of a ready-to-use spot-on formulation. One properly timed application per season provides effective control of cattle grubs and aids in the control of lice.

D. Controlled-Release Systems, Patches

While controlled-release transdermal systems as veterinary dosage forms have a wide spectrum of possible uses, actual use has been limited. Properties of a potential drug candidate for this dosage form include poor oral availability, intestinal upset as a side effect, and significant first-pass metabolism. Ensuring adhesion of a transdermal system over a therapeutically relevant time to either a companion animal or a production animal poses some challenges: the presence of fur, animal-to-animal interaction, scratching and rubbing, and irritation due to adhesives or drug or both. Sensitization to the system components and skin reaction due to occlusion are also issues.

Controlled-release systems have been used to deliver fentanyl for postoperative analgesia in dogs and appear to have promise (93). Two additional examples of the use of the transdermal route of drug delivery are an adhesive diffusion-controlled matrix, or transdermal tape, described in a paper by Gupta et al. (94), and transdermal delivery of ivermectin to treat mange in rabbits. The class of compound being delivered in both cases is an anthelmintic. Transdermal technology is described in more detail in Section VII.D.

VII. CONTROLLED-RELEASE TECHNOLOGIES

A. Special Considerations

1. Benefits

A drug delivery system utilizing controlled-release technology is a dosage form that releases one drug or more in a programmed pattern for an extended period, which can range from days to months. The drug is delivered systemically or to a targeted organ. By this definition, most pastes, powders, injectable solutions, or pour-on formulations would not be considered controlled-release drug delivery systems, although some of these preparations are administered several days apart. The therapeutic efficacy of many drugs can be enhanced by innovative drug delivery processes. Some benefits of controlled-release technologies follow:

- Dosing intervals are minimized. Frequent dosing, restraint, or handling causes stress to animals. A single administration of a controlled-release dosage form minimizes stress and economic loss from treatment compared with conventional dosage forms which require frequent dosing.
- A single administration of a controlled-release product is convenient to the animal owner or producer by eliminating the need for procedures that are often costly, difficult, and labor-intensive, and that need to be repeated intermittently.

- Controlled-release of short-half-life compounds is more efficient than intermittent administration. Because of biological clearance processes, large doses of expensive compounds must often be given frequently using conventional methods of administration.
- The exact dose the animal receives is known. Because administration of compounds in food and drinking water is imprecise, it is difficult to know the exact dose the animal has received. Also, constant administration is not possible by this method.
- Because many potentially useful veterinary compounds cannot be effectively administered with feed, the range of potentially useful veterinary compounds can be expanded with controlled-release technology.
- Constant, controlled administration may provide dose-sparing effects; lower effective doses often minimize unwanted side effects.
- Controlled-release dosage forms minimize human exposure. Some veterinary compounds are unsafe for humans to handle or may have untoward side effects in humans.

2. Economics and Animal Health Management Practices

The economics of medicating animals changes with each class of animal. Price sensitivity is less in the companion animal market than in the production animal market. Owners of domestic production animals constantly compare cost of treatment with results. The cost of treatment with a controlled-release product may be higher than conventional therapies for a number of reasons, including higher development costs and higher unit cost per administration. Cost:benefit ratio then comes into play. Controlled-release products often offer more benefits than conventional therapies, such that the cost:benefit ratio is attractive enough to make these products and technologies competitive.

There are a number of factors an animal owner or producer must evaluate to determine whether or not to purchase and use a controlled-release product compared with using a conventional therapy or no therapy at all:

- Is the incidence of the disease state high enough to warrant treatment?
- Is the species in which the disease occurs of high value? High value per head in livestock refers to the dollar value of sale or show of the animal. High value in domestic companion animals depends on the emotional value and financial resources an owner has for his or her pet.
- Does the producer or owner have confidence in the stated return from use of the product? For example, how many extra kilograms at how much faster a rate of gain could a livestock producer expect to realize after the use of growth enhancer "X" administered as a controlled-release product compared with that administered as a conventional product or no product? Is this return worth the price the manufacturer is ask-

ing? Pricing in the food-producing animal market has historically commanded only a \$1 price for a purported \$5 (or greater) return.

• Are there drawbacks to the current therapy which the controlled-release product addresses?

• If there is no alternative to the controlled-release product, is the potential outcome of no treatment acceptable?

The use of controlled-release products makes more sense in fields where the value per head is higher, and where treatment time is longer. This can be seen in comparing the economics of production between the cattle and poultry industries. Use of such products is more prevalent in the cattle industry relative to poultry due to the higher per-animal value, and potential for longer durations of treatment.

B. Oral

1. Ruminal Boluses

Because of the high value on an individual-animal basis of cattle, the economics of using a controlled-release dosage form are generally acceptable. Additionally, in comparison to other routes of providing sustained delivery, the rumen is a convenient, somewhat robust environment. All that is needed to retain the delivery system in place is high enough density or a geometry sufficient to prevent passage out of the reticulorumen. The size and drug content of a ruminal bolus determine its suitability for use in cattle within a defined weight range and in sheep. Examples of ruminal boluses either reported in the literature as under development or currently marketed in the U.S. are shown in Table 16.

The three technologies employed are described in more detail in the following paragraphs: the use of corrosion and degradable materials to control release of fenbendazole (Pancur), osmotic control of release of ivermectin (Ivomec SR), and a combination of chemical and mechanical controls of release of albendazole (Captec). The bolus is maintained in the rumen by the use of density for the Ivomec SR bolus and the Pancur bolus and by geometry for the Captec bolus.

The rate of release of 12 g of fenbendazole over 4 to 5 months from the Pancur SR bolus is controlled by the corrosion of two magnesium alloy tubes; each tube encloses five tight-fitting cylindrical tablets, each containing 1.2 g fenbendazole. The exterior of the magnesium alloy tubes is protected from the rumen contents by a series of close-fitting rigid plastic rings; the central ring connects both tubes. This particular bolus is completely degradable and leaves no residue in the forestomach (95).

An example of a sustained-release bolus is the Ivomec SR bolus containing ivermectin (Fig. 24). This bolus is a cylindrical device with an outer semipermeable membrane enclosing a metal density element at one end, an osmotic

Table 16 Ruminal Boluses

Drug	Bolus type	Indication(s)
Levamisol	"Chronomintic-Virbac"	Grazing heifers; weight gain
	Levasol	Cattle and sheep anthelmintic
	Tramisol Oblets	Sheep wormer
Morantel tartrate	Tri-laminate; Paratect-Flex bolus	Anthelmintic for cattle
Oxfendazole	Castex	Anthelmintic for sheep and cattle
	Autoworm (pulsed delivery)	Anthelmintic
Ivermectin	Ivomec-SR osmoticallly trolled	Anthelmintic for cattle
Albendazole	Captec device	Anthelmintic for sheep
Moxidectin		Antiparasiticide
Fenbendazole	Pancur	Antiparasiticide
Sulfamethazine	Sulka-S Bolus; Spanbolet II Tablets	Antibacterial
Methoprene	Inhibitor	Antiparasiticide

energy source at the other end, and a drug formulation in the center. The osmotic energy source consists of a tablet of a polymeric salt mixture. As water is absorbed through the semipermeable membrane, the tablet expands to drive the ivermectin-impregnated wax formulation through a central delivery orifice in the density element (96).

The Captec device (97) design consists of a polypropylene barrel with a plunger and spring in one closed end (Fig. 25). The spring is compressed and the barrel is filled with seven tablets containing a drug—albendazole in this example. Drug is then extruded when contact is made between ruminal fluid and the albendazole tablet at the orifice at the opposite end of the barrel. The primary factors governing drug release rate are the diameter of the orifice, spring strength, and chemical properties of the gel that forms between the ruminal fluid and the tablets at the orifice. The Captec device includes a set of "wings" secured to the barrel with water-soluble tape prior to administration. On contact with the ruminal environment, the tape dissolves and the wings expand to a predetermined angle, preventing regurgitation of the device.

2. Tablet Forms

The use of sustained-release tablet forms without the retention ability utilized by ruminal boluses naturally limits the duration of the potential delivery duration to the transit time in the gastrointestinal tract. Transit time through the gastrointestinal tract is affected by numerous variables including food, activity, time of day, and species differences. The duration of such a therapy is there-

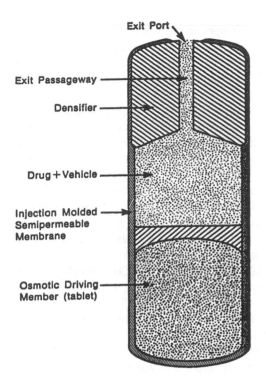

Figure 24 IVOMEC schematic.

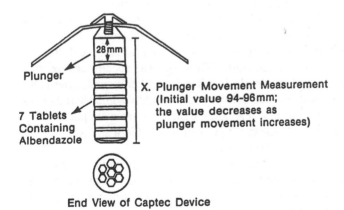

End View of Captec Device

Figure 25 Captec schematic.

fore limited to approximately a day, as opposed to weeks or even months for other controlled-release devices. If the treatment requires extended duration, then dosing must be repeated frequently. Benefits of sustained-release tablets would be found in therapies that would otherwise require multiple daily doses to achieve efficacy.

Although there is a wide array of animal health products available in oral tablet form, there are relatively few sustained-release tablets. The mechanisms of drug release from most oral controlled-release systems rely primarily on diffusion or erosion of a drug-polymer matrix or on osmotic pumping from a tablet coated with a nonerodible film or membrane. With matrix-designed tablets, a key variable in the release pattern is surface area of the tablet; hydrogels are often used as the matrix material. Administering tablets to animals is difficult at best, and one cannot always be sure that the tablet has not been bitten or chewed before swallowing. This uncertainty leads to potential variability in the controlled-release delivery profile, thus mitigating the potential value of the technology. In the case of coated tablets, such as the OROS technology, any breech of the membrane integrity by biting or chewing alters the performance of the dosage form.

Oral dosing of intact tablets can, however, be achieved by proper technique or by novel dispensers that deposit the tablet to the back of the throat without damaging tissue. With confidence in dosing technique, there are oral sustained-release technologies that could be usefully applied to veterinary dosage forms.

3. Osmotic Tablets

In addition to the conventional hydrogel-based matrix tablet forms of sustained oral delivery, there exists technology which utilizes osmotic energy to effect a constant, controlled rate of drug release into the gastrointestinal tract. The OROS® technology employs various mechanisms in an effort to attain predictable kinetic patterns of drug release in the face of the gastrointestinal tract's seven-order-of-magnitude variation in hydrogen ion concentration and its wide variations in motility and luminal contents. Thus, one can specify desired rate of in vivo drug release based on simple in vitro tests. Additionally, the technology can be used to provide desired delivery rate of orally administered drugs of different solubilities for a range of relevant durations, to maintain plasma concentrations of a drug within a narrower range than is possible with dosage forms that are specified only by quantity (total dose).

In its simplest form, the Elementary Osmotic Pump, the solid drug (alone or in combination with an osmotic driving agent) is surrounded by a semipermeable membrane having one delivery orifice (Fig. 26). During pump operation, water from the environment is continuously imbibed across the semipermeable membrane by osmosis to produce the fluid-drug formulation. The membrane's structure does not allow expansion of the tablet; thus, fluid must leave the interior of the tablet at the same rate as water enters by osmosis. As

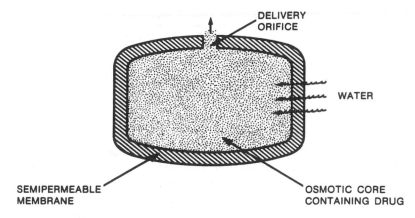

Figure 26 Elementary osmotic pump schematic.

the system moves through the gastrointestinal tract, fluid (i.e., drug in solution) flows out through the single, small orifice at a constant rate until the last of the solid drug in the core has dissolved. The membrane is excreted intact (98). There are currently no products on the market utilizing this technology for animal health applications, but there are many successful human pharmaceuticals based on OROS® technology. With the development of proper dosing techniques, this technology could have a number of applications in animal health.

C. Parenteral

1. Nonretrievable Implants

Implants have long been used in animals, mainly for production enhancement purposes. Since the early 1950s, compressed tablet implants containing estrogenic anabolic steroids have been administered for 2- to 3-month periods to improve the rate of weight gain and feed conversion in beef cattle. The economic benefits of growth enhancers have been well documented. For example, in cattle production, few if any management options equal the economic return to producers that these materials can provide. The effectiveness of the various products available is difficult to rank since conditions of use and management systems differ, causing wide variability in results.

Commercially available implants used in cattle production are inserted under the skin of the back side of the ear where they slowly release a very small amount of drug into the animal's system. The size and number of implants to be inserted vary among products, as do placement and the instrument used for implanting. Current products contain similar types of ingredients—an estrogen or estrogenlike substance that stimulates growth hormones in the animal's pi-

Table 17 Examples of Implantable Anabolic Growth Promoters

Commercial name	Active agent(s)
Compudose	Estradiol-17β
Synovex family of products	Various combinations of progesterone, estradiol ate, and testosterone propionate
Ralgro	Zeranol

tuitary gland. These weight gain responses are typical of those found in the literature: 8% in suckling calves, 15% in growing cattle, and 10% in finishing cattle (99). Throughout the literature, it is agreed that proper implanting technique is the key to obtaining consistent benefits. Examples of some commercially available products using implant technology are shown in Table 17. Other combinations of the compounds in Table 17, as well as trenbolone acetate (Finaplix), are used in other products on the market or under development.

The mechanism of action of the compressed tablet technologies is slow erosion. At least one product, Compudose, utilizes a matrix diffusion-controlled process for estradiol release. This implant is made in a continuous process by coating a nonmedicated silicone rubber core with a thin layer of silicone rubber which contains micronized crystalline estradiol-17β (Fig. 27). The strand is heat-cured, then cut into 3-cm lengths. The dimensions of the controlled-release implant are monitored so that the curved surface area for estradiol release is within target limits (99).

The use of implants for delivery of antiparasiticides has been reported in the literature (100). Ivermectin release from a crosslinked poly (ortho ester) matrix over a 6- to 12-month period has been shown to be effective in canine heartworm prophylaxis.

A slow-release device containing homidium bromide implanted in rabbits infected with *Trypanosoma congolense* provided prophylaxis against parasites. The

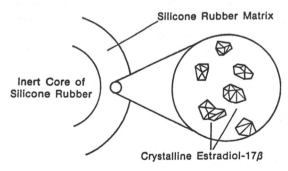

Figure 27 Compudose schematic.

implants were prepared by extrusion of a polyester composed of a copolymer of ε-caprolactone and L-lactide and homidium bromide as cylindrical fibers. The fibers were coated by dipping them in a solution of the polymer alone. The length of the rods was adjusted to achieve the target dose (101).

Bioresorbable implants have been evaluated over the years, primarily based on the lactide and glycoside family of bioerodible polymers. Implantable subdermic rods of polylactide have been shown to provide a viable mechanism of long-term drug delivery using pyrimethamine as a drug model (102).

2. Retrievable Implants

Historically, marketed implants for animal health applications have utilized erosion mechanisms (compressed tablets) or matrix diffusion control (Compudose) to effect the desired delivery of the active agent. These technologies have been suitable for the stable steroid-based products used for growth promotion purposes in food-producing animals. However, with the advent of availability of biotechnology products, many of which are peptide- or protein-based, these delivery technologies have proven to be less than optimal. The mechanism of action of these devices allows direct body fluid access to the active agent. In the case of stable steroidlike compounds, this is not a problem. However, peptides and proteins are extremely water-sensitive, often degrading in the presence of water or body fluids. Technologies under development provide protection of these water-sensitive compounds and allow for extended delivery of bioactive agents. An example of one of these technologies is the Veterinary Implantable Therapeutic System, or VITS, under development at ALZA Corporation (Fig. 28).

It is composed of an impermeable drug reservoir, into which the drug formulation is filled, and a rate-controlling semipermeable membrane cup, into which osmotic tablets are inserted with a piston stacked on top of the tablets. The filled membrane cup subassembly is then coupled with the drug reservoir. The piston isolates the osmotic energy source from the drug formulation. After implantation, water from the animal's body tissues moves across the membrane wall at a constant rate governed by the thickness and composition of the wall, the osmotic material swells, which then displaces the piston forward into the drug reservoir. The movement of the piston pumps drug formulation through the orifice into the animal. References in the literature confirm delivery of bioactive proteins for prolonged periods of time using this type of technology (103–107).

D. Transdermal

1. Passive

Transdermal technologies allow drug delivery through intact skin rather than by invasive methods, such as injection. Administering a drug transdermally is

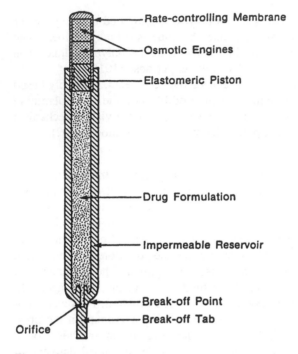

- Rate-controlling Membrane
- Osmotic Engines
- Elastomeric Piston
- Drug Formulation
- Impermeable Reservoir
- Break-off Point
- Break-off Tab
- Orifice

Figure 28 VITS schematic.

similar to giving a closely monitored intravenous infusion, but is relatively noninvasive. Many transdermal systems, commonly referred to as "patches," resemble a small adhesive bandage. The drug is usually contained in the drug reservoir compartment, and the rate of delivery is determined by a rate-controlling membrane (Fig. 29). The drug permeates the skin and enters the bloodstream at a regulated rate. Rates of drug delivery are programmed to target drug concentrations at levels that maximize therapeutic effects and minimize side effects.

At peak concentrations, drugs may cause side effects; at low concentrations they may not be effective. Rate-controlled transdermal products produce much steadier concentrations of drug, often reducing side effects. With some oral medications, the drug may be metabolized or inactivated in the stomach, intestine, or liver before reaching the bloodstream. However, when drugs are delivered transdermally, presystemic metabolism is avoided, and lower doses may be effective. An additional benefit is that treatment can be easily discontinued simply by removing the transdermal patch (108).

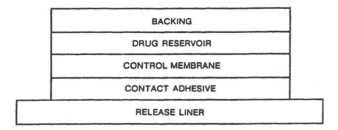

BACKING
DRUG RESERVOIR
CONTROL MEMBRANE
CONTACT ADHESIVE
RELEASE LINER

Figure 29 Transdermal system schematic.

2. Electrically Assisted

Drug delivery technologies are being developed that use low-level electrical energy to assist the transport of drugs across the skin (Fig. 30). This type of technology, electrotransport, could be used to treat animals with drugs or polypeptides that cannot be administered in oral forms or that are not suitable for passive transdermal delivery. Using this concept, higher permeation rates through the skin can be achieved than with passive transdermal drug delivery. Another benefit of the technology is that the drug absorption rate is very rapid compared with passive transdermal delivery—minutes rather than hours.

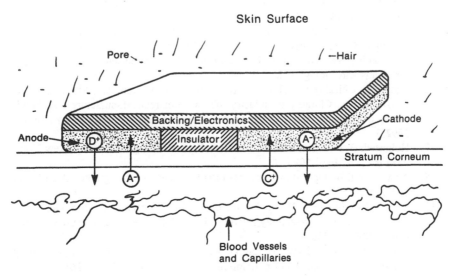

Figure 30 Electrotransport system schematic.

REFERENCES

1. Wells JI. Pharmaceutical Preformulation: The Physicochemical Properties of Drug Substances. Chichester, England: Ellis Horwood Limited, 1988.
2. Akers MJ. Personal communication.
3. Pope DG. Physico-chemical and formulation-induced veterinary drug-product bioinequivalencies. J Vet Pharmacol Ther 1984; 7:85.
4. Tomaszewski J, Rumore MM. Stereoisomeric drugs: FDA's policy statement and the impact on drug development. Drug Dev Indust Pharm 1994; 20:119.
5. FDA. Drug Stability Guidelines. Center for Veterinary Medicine, Dec. 1, 1990.
6. Young WR. Accelerated temperature pharmaceutical product stability determinations. Drug Dev Indust Pharm 1990; 16:551.
7. Yoshioka S, Carstensen JT. Rational storage conditions for accelerated testing of stability of solid pharmaceuticals. J Pharm Sci 1990; 79:943 (1990).
8. Nema S, Washkuhn RJ, Beussink DR. Photostability testing: An overview. Pharm Technol 1995; 170.
9. Carstensen JT, Rhodes CT. Cyclic temperature stress testing of pharmaceuticals. Drug Dev Indust Pharm 1993; 19:401.
10. Carstensen JT, Franchini M, Ertel K. Statistical approaches to stability protocol design. J Pharm Sci 1992; 81:303.
11. Bohidar NR. Short-term stability determination using SAS. Drug Dev Indust Pharm 1991; 17:39.
12. Golden MH, Cooper DC, Riebe MT, Carswell KE. A matrixed approach to long-term stability testing of pharmaceutical products. J Pharm Sci 1996; 85:240.
13. Patel I, Marshall KRB, Williams C, Othman H, Crosby NT. Investigation of the stability of medicinal additives in animal feedingstuffs to prepare reference feeds. Analyst 1994; 119:1483.
14. Mutton IM. In: Subramanian G, ed. A Practical Approach to Chiral Separations by Liquid Chromatography. New York: VCH Publishers, 1994:329.
15. FDA. Policy Statement for the Development of New Stereoisomeric Drugs. Center for Drugs and Biologics, May 1, 1992.
16. Crosby J. Chirality in industry - An overview in: Collins AN, Sheldrake GN, Crosby J, eds. Chirality in Industry: The Commercial Manufacture and Applications of Optically Active Compounds. Chichester, England: John Wiley & Sons, 1992:3.
17. Holmstedt B, Frank H, Tests B, eds. Chirality and Biological Activity. New York: Alan R. Liss, Inc., 1990.
18. Armstrong DW. Optical isomer separation by liquid chromatography. Anal Chem 1987; 59:84A.
19. Ruffolo RR, Bondinell W, Hieble JP. α- and β-adrenoceptors: From the gene to the clinic. 2. Structure-activity relationships and therapeutic applications. J Med Chem 1995; 38:3681.
20. Midha KK, Hubbard JW, McKay G, Rawson M, Schwede R. Stereoselectivity in bioequivalence studies of nortriptyline. J Pharm Sci 1995; 84:1265.
21. Roy SD, Chatterjee DJ, Manoukian E, Divor A. Permeability of pure enantiomers of ketorolac through human cadaver skin. J Pharm Sci 1995; 84:987.

22. Hammer RE, Brinster RL, Palmiter RD. Use of gene transfer to increase animal growth. Cold Spring Harb Symp Quant Biol 1985; 50:379.
23. Bolt DJ, Pursel VG, Rexroad CE, Wall RJ. Genetic engineering of livestock: Challenges for the delivery scientist. Proc Int Symp Control Rel Bioact Mater 1990; 17:47.
24. Couvreur P, Huang L, eds. Pharmaceutical Approaches to Oligonucleotide and Gene Therapy, Controlled Release Society Workshop, Seattle, Wash., Aug. 3–4, 1995.
25. Akers MJ. Preformulation testing of solid oral dosage form drugs methodology management and evaluation. Can J Pharm Sci 1976; 11:1.
26. Jacobs AL. Determining optimum drug/excipient compatibility through pre-formulation testing. Pharm Manuf 1985; 2:42.
27. Brittain HG, et al. Physical characterization of pharmaceutical solids. Pharm Res 1991; 8:963.
28. Chowhan ZT. Excipients and their functionality in drug product development. Pharm Technol 1993; 17:72.
29. Chowhan ZT. Drug sunstance physical properties and their relationship to the performance of solid dosage forms. Pharm Technol 1994; 18:45.
30. Blodinger J. Formulation of drug dosage forms for animals. In: Blodinger J. ed. Formulation of Veterinary Dosage Forms. New York: Marcel Dekker, Inc., 1983:135–173.
31. Taylor K. Personal communication.
32. Cardinal JR. Veterinary Applications of Controlled Drug Delivery. Controlled Release Society Workshop, Orlando, Fl., July 30–31, 1992.
33. Ascoli G, Bertucci C, Salvadori P. Stereospecific and competitive stereospecific and competitive binding of drugs to human serum albumin. J Pharm Sci 1995; 84:737.
34. Rahim S, Aubry AF. Location of binding sites on immobilized human serum albumin for some nonsteroidal anti-inflammatory drugs. J Pharm Sci 1995; 84:949.
35. Pope DG, Baggot JD. The basis for selection of the dosage form. In: Blodinger J. ed. Formulation of Veterinary Dosage Forms. New York: Marcel Dekker, Inc., 1983:1–70.
36. Hieble JP, Bondinell WE, Ruffolo RR. α- and β-sdrenoceptors: From the gene to the clinic. 1. Molecular biology and adrenoceptor subclassification. J Med Chem 1995; 38:3415.
37. Ruffolo RR, Bondinell WE, Hieble JP. α- and β-sdrenoceptors: From the gene to the clinic. 2. Structure-activity relationships and therapeutic applications. J Med Chem 1995; 38:3681.
38. Smith DJ, Feil WVJ, Huwe JK, Paulson GD. Metabolism and dispositon of ractopamine hydrochloride by turkey poults. Drug Metab Dispos 1993; 21:624.
39. Mersmann HJ. Evidence of classic β_3-adrenergic receptors in porcine adipocytes. J Anim Sci 1996; 74:984.
40. Code of Federal Regulations, U.S. Government Printing Office, Washington, D.C., 1996.

41. Laing N. Reducing drug potency loss. Feed Management 1994; 45(12):17.
42. Ogden TL, ed. British Occupational Hygiene Society Technical Guide No. 4., Dustiness Estimation Methods for Dry Methods. Northwood, U.K.: Science Reviews Ltd., 1995.
43. Stauber D, Beutel R. Determination and control of the dusting potential of feed premixes. Fresenius Z Anal Chem 1984; 318:522.
44. Carlson KH, Herman DR, Markey TF, Wolff RK, Dorato MA. A comparison of two dustiness evaluation methods. Am Indust Hyg Assoc J 1992; 54(7):448.
45. Kniep H, Wolfe WA, Zarow AI. Pellet Mill Operators Manual. Arlington, VA.: American Feed Manufacturers Association, 1982:1-40.
46. Hicks DC, Freese HL. Extrusion and spheronizing equipment. In: Ghebre-Sellassie I, ed. Pharmaceutical Pelletization Technology. New York: Marcel Dekker, Inc., 1989:71-100.
47. O'Connor RE, Schwartz JB. Extrusion and spheronization technology. In: Ghebre-Sellassi I, ed. Pharmaceutical Pelletization Technology. New York: Marcel Dekker, Inc. 1989:187-216.
48. Klothen I. Process for the Preparation of Medicated Animal Feed Supplement, U.S. Patent 4,447,421, May 8, 1984.
49. Official Publication of the Association of American Feed Control Officials, Inc., Boulder, Colo.: Johnson Printing, 1995:247-248.
50. Shryock TR, Klink PR, Readnour RS, Tonkinson LV. Effect of Bentonite incorporated in a feed ration with Tilmicosin in the prevention of induced *Mycoplasma gallisepticum* airsacculitis in broiler chickens. Avian Dis 1994; 38:501-505.
51. Lilly Research Laboratories. Unpublished data.
52. Sparks RE, Mason NS, Jacobs IC. New Process for Microencapsulation and Granulation. St. Louis, Mo.: Washington University Technology Associates, no date, pp. 16-17.
53. Beuchat LR. Microbial stability as affected by water activity. Cereal Foods World 1981; 26(7):345.
54. Federal Register, Vol. 44(53). U.S. Washington, D.C.: Government Printing Office, March 16, 1979.
55. Wicker DL, Poole DR. How is your mixer performing? Feed Management 1991; 42(9):40.
56. Wilcox RA, Unruh DL. Feed Mixing Times and Feed Mixers. Kansas State Extension Service Pub. MF-829.
57. Yalkowsky SH. Techniques of Solubilization of Drugs. New York: Marcel Dekker, 1981:91.
58. Tingstad JE. Physical stability testing of pharmaceuticals. J Pharm Sci 1964; 53:955.
59. Ansel HC, Popovich NG, Allen LV Jr. Pharmaceutical Dosage Forms and Drug Delivery Systems. Malvern Pa.: Williams and Wilkins, 1995:253.
60. McEllhiney RR, Bitel SM, Lierz TR. Towards rapid mixing. Feed Management 1990; 41(8):27.
61. Code of Federal Regulations 21 § 558.311. Washington, D.C.: U.S. Government Printing Office, 1996.

62. McKinzie CO. U.S. Patent 3,961,081 (1976).
63. Schroeder JJ, Findley JE. U.S. Patent 4,431,675 (1984).
64. Rawlings RT, Bowman K, Holton BR. U.S. Patent 5,335,625 (1994).
65. Bernard B, Lazarus J. Semisolids. In: Lachman L, Lieberman HA, Kanig JL, eds. The Theory and Practice of Industrial Pharmacy. 3rd ed. Philadelphia: Lea & Febiger, 1986:534-563.
66. Office of New Animal Drug Evaluation. Guideline for target animal and human food safety, drug efficacy, environmental and manufacturing studies. For: Anti-Infective Bovine Mastitis Products. Draft. Feb. 1, 1993. Rockville, Md: CVM.
67. Bodie RL, Nickerson SC. Dry cow therapy: effects of drug administration on occurrence of intramammary infection. J Dairy Sci 1986; 69:253-257.
68. Banker GS, Anderson NR. Tablets. In: Lachman L, Liberman HA, Kanig JL, eds. The Theory and Practice of Industrial Pharmacy. 3rd ed. Philadelphia PA: Lea & Febiger, 1986:293-345.
69. Ansel HC, Popovich NG, Allen LV Jr. Peroral solids, capsules, tablets and controlled-release dosage forms. In: Pharmaceutical Dosage Forms and Drug Delivery Systems. 6th ed. Malvern Pa.: Williams & Wilkins, 1995:155-225.
70. Rudnic EM, Schwartz JB. Oral solid dosage forms. In: Remington, ed. The Science and Practice of Pharmacy. 19th ed. Easton Pa.: Mack Publishing Company, 1995; pp. 1615-1649.
71. Rudnic EM, Kottke MK. Tablet dosage forms. In: Banker GS, Rhodes CT, eds. Modern Pharmaceutics. 3rd ed. New York: Marcel Dekker, 1996:333-394.
72. Kaneniwa N, Imagawa K, Ichikawa JI. The effect of particle size on the compaction properties and compaction mechanism of sulfadimethoxine and sulfaphenazole. Chem Pharm Bull 1988; 36:2531-2537.
73. Augsburger LL. Hard and soft shell capsules. In: Banker GS, Rhodes CT, eds. Modern Pharmaceutics. 3rd ed. New York: Marcel Dekker, 1996:333-394.
74. Use of aseptic processing and terminal sterilization in the preparation of sterile pharmaceuticals for human and veterinary use. Fed Reg 1991; 56:51354.
75. Chew NJ. Sterilization process validation. BioPharm 1993; 20(Oct.).
76. Forcinio H. Increasing confidence in aseptic filling options. Pharm Technol 1996; 332 (March).
77. Sterility and pyrogen requirements for injectable drug products. In: Staff Manual Guide. Food and Drug Administration, Center for Veterinary Medicine Policy and Procedures Manual. Guide 1240.4122, 1995.
78. Ferguson TH. Sterilization of controlled release systems. In: Hsieh DST, ed. Controlled Release Systems: Fabrication Technology, Vol. II. Boca Raton, Fla.: CRC Press, Inc., 1988:163.
79. PMA's Sterilization Steering Committee. Terminal sterilization and aseptic processing. Pharm Technol 1992; 52(April).
80. Yoshioka S, Aso Y, Otsuka T, Kojima S. The effect of γ-irradiation on drug release from poly(lactide) microspheres. Radiat Phys Chem 1995; 46:281-285.
81. Merkli A, Heller J, Tabatabay C, Gurny R. Gamma sterilization of a semisolid poly(ortho ester) designed for controlled drug delivery—validation and radiation effects. Pharm Res 1994; 11:1485-1491.

82. Ruiz JM, Busnel JP, Benoit JP. Influence of average molecular weights of poly(DL-lactic acid-co-glycolic acid) copolymers 50/50 on phase separation and in vitro drug release from microspheres. Pharm Res 1990; 7:928–934.

83. Machi S. New trends of radiation processing applications. Radiat Phys Chem 1996; 47:333–336.

84. Sebert P, Bourny E, Rollet M. Gamma radiation of carboxymethylcellulose: technological and pharmaceutical aspects. Int J Pharm 1994; 106:103–108.

85. El-Bagory I, Reid BD, Mitchell AG. The effect of gamma radiation on the tableting properties of some pharmaceutical excipients. Int J Pharm 1994; 105:255–258.

86. Yoshioka S, Aaso Y, Kojima S. Drug release from poly(dl-lactide) microspheres controlled by γ-irradiation. J Controlled Release 1995; 37:263–267.

87. Volland C, Wolff M, Kissel T. The influence of terminal gamma-sterilization on captopril containing poly(D,L-lactide-co-glycolide) microspheres. J Controlled Release 1994; 31:293–305.

88. Bawa R, Mahendra N. Physicochemical considerations in the development of an ocular polymeric drug delivery system. Biomaterials 1990; 11:724–728.

89. Schwarz C, Mehnert W, Lucks JS, Mueller RH. Solid lipid nanoparticles (SLN) for controlled drug delivery. I. Production, characterization and sterilization. J Controlled Release 1994; 30:83–96.

90. Code of Federal Regulations. Washington, D.C.: U.S. Government Printing Office, 1995.

91. Sanders LM, Kent JS, McRae GI, Vickery BH, Tice TR, Lewis DH. Controlled release of a luteinizing hormone-releasing hormone analogue from poly(d,l-lactide-co-glycolide) microspheres. J Pharm Sci 1984; 73:1294.

92. Pitman IH, Rostas SJ. Topical drug delivery to cattle and sheep. J Pharm Sci 1981; 70:1181–1194.

93. Kyles AE, Papich M, Hardie EM. Evaluation of the pharmacokinetics of transdermal fentanyl in the dog. Vet Surg 1994; 23:407.

94. Gupta S, Srivastava JK, Katiyar JC, Jain GK, Singh S, Sarin JPS. Transdermal device of a substituted benzimaidazole carbamate and its efficacy against helminth parasites. Trop Med 1992; 34:113–119.

95. Berghen P, Hilderson H, Vercruysse J, Claerebout E, Dorny P. Field evaluation of the efficacy of the fenbendazole slow-release bolus in the control of gastrointestinal nematodes of first-season grazing cattle. Vet Q 1994; 16:161–164.

96. Baggott DG, Ross DB, Preston JM, Gross SJ. Nematode burdens and productivity of grazing cattle treated with a prototype sustained-release bolus containing ivermectin. Vet Rec 1994; 135:503–506.

97. Ho MYK, Gottschall DW, Wang R. Albendazole in cattle administered via a sustained-release "Captec" device. In: Hutson DH, ed. Xenobiotics and Food-Producing Animals: Metabolism and Residues. New York: American Chemical Society, 1992:149–157.

98. ALZA Corporation. ALZA Technology. Palo Alto, Calif.: ALZA Corporation, 1992:8–9.

99. Doane Information Services. Growth stimulants for beef cattle. In: Doane's Agricultural Report. St. Louis: Doane Information Services 1988:266.1–266.2.

100. Ferguson TH, Needham GF, Wagner JF. Compudose: an implant system for growth promotion and feed efficiency in cattle. J Controlled Release 1988; 8:45–54.
101. Shih C, Fix J, Seward RL. In vivo and in vitro release of ivermectin from poly(ortho ester) matrices. I. Crosslinked matrix prepared from ketene acetal end-capped prepolymer. J Controlled Release 1993; 25:155–162.
102. Geerts S, De Deken R, Kageruka P, Lootens K, Schacht E. Evaluation of the efficacy of a slow release device containing homidium bromide in rabbits infected with *Trypanosoma congolense*. Vet Parasitol 1993; 50:15–21.
103. Azain MJ, Bullock KD, Kasser TR, Veenhuizen JJ. Relationship of mode of porcine somatotropin administration and dietary fat to the growth performance and carcass characteristics of finishing pigs. J Anim Sci 1992; 70:3086–3095.
104. Baile CA, Kasser TR, Hampton TR, et al. The bST delivery profile and efficacy in Holstein heifers receiving intraperitoneal 84-day sustained delivery systems. J Anim Sci 1993; 71(suppl 1):131.
105. Kasser TR, Day JW, Hale RL, Hampton TR, Hartnell GF, Baile CA. Continuous intraperitoneal bST delivery and improved performance in feedlot steers also treated with steroids, tylosin and monensin. J Anim Sci 71(suppl 1):131.
106. Armstrong JD, Harvey RW, Poore MH, et al. Effect of sometribove or immunization against growth hormone releasing factor (GRFi) on milk yield and composition, calf gain, insulin, and metabolites in multiparous beef cows. J Anim Sci 1994; 72(suppl 1):182.
107 Stine CA, Althen TG, Essig HW, et al. Effects of rBST and protein supplementation on growth performance and IGF-I of stocker steers. J Anim Sci 1994; 72(supply 1):182.
108. Eckenhoff JB, Theeuwes F, Urquhart J. Osmotically actuated dosage forms for rate-controlled drug delivery. Pharm Technol 1987; 96–105.

3

Protein/Peptide Veterinary Formulations

TODD P. FOSTER
Pharmacia & Upjohn, Inc., Kalamazoo, Michigan

I. INTRODUCTION

The Russian scientists Asimov and Krouze were the first to show the usefulness in animals of exogenous delivered proteins when they observed galactopoietic efficacy of a crude extract of ox anterior-pituitary administered to cows (Asimov and Krouze, 1937). One year later, in 1938, Folley and Young published similar results showing a single crude extract of ox anterior-pituitary extract injection yielded a mean 10% daily rise in milk over 5 or 6 days (Young, 1947). With the advent of World War II, the interest in significantly increasing Britian's milk supply using hormones intensified (Young, 1947). More than 50 years later a highly purified form of those first crude anterior-pituitary extracts was given regulatory approval for increasing milk yield in lactating cows. In 1988 the first approval was granted in South Africa, which was followed by approval in 1993 in the U.S. The large market in the U.S. has resulted in bovine somatotropin (bSt) being the first protein administered to animals with potential to become a commercial success. Agricultural and pharmaceutical companies hope to capitalize on the 1940 early market research showing the gross profit per cow per lactation cycle increased 4 shillings when accounting for the 11-shilling cost per animal for the extract and the extra food the animals consumed (Young, 1947).

The 50-year delay from concept to market introduction for bovine growth hormone will not be typical for other proteins/peptides approved for use in food-producing animals. One major challenge in gaining regulatory approval is discovering a marketable dosage form. The protein/peptide formulation must be chemically and physically stable. The compound must be compatible with its packaging and maintain stability during storage to ensure adequate biological activity. The formulation must be manufacturable, be easy to administer, leave no harmful residues, be safe to the animal, and be economically viable when used. Like nonprotein/peptide molecules, the physicochemical properties need to be elucidated so that a successful formulation may be discovered. Additionally, the larger molecular weight of proteins, compared to nonprotein/peptide molecules, requires additional analytical methods to adequately characterize the molecules.

This chapter describes the analytical methods used to characterize protein/peptides, and presents a comprehensive review of protein/peptide veterinary formulations. Examples of proteins or peptides being developed for use in food-producing animals are given. These include bSt for increased milk lactation and growth, porcine somatotropin (pSt) for growth, and bovine growth hormone-releasing factor (bGRF) for growth and lactation in bovines. Other proteins/peptides mentioned are gonadotropin for reproduction synchronization and lysostaphin for mastitis control. Several unique indications and formulations, such as bSt for growth in salmon and chickens, are described. The chapter has been organized with respect to proteins/peptides to provide individuals interested in a particular molecule with a review of the molecular characterization and formulations.

II. CHARACTERIZATION

Although the ultimate goal is a marketable formulation, an adequate understanding of the available analytical tools and methods to characterize the molecule and formulation must be appreciated. These analytical techniques are used to discover a formulation and to ensure quality of the bulk and formulated protein/peptide. Selection of the most appropriate formulation is based on the physicochemical properties of the protein/peptide and the required product attributes. The major molecular properties evaluated, the reasons for examining each property and the analytical techniques used are shown in Table 1. If further details on specific analytical techniques for characterizing proteins/peptides are required, the reader is referred to published review papers (Randall et al., 1991; Pearlman and Nguyen, 1991; Hanson and Rouan, 1992; Wang and Hanson, 1988; Chen, 1992).

Table 1 Protein/Peptide Physicochemical Properties and Characterization Techniques

Physicochemical Properties	Reason for characterizing	Techniques used
Solubility	Select formulation dosage form/vehicle, help understand ADME[a]	Saturated solubility[b]
Dissolution	Understand in vivo release/ADME, time for powder reconstitution	Rotating disk[b]
Aggregation	Affects ADME, potency, safety, pharmaceutical elegance	SEC, turbidity measurements
Particle size	Influence suspension stability, in vivo dissolution, powder flow	Microscopy, laser light scattering, analytical ultracentrifugation
Molecule size	Defines the molecular entity	SDS-PAGE, SEC, dynamic light scattering
Primary structure	Defines the molecular entity	Peptide mapping, HPLC
Secondary structure	Alpha helical, beta sheets and random coil will influence other physico-chemical perties	FTIR, x-ray crystallography far UV-CD
Tertiary structure	Defines stability and biological activity	Near UV-CD, x-ray crystallography, Raman spectroscopy
Quaternary structure	Defines stability and biological activity	SEC, x-ray crystallography
Solid state stability	Helps defines storage/handling, formulation and manufacturing strategy	Stability indicating assay[c]
Solution stability	Helps define formulation and manufacturing strategy	Stability indicating assay[c]
Hygroscopicity	Helps define drug handling/storage, useful when examining stability, helps define formulation strategy	Moisture isotherms
Flowability	Useful for some manufacturing methods	Compressibility, shear cell, angle of repose
Isoelectric point	Helps predict aqueous solubility and stability conditions	Calculated, isoelectric focusing

[a]Abbreviations: (ADME = absorption, distribution, metabolism, and excretion; SDS-PAGE = sodium dodecyl sulfate polyacrylamide gel electrophoresis; SEC = size exclusion chromatography; CD = circular dichroism; RP-HPLC = reversed-phase high-performance chromatography; FTIR = Fourier transform infrared spectroscopy.
[b]Measured by RP-HPLC, UV/fluorescence spectroscopy, colorimetric methods, immunoassays, others.
[c]The stability-indicating assay is unique to the protein/peptide and is correlated with efficacy. It could be a biological potency assay (e.g., rat growth) or a chemical assay (e.g., RP-HPLC).

A. Somatotropins

Native bovine somatotropin (also called bovine growth hormone, bSt, BGH) is a 191-amino acid (AA) globular protein produced in the pituitary gland. Four companies have used the native bSt, or modified the NH_2-terminal sequence with one to eight amino acids to create a molecule that can be formulated and used to increase milk production in dairy cattle (Davio and Hageman, 1993). Porcine somatotropin is a 191-AA globular protein with a high degree of homology (> 90%) to bSt (Carlacci et al., 1991). It is being used to increase both feed efficiency and lean/fat ratio in swine (Etherton et al., 1986, McNamara et al., 1991).

1. Structural Characterization

An FDA-approved product must be safe and efficacious. To ensure safety and efficacy, adequate characterization of the molecule and formulation must be made. The process of recombinantly synthesizing, isolating, and purifying bovine somatotropin can result in undesirable structural and chemical modifications if inappropriate conditions are utilized (Wingfield et al., 1987a; Dellacha et al., 1968). Recombinantly produced bSt, nearly identical to pituitary bSt, is produced when appropriate and controlled conditions are used (Langley et al., 1987a,b; Wood et al., 1989; Leung et al., 1986). Formulating the product as a solution, lyophilizing it to a powder, or compressing it into an implant may alter the structure and render the molecule biologically inactive. To ensure that deleterious structural changes do not occur, the primary, secondary, tertiary, and quaternary structures are examined.

   a. *Primary Structure.* The primary structure is the unique amino acid sequence of the molecule. The method for determining this sequence involves digesting thermally, chemically, or enzymatically the protein/peptide followed by identification via amino acid composition analysis, N- and C-terminal analysis, or mass spectrometry. Separation of the fragments can be done using chromatographic methods such as gas, thin-layer, ion-exchange, or reversed-phase high performance liquid (RP-HPLC) chromatography. Hartman et al. (1986) have used trypsin to cleave bSt, RP-HPLC to separate the tryptic peptides and amino acid analysis to identify the bSt components. Trypsin was selected as the cleaving agent because of its specificity for the carboxyl side of the peptide bonds in lysine and arginine. Reversed-phase HPLC using wide-pore silica and perfluorinated acids as solvent modifiers was utilized because of its resolving power and speed. Amino acid analysis using phenylthiocarbamyl derivatives was selected because of its speed and sensitivity. Tryptic mapping bSt has also been studied by Hara et al. (1978), Dougherty et al. (1990), and Yamasaki et al. (1970). An RP-HPLC method for analyzing tryptic digests of pSt has been developed and validated (Charman et al., 1993).

The primary sequences of pituitary bSt and pSt have been reported (Wallis, 1973; Santome et al., 1973; Graf and Li, 1974; Seeburg et al., 1983). In studying bSt, Wallis (1973) used trypsin, α-chymotrypsin, pepsin, carboxypeptidases A and B, and cyanogen bromide to cleave the protein. The soluble peptides were fractionated using either gel filtration followed by paper electrophoresis and chromatography, or a peptide mapping technique. Further details on the primary structure and differences between the four recombinant bSt molecules on the market or in development (somavubove, sometribove, somidobove, and somagrebove) are discussed elsewhere (Davio and Hageman, 1993). Seeburg and co-workers (1983) described the construction of the bacterial vector used to produce pSt and the AA sequence of pSt. Homology of 90% between pSt and bSt was observed.

Changing the amino acid sequence is one method of producing a somatotropin with advantageous formulation properties. Mutations in helices 1, 2, or 3 of rpSt were attempted to decrease the tendency for protein aggregation at high protein concentrations (Fischer et al., 1991). For example, one focus by Fischer and co-workers was the amino acid residues 112 to 129 in the hydrophobic face of the α helix 3 region in rpSt. An isoleucine at position 122 was replaced with leucine, along with changing the cysteines to alanines at positions 181 and 183. It was claimed that enhanced physical solution stability was obtained, although no supporting data were presented.

b. Secondary Structure. Proteins and peptides rarely exist as linear molecules; instead, they adopt different folded patterns or conformations. The first level of folding is referred to as secondary structure and consists of amino acids arranged along a polypeptide backbone. They may exist as well-ordered structures such as α helices, β sheets, loops, or unordered random coils. A single crystal x-ray diffraction technique was used with methionyl porcine somatotropin to show it is mainly helical (54% of the AA) containing four tightly packed antiparallel α helices (Abdel-Meguid et al., 1987). They determined that the four helices consisted of residues 7–34, 75–87, 106–127, and 152–183.

Fourier transform infrared (FT-IR) spectroscopy is often used to characterize the secondary structure of proteins. The amide linkages of the protein produce vibrations in distinct regions termed amide I (1620 to 1700 cm^{-1}), amide II (1520 to 1580 cm^{-1}), and amide III (1220 to 1350 cm^{-1}) (Havel et al., 1989). The amide I region of the IR absorption spectra for bSt is shown in Figure 1. Various amounts of α helical, β sheet, β turn, and disordered structure are observed depending on how the bSt was treated (Havel et al., 1989).

Far-UV circular dichroism spectroscopy can also be used to determine secondary structure. This technique, which uses elliptically polarized light to characterize amide-bond absorption, showed 45% to 50% α helicity in bSt (Holladay et al., 1974). Sonenberg and Beychok (1971) stated that bSt has 10% β struc-

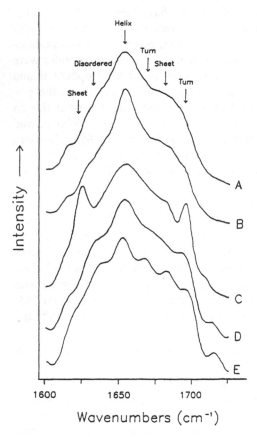

Figure 1   The amide I region of the IR spectra of solid-state bSt. Band resolution was enhanced by Fourier self-deconvolution. The positions of vibrations due to α helix, β sheet, β turn, and disordered secondary structure are shown with arrows. (A) Reduction of the dM. (D) bSt γ-irradiated in nitrogen environment. (E) bSt γ-irradiated in air. (Reprinted with permission from Havel *et al.*, 1989. Copyright 1989, American Chemical Society.)

ture and 55% α-helical content based on examination of a far-UV CD spectrum. Far-UV CD spectroscopy was used to show that pituitary bSt and recombinant bSt are identical (Langley et al., 1987a). In the far-UV spectra, negative bands at 208 and 221 nm existed when both compounds were examined. The molar ellipticities at 221 nm were approximately 13,500, which was consistent with the published 45% to 55% α-helical content. For more data on far-UV CD spectroscopy, Havel and co-workers (1989) have reviewed far-UV CD spectrums obtained with bSt.

c. Tertiary Structure. The tertiary structure refers to how the secondary structure orients to produce some three-dimensional structure. Understanding tertiary structure assists in identifying amino acid residues involved in binding of the somatotropin to its receptor. Using pST and x-ray diffraction, Abdel-Meguid et al. (1987) showed that the antiparallel α helices were connected differently than other proteins. They discovered an up-up-down-down connectivity (helix A connected to C, C to B, and B to D) that has never been reported for other proteins.

Carlacci et al. (1991) used the pSt three-dimensional structure and a combination of heuristic approach and energy minimization to predict the tertiary structure of bSt. This molecular modeling technique uses the amino acid sequence, rotational constraints on the helices, constraints on distance between atoms due to the disulfide bonds, and similarity of loops to arrive at stereoribbon, or space-filling drawings.

d. Quaternary Structure. Quaternary structure refers to how aggregates of simple subunits of proteins are constructed. The somatotropins do not have identified subunits like hemoglobin. The quaternary structure has been defined as the noncovalent interaction or spatial arrangement of individual protein subunits (Oeswein and Shire, 1991). Using this definition, somatotropins would not generally have quaternary structure. However, they do aggregate to create higher-molecular-weight entities (e.g., dimers, trimers, and hither oligomers). Hageman et al. (1992) showed the rate of dimer formation occurring when incubating 10 mg/ml pSt and bSt in pH 9.8 bicarbonate buffer at 30°C. Their high-performance size exclusion chromatography method (HPSEC) used SDS in the mobile phase, so only covalently dimerized or higher oligomers were detected. Further details on HPSEC used for analyzing bSt have been reported (Stodola et al., 1986). Their attempt to use a simple, nondenaturing buffer mobile phase to study bSt in its native conformation, instead of using SDS, was not successful as the bSt did not dissolve in the buffer. Chang et al. (1994) were successful in developing a nondenaturing HPSEC assay to measure the potency of somidobove. A rat mass-gain assay and radioreceptor assay showed that somidobove was nondenatured when exposed to the Spherogel TSK 3000 PW column and the ammonium hydrogencarbonate buffer (pH 9.0) as mobile phase.

When studying the solution thermostability of pSt, covalent dimeric species have been identified (Buckwalter et al., 1992). Phosphate-buffered solutions containing up to 250 mg/ml pSt were incubated for 14 days at 39°C. The solutions were initially clear but by day 14 were thixotropic gels. Analysis with SDS-PAGE showed that a covalent dimeric species with a molecular weight (MW) of 44,000 was formed.

2. Chemical Stability

Information on the chemical stability of somatotropins helps assist in defining methods to recombinantly produce somatotropins (Violand et al., 1989, Lewis et al., 1970; Liberti et al., 1969; Glaser and Li, 1974; Cascone et al., 1980; Wolfenstein-Todel et al., 1983; Delfino et al., 1986; Graf et al., 1975). The temperature, ionic strength, solution holding times, somatotropin concentrations, buffer, surfactant, and detergent types and concentrations all influence the stability of the somatotropin. The use of the optimal somatotropin solution stability conditions will help ensure that a high-quality somatotropin is produced. Knowledge of the chemical stability is also important in selecting a formulation (Davio and Hageman, 1993; Buckwalter et al., 1992; Ferguson et al., 1988, Hageman et al., 1992). Good reviews on the chemical stability of protein pharmaceuticals have been published (Manning et al., 1989, Clarke et al., 1992, Kosen, 1992). Protein chemical instability includes decomposition via deamidation, racemization, hydrolysis, oxidation, beta elimination and disulfide exchange (Manning et al., 1989).

Somatotropins undergo a deamidation reaction where side chain amide linkages are hydrolyzed to form a carboxylic acid. Lewis et al. (1970) proposed that deamidation of glutamyl and asparaginyl residues in bSt occurred. They observed faster-migrating compounds during gel electrophoresis and NH_3 liberation. Increased temperature, increased pH (above pH 7.5), increased ionic strength, and the addition of urea all caused a greater rate of conversion. Violand et al. (1990) was able to isolate asparagine 99 as one amino acid undergoing deamidation in pSt and bSt. They proposed that isoaspartate residue was formed via a succinimide rearrangement. Peptide bond cleavage at position 99 also occurred, but to a lesser degree.

Oxidation of somatotropins may occur. Glaser and Li (1974) used hydrogen peroxide to oxidize three of the four methionine residues of pituitary bSt. They showed that both structural and biological activity were maintained after oxidation. Cascone et al. (1980) showed that methionine 4 was the most reactive site followed in decreasing order by methionine at positions 148, 123, and 178. They also showed, using a rat growth bioassay, that biological activity was maintained in the oxidized bSt.

Modification of the somatotropin through reactions with the carboxylate groups can occur (Liberti et al., 1969; Delfino et al., 1986). When methylated in acid methanol, bovine somatotropin retained biological activity if less than 15% esterification occurred (Liberti et al., 1969). However, biological activity was progressively lost once between 15% and 50% esterification occurred. Once 50% esterified, no growth-promoting activity remained in the preparation. Other methods to stabilize somatotropins include modification of arginines in bSt (Wolfenstein-Todel et al., 1983) and the COOH-terminal disulfide bond (Graf et al., 1975).

Usually a marketed formulation requires at least a 2-year shelf life when stored using the labeled conditions. Two years allows enough time for adequate inventory and product distribution. Additionally, upon injection the protein/peptide must be stable for long enough to allow adequate absorption. If inadequate stability exists upon administration, the formulation vehicle may be used to protect the molecule. Studying somatotropin stability in solution and in the solid state assists in selecting a good formulation. With respect to solution stability, Hageman et al. (1992) showed that bSt and pSt at 30°C in a bicarbonate buffer (pH 9.8) had half-lives of less than 14 days when using a RP-HPLC assay. Deamidation and chain clipping at position 99 were thought to be the major processes of degradation in solution. Modification of the somatotropin may help resolve solution instability. Buckwalter and co-workers (1992) were able to increase the solution stability of pSt by carboxymethylating the protein. Figure 2 shows the increased solution stability obtained for carboxymethylated pSt when the initial protein concentration was greater than 25 mg/ml.

The major degradation pathway in the solid state for bSt and pSt is covalent crosslinking instead of deamidation and chain clipping observed in solutions (Hageman et al., 1992). This is presumably due to intermolecular interaction occurring with the close-proximity somatotropin molecules. The effect of oxy-

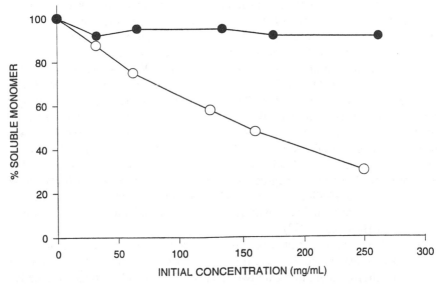

Figure 2 Concentration-dependent solution stability of carboxymethylated pSt (●, CM-pSt) and (○, pSt). The proteins were dissolved in PBS at various concentrations and incubated at 39°C in duplicate for 14 days, and the soluble monomer determined using SE-HPLC. (Reprinted with permission from Buckwalter et al., 1992. Copyright 1992, American Chemical Society.)

gen, heat, and moisture on the solid-state stability of bSt has been studied (Ferguson et al., 1988). By replacing air with nitrogen in sealed vials, they concluded that oxidation was not a major degradation pathway for powder bSt. Powder stored with an unspecified water content at 4°C had a 2% loss of potency defined by RP-HPLC during 12 months. When stored at 25°C or 37°C, loss was 12% and at least 19%, respectively.

Moisture is a key factor in the stability of somatotropins. Stability of protein powders containing less than a monolayer of water normally is at a maximum (Hageman, 1988). Decomposition increases above a monolayer of water because of increased protein conformational flexibility. Sigmoidal-shaped isotherms were observed during water sorption studies of bSt (Hageman et al., 1992). A monolayer of 5 to 8 g water/100 g bSt was shown using either the Brunauer, Emmett, and Teller equation or the improved equation by Guggenheim, deBoer, and Anderson. As the moisture content changes, the decomposition pathway changes for bSt (Bell et al., 1995). Hydrophobic aggregation is the major degradation process in the presence of water; covalent modifications (e.g. dimers, cleaved bSt) occur when moisture contents are low.

3. Physical Stability

A protein may exhibit good chemical stability, but if physical instability (i.e., aggregation) occurs, synthesis, formulating, and obtaining adequate biological activity may be extremely difficult. One of the first indications that somatotropins aggregate is the variety of molecular weights that have been reported. In the 1940s the reported MW for bSt varied between 39,300 and 45,000 daltons depending on the buffer solutions (Dellacha et al., 1966). Andrews and Folley (1963) reported the MW to be 20,000 daltons. The monomer bSt MW was reported to be approximately 22,000 daltons at either acidic (pH 3.6) or strongly alkaline (pH 11.5) conditions (Bewley and Li, 1972). Working with glycine/hydrochloric acid solutions at a pH of 3.6, a MW of around 21,000 daltons was proposed for bSt (Dellacha et al., 1968). They showed that at pH 9.4 the inclusion of sodium dodecyl sulfate (SDS) would result in a similar MW. The SDS presumably prevented aggregation, which led to the erroneous higher-MW measurements in alkaline solutions (pH 9.4). With the exact sequence determined, the pituitary bovine somatotropin MW was stated to be 21,812 daltons (Davio and Hageman, 1993).

Recombinant protein synthesized from *E. coli* as inclusion bodies produces somatotropin (St) without the disulfide bonds formed (Violand et al., 1989). A step of refolding occurs to create two disulfide bonds and introduce the native bSt and pSt structure. Studies have been conducted both to determine the optimal conditions for this folding to occur and to characterize the folding process for bSt (Brems et al., 1985, 1986, 1987; Havel et al., 1986; Edelhoch et al.,

1966; Burger et al., 1966; Holladay et al., 1974; Holzman et al., 1990) and pSt (Bastiras and Wallace, 1992; Puri, 1991; Puri and Cardamone, 1992; Cardamone et al., 1994). Guanidine hydrochloride, urea and acid were used in equilibrium denaturation studies to show that the process is multistaged. Four protein species have been identified: the native; a monomeric folded intermediate; an associated folded intermediate; and the unfolded bSt. Porcine somatotropin and bSt have similar equilibrium denaturation while the more conformationally stable human somatotropin follows a two-step process (Bastiras and Wallace, 1992).

Certain secondary structures—e.g., the third helix in bSt—are responsible for the association of partially unfolded bSt (Brems et al., 1986). Partially exposed lipophilic faces of the helices associate through hydrophobic interaction. To decrease aggregation, Lehrman et al. (1991) used this information to justify site-directed mutagenesis in the third helix of bSt. They substituted the human St sequence between amino acid residues 109 and 133 and then utilized near-UV CD spectroscopy, kinetic folding, size-exclusion chromatography, and dynamic light-scattering techniques to show decreased aggregation of the mutant bSt at higher protein concentrations ($> 2 \mu M$). Brems (1988) used site-directed mutagenesis to modify the refolding and precipitation of bSt.

Dimers can exist as reversible (soluble) or irreversible (aggregates) forms. The process of forming dimers has been explored (Violand et al., 1989; Mao, 1990; Oppezzo and Fernandez, 1991). Violand et al. (1991) claim that a concatenated dimer of bSt exists. They proposed that interlocking of the disulfide loops could create a bSt dimer and justified their hypothesis with trypsin peptide mapping and thrombin bSt digestion. Mao (1990) disagreed with their hypothesis, stating that thrombin digestion experiments cannot eliminate the possibility of disymmetric disulfide-linked dimers. Oppezzo and Fernandez (1991) showed the involvement of tyrosine 142 in forming bSt dimers. A loss of self-association was observed when radioiodinated Tyr-142 was introduced into the molecule.

Irreversible aggregation reduces the amount of St available for absorption in an animal. The use of excipients to prevent aggregation is one formulation technique. The method for inducing the aggregation to examine the effectiveness of the excipients is important, as was shown by Charman et al. (1993) when studying pSt. They observed irreversible aggregation of pSt when causing denaturation thermally, interfacially, or with guanidine HCl (Gdn). Using Tween 20, hydroxypropyl-β-cyclodextrin (HPCD), and sorbitol, they obtained varying degrees of success in preventing aggregation. For example, HPCD was effective in reducing precipitation when induced by thermal and interfacial methods but was not effective against Gdn-induced aggregation.

4. Solubility

Since changes to conformational structure are highly influenced by the surrounding environment, the reported solubilities of somatotropins vary depending on the conditions used. The solubility of pSt at 37°C varied from 20 mg/ml in 0.05 M Tris buffer (pH 6.5, ionic strength 0.12) to 150 mg/ml in the same buffer at pH 8 (Hageman et al., 1992). The solubility of bSt was lower with a range of 7.5 mg/ml (pH 6.5) to 19 mg/ml (pH 8) in that same Tris buffer. Decreasing the temperature to 25°C had little effect on the solubility (Fig. 3). The solubility of somidobove depended on the pH, ionic strength and salt content (Ferguson et al., 1988). Examples of solubility at 25°C included 0.49 to 1.36 mg/ml when using phosphate buffers of different pH, to 49.09 mg/ml for a 0.1 M phosphoric acid solution. The solubility will depend on the purity of the protein used and the technique employed. Techniques utilized to determine solubility have included dialysis, ultrafiltration, dissolution to saturation, and pH jumps (Davio and Hageman, 1993).

5. Pharmacokinetics

To design correct dosage regimens, an understanding of the mechanisms of drug absorption, distribution, and elimination and the kinetics of these processes is required. Studying the pharmacokinetics of somatotropins is difficult because

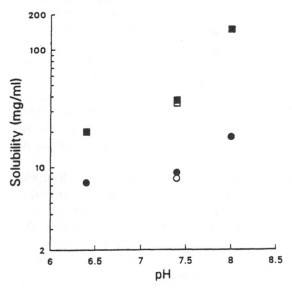

Figure 3 Solubility of rbSt (●) and rpSt (■) at 37°C as a function of pH. Solubility at pH 7.4, 25°C, is shown by the open symbols. Solubility measured in 0.05 M Tris buffer with an ionic strength of 0.12. (Reprinted with permission from Hageman et al., 1992. Copyright 1992, American Chemical Society.)

endogenous St release is pulsatile. Furthermore, the size and sex of the animal may influence the St concentrations. Arbona et al. (1988) showed that during a 6-h period approximately two secretory episodes occurred in pigs, but they were spontaneous and random. The metabolic clearance rate (MCR) of exogenous pSt in boars and gilts was 83 ml/min. The mean half-life, MCR, and secretion rate (SR) in obese, control, and lean swine were shown to be similar (Althen and Gerrits, 1986). At 15 weeks of age (mean weight 33 kg), the mean half-lives for the three groups ranged from 7.4 to 9.8 min, the MCR was 158 to 341 ml/min, and the SR was 520 to 907 ng/min. Swine at age 15 weeks had a shorter St half-life and secreted and cleared more St/kg bodyweight than 30-week-old pigs (mean weight 90 kg). Hu et al. (1995) administered a single IV dose of 250 μg rpGH/kg to six gilts. A radioimmunoassay sensitivity of 50 pg/ml was used to measure plasma St concentrations. Fitting the plasma concentration-time data to a three-compartment open model provided a terminal half-life of 40.2 ± 5.6 min, a volume of distribution of 0.029 ± 0.003 L/kg, and a total plasma clearance of 0.53 ± 0.05 ml/min/kg.

A study designed to examine the disposition of three molecular variants of recombinant bSt in Holstein cows was conducted by Eppard and co-workers (1993). The serum GH concentrations obtained for 12 h after a 25-mg rbSt bolus infusion were fitted to a two-compartment open model. No differences in pharmacokinetic parameters were observed among the three analogs. With no differences observed, average pharmacokinetic parameters were calculated by combining data from the three analogs. A clearance of 0.15 L/min/100 kg, a volume of distribution of the central compartment of 2.59 L/kg, and half-lives of 8.2 and 29.1 min in the two compartments were found when averaging the three analogs. Differences in pharmacokinetic parameters have been observed between pituitary and recombinant bSt. Toutain et al. (1993) observed lower clearance (0.119 ± 0.012 vs. 0.143 ± 0.011 L/h/kg) and volume of distribution (0.10 ± 0.018 vs. 0.12 ± 0.015 L/kg) for the pituitary bSt compared to somidobove. The half-lives of the pituitary and recombinant bSt were 61.8 ± 5.5 min and 54.8 ± 5.5 min, respectively. Somidobove has eight additional amino acids at the NH_2 terminus. Possibly reasons for these observed differences, such as the analytical techniques used, solubility of the two bSt, and protein binding are discussed, but no final explanation for the observed differences in pharmacokinetic parameters was provided.

6. Assays

Reference has been made to various assay methods. Other assays that are used to characterize bovine and porcine somatotropins include capillary electrophoresis (Tsuji, 1993), radioreceptor (Haro et al., 1984), enzyme-linked immunosorbent assay (ELISA) (Secchi et al., 1988), radioimmunoassay (Reynaert and Franchimont, 1974), optical spectroscopy (Havel et al., 1989), nuclear magnetic resonance (MacKenzie et al., 1989; Gooley et al., 1988), and chromato-focusing (Wingfield et al., 1987b).

The SDS nonacrylamide gel-filled capillary electrophoresis assay was suggested as an alternative to HPSEC to determine the composition of bSt (Tsuji, 1993). It had good peak resolution, peak time remained relatively constant, and molecular mass relative standard deviation was approximately 2% to 3%. The technique compared well to HPSEC with adequate resolution of monomer, dimer, trimer, and tetramer. A useful review of UV absorption, circular dichroism, fluorescence, Raman and infrared spectroscopy, as it applies to bSt, has been published (Havel et al., 1989).

B. Growth Hormone-Releasing Factor

Several human growth hormone-releasing factors (AA residues of 37 to 44) have been discovered (Rivier et al., 1982, Guillemin et al., 1982). By 1984 the major endogenous peptide was isolated in hypothalamic tissue as $GRF(1-44)-NH_2$ (Bohlen et al., 1983a; Ling et al., 1984a). Growth hormone-releasing factor (GRF) has been isolated in cattle, sheep, goats, pigs, rats, mice, carp, and salmon (Campbell et al., 1991; Esch et al., 1983; Brazeau et al., 1984; Bohlen et al., 1983b; Suhr et al., 1989; Frohman et al., 1989a; Vaughan et al., 1992, Parker et al., 1993). In a comprehensive review by Campbell et al. (1995), a sequence homology of 86% to 93% among human, cattle, sheep, goat, and pig GRF was reported. GRF stimulates the synthesis and release of growth hormone from the anterior pituitary gland (Esch et al., 1983; Brazeau et al., 1984). The enhanced synthesis and release of GH means that GRF could be used to increase milk production, increase growth rate, or produce a leaner carcass in food-producing animals (Dahl et al., 1990; Enright et al., 1993; Pommier et al., 1990). Because of the high sequence homology between human and bovine/swine GRFs and because relatively little information pertaining to animal GRFs has been published, both animal and human GRFs will be discussed.

1. Structural Characterization

The primary sequence of human, pig, cattle, goat, sheep, rat, mouse, carp, and salmon GRFs was reviewed by Campbell et al. (1995). It has been determined that the biologically active core is the one to 29 amino acids from the amino terminus (Coy et al., 1993; Ling et al., 1984b; Lance et al., 1984). Ling et al. (1984b) systematically deleted residues to create human GRF analogs that had 1–34, 1–31, 1–30, 1–29, 1–23, 1–22, and 1–21 amino acids. As amino acids were removed, they observed a decrease in biological activity when using a rat pituitary cell assay. In other research to define the sequence responsible for biological activity, Petitclerc and co-workers (1987) infused hGRF $(1-44)NH_2$ and $hGRF(1-29)NH_2$ into dairy heifers and pigs to show that the two compounds were biologically equivalent. The average peak concentrations of growth hormone and the area under the growth hormone response curves were not significantly different ($P > .05$) after injecting the two compounds. In a similar study, hGRF (1–26)NH_2 was shown to have lower growth hormone-releasing activity than hGRF (1–29)NH_2 and hGRF (1–44)NH_2, while the 1–29 GRF analog and 1-44 GRF had similar activity (Hodate et al., 1986). Numerous analogs of the native GRF have been synthesized to enhance the peptide's stability. These will be discussed in the section dealing with stability.

One strategy to improve the biological potency of GRF is to create analogs that have extended or stabilized α-helical regions as these analogs may have increased affinity for the receptor (Campell et al., 1991). Using circular dichroism and two-dimensional NMR spectroscopy, two human GRF analogs (1-45 and 1-29) were shown to have 23 to 25 residues in a helical state (Clore et al., 1986). Two distinct α-helical regions extend from residues 6 to 13 and 16 to 19. In water, no ordered structure of the protein was observed. A 30% trifluoroethanol solution was required to obtain helical structure. Using the same technique, a water solution containing hGRF contained helical structure between residues 9–14 and 24–28 while a 75% methanol/water solution had α-helical character between residues 4–29 (Campell et al., 1991). Analogs designed with enhanced amphiphilic, α-helical character have greater in vitro and in vivo GH releasing activity (Felix et al., 1986, 1988; Dubreuil et al., 1990). For example, the replacement of glycine at position 15 with other hydrophobic amino acids (Ala, Leu, Val) increased the extent of α-helicity and the biological potency as measured by a rat pituitary bioassay by up to fourfold (Felix et al., 1986).

2. *Stability*

GRF can decompose enzymatically or chemically. The primary enzyme for degrading bovine and porcine GRFs is dipeptidylpeptidase IV enzyme (DPP-IV) (Martin et al., 1993; Kubiak et al., 1989, 1992a,b, 1993; Su et al., 1991; Frohman et al., 1986, 1989b; Mentlein et al., 1993; Campbell et al., 1995). This enzyme hydrolyzes the peptide bond between the Ala^2 and Asp^3 residues. Frohman and co-workers (1986) were the first to suggest that inactivation of human GRF(1-44)-NH_2 was due to DPP-IV enzyme. Using HPLC, RIA, and a pituitary cell bioassay, they observed a rapid loss ($t_{1/2} = 17$ min by HPLC) of hGRF in plasma. In subsequent research they confirmed that DPP-IV was causing proteolytic cleavage between the second and third amino acid residues (Frohman et al., 1989b). Native GRF and terminally shortened analogs (1-32 or 1–29) were rapidly cleaved while a 2-32 analog was not degraded. Kubiak et al. (1989) observed DPP-IV hydrolysis when incubating Leu27-bGRF(1-29)NH_2 in bovine and porcine plasma. They were able to extend the half-life in bovine plasma from 22.1 min to 83.3 min by including diprotin A, a competitive DPP-IV inhibitor.

Modifications were made to residue 2 in GRF to prevent enzymatic degradation (Kubiak et al., 1992a,b, 1993; Martin et al., 1993; Su et al., 1991). For example, Ala^2 was replaced with Ser, Thr, or Gly in Leu27-bGRF(1–29)NH_2 and the peptide stability tested in bovine plasma (Kubiak et al., 1993). The peptide half-lives were increased from threefold to eightfold depending on the substitution. However, when using a bovine pituitary cell culture, the substituted analogs had growth hormone-releasing potencies of 4.7% to 24% of Leu27-bGRF(1–29)NH_2 or bGRF(1–44)NH_2. Complete prevention of the cleavage at position 2 occurred by removing the free amino group at the N terminus and/or substituting with a D-amino acid residue (Su et al., 1991). Martin et al. (1993) have shown the improved enzymatic stability of Leu27-bGRF(1–29)NH_2 that is achieved by replacing position 2 with Ser or position 15 with Ala. A Ser^2Ala^{15} substitution in the GRF resulted in protection from DDP-IV and a 28-fold increase in the in vitro half-life (Table 2).

Besides enzymatic degradation, deamidation occurs with GRFs (Stevenson et al, 1993; Bongers et al. 1992; Friedman et al., 1991, 1992). The incubation of $Leu^{27}hGRF(1-32)NH_2$ in an aqueous phosphate buffer (pH 7.4, 37°C) caused deamidation at Asn^8 residue to produce β-Asp^8,$Leu^{27}hGRF(1-32)NH_2$ and α-Asp^8,$Leu^{27}hGRF(1-32)NH_2$ (Friedman et al., 1991). In a bovine pituitary cell bioassay the potency was reduced for β-Asp^8,$Leu^{27}hGRF(1-32)NH_2$ by 400- to 500-fold and 25-fold for α-Asp^8,$Leu^{27}hGRF(1-32)NH_2$ when compared to the parent peptide. Potency reductions were observed by Bongers and co-workers (1992) when they generated deamidated products from human GRF(1–44)NH_2 and GRF(1–29)NH_2. They observed that GRF was most stable at pH 4 to 5. Below pH 4 (the pK_a of the Asp^3 side chain was 4), cleavage at Asp^3-Ala^4 was

Table 2 Stability of Leu^{27} bGRF(1–29)NH_2 and Several Analogs from Enzyme (DPP-IV) Degradation

Analog substitution	Mean k values (min^{-1})	$t_{1/2}$ (min)
None	0.0579 ± 0.00685[a]	12.0
Ala^{15}	0.0497 ± 0.00064[a]	13.9
Ser^2	0.0029 ± 0.00019[b]	239
Ser^2Ala^{15}	0.0020 ± 0.00019[c]	341

Source: Reprinted from Martin et al., 1993, with kind permission from Elsevier Science-NL, Sara Burgerhartstraat 25, 1055 KV Amsterdam, The Netherlands.
[a,b,c]Mean k values (n = 3) with different superscripts are different at $P < .05$.
Peptides at 30 µM were incubated in DPP-IV/PBS at 37°C. Peptide concentration monitored at specified time intervals with reversed-phase HPLC.

the predominant degradation route. At a pH > 4, the isomerization of Asp^3 to β-Asp^3 (iso-aspartic) was the major degradation pathway. Deamidation of Asn^8 to β-Asn^8 and Asp^8 occurred at pHs equal to or above 5.

III. FORMULATIONS

The decision criteria for selecting a protein/peptide formulation are the same as those used for selecting nonprotein/peptide formulations. Desired product attributes for the best marketable formulation are balanced with the molecule's physicochemical properties. Product attributes which define a formulation's in vivo duration of action, product storage conditions, "ease of use," packaging, and site of administration are combined with chemical, physical, and structural stability, excipient compatibility, and solubility of the molecule to produce the ideal formulation. Numerous attempts have been made to produce efficacious, stable, manufacturable, and marketable formulations for bovine somatotropin, porcine somatotropin, and GRF. Formulation discovery has focused on parenteral administration because proteins are degraded by enzymes, water, extreme pH and heat, and not sufficiently absorbed orally. Although immediate-release parenteral formulations are probably easier to discover and develop, prolonged-release formulations have received much attention because of the inconvenience associated with frequent administration to farm animals. Formulation types used with example proteins/peptides are shown in Table 3.

A. Bovine Somatotropin

Bovine somatotropin is used to increase milk production in dairy cows (Fronk et al., 1983; Jenny et al., 1992; Hartnell et al., 1991; Eppard et al., 1991). It has also been tested for improving average daily gain, feed efficiency, and carcass composition in finishing beef steers (Moseley et al., 1992). It is the only

Table 3 Formulation Types Used to Administer Proteins/Peptides to Animals

Formulation Types	Protein/ peptide	Comments	Reference (year)
Aqueous solution	GRF	GRF in sterile water	Enright et al. (1988)
Oil suspension	bSt	15–50% suspensions	Mitchell (1991)
Microspheres	bSt	20% increase in milk	Cady (1989)
Coated beads	pSt	Coated on nonpareil seeds	Raman et al. (1994)
Solid implants	GRF	Cholesterol matrix	Leonard et al. (1991)
Coated implants	bSt	Polyvinyl alcohol coating	Pitt et al. (1992)
Osmotic pumps	GRF	7-day delivery	Wheaton et al. (1988)

approved purified protein molecule indicated for use in food-producing animals that is currently being marketed. Bovine somatotropin is administered from approximately 10 to 15 weeks postpartum until the dry period. A typical milk lactation curve showing enhanced milk production when using bSt is presented in Figure 4. It can increase milk production by 10% or more depending on the dose, breed of cattle, formulation, and study design.

1. Solutions

Over 50 different formulations have been reported in journals and patents. Most formulation research has focused on providing bSt in an easy-to-use, conveniently stored, sustained-release formulation. During early formulation development, when little bSt was available, or the formulation variable was purposely eliminated from the study, bSt was administered in solution. An early Monsanto study (1973) used bSt dissolved in 0.01 M NaOH in saline and the pH adjusted to 9.0 with 0.01 M HCl (Machlin, 1973). Peel et al. (1981) solubilized 1.3 mg bSt/ml in 0.1 M $(NH_4)_2CO_3$ with the pH adjusted to 12.0 with 0.1 M NaOH.

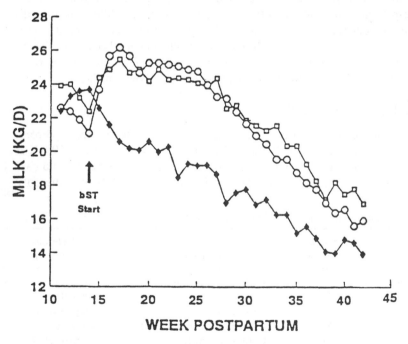

Figure 4 Typical milk lactation curve. Mean weekly milk yields of cows administered either 0 (◆) or 15.5 mg (○) of bST/d or 310 mg of bST/14 days in a sustained-release vehicle (□) from week 14 through 42 weeks postpartum. (Reprinted with permission from Jenny et al., 1992.)

The pH was then dropped to 9.5 with 0.1 M HCl before subcutaneous administration in the shoulder region. They subsequently modified the solution to include a carbonate buffer composed of 0.025 M $NaHCO_3$ and 0.025 M Na_2CO_3 at pH 9.4. This carbonate buffer maintained bSt in solution at about 2 pH units above the isoelectric point. Solutions were freshly made every 2 to 3 days and were stored at 4°C when not being used. Growth hormone in solution can readily degrade via deamidation or oxidation and may also aggregate to form dimers or other oligomers. Fronk et al. (1983) used the carbonate buffer formulation to administer bSt to Holstein cows. They showed similar milk production increases (31% increase) whether a 39.6 mg/day dose was injected subcutaneously, pulsed IV four times per day, or administered by constant infusion. These results showed that some flexibility in formulation design could be tolerated. Other scientists reported using bSt solutions at concentrations of 10 to 20 mg/ml in unspecified buffered saline (Moseley et al., 1992; Jenny et al., 1992).

2. Lyophilized Powders

Because most proteins are originally isolated as lyophilized powders, and since long-term storage as a solution is not possible due to bSt degradation, an obvious formulation strategy is to create lyophilized bSt formulations. These formulations maintain low water contents and when properly formulated allow reconstitution to precise concentrations. Bauman et al. (1985) used pituitary-derived bovine somatotropin and N-terminus methionine bovine somatotropin prepared with sodium bicarbonate as lyophilized powders. They were reconstituted to 9 mg/ml with a 0.05 M bicarbonate buffer and administered intramuscularly in the gluteal or thigh muscle. A lyophilized formulation was also used in a large, nine-herd study (n = 598 cattle) (Stanisiewski et al., 1994). No excipients were mentioned. The lyophilized powder was reconstituted to 8.6 mg/ml with Sterile Water for Injection, USP, and administered intramuscularly. Effective doses per day to increase 3.5% fat corrected milk were 5.0 to 16.7 mg/day for heifers (primiparous) and 4.3 to 13.2 mg/day for cows (multiparous).

More details on lyophilization excipients used with bSt are given in several patents (Arendt, 1992; Hamilton and Burleigh, 1987). Arendt (1992) discusses appropriate concentrations of sodium carbonate (0.1% to 7.6% w/w) and sodium bicarbonate (7.6% to 15.1% w/w) with bSt to prevent vial breakage during lyophilization. The product is to be reconstituted with saline such that the final concentrations are 1.26% bSt, 0.13% sodium carbonate, 0.33% sodium bicarbonate, and 0.85% sodium chloride. Hamilton and Burleigh (1987) patented animal growth hormones in powder form with stabilizers such as polyols, amino acids, choline derivatives, and amino acid polymers having charged side groups at physiological pH. These excipients prevent growth hormone in solution from forming insoluble aggregates and, in doing so, preserve biological activity.

3. Implants

Solution formulations do not normally provide sustained release. One method to obtain longer durations is to administer the drug formulated as a powder. Solid implants formed from powders can be administered using a trochar. The longer in vivo duration results from bSt dissolution and diffusion from the solid compact. The release rate may be further extended by including a retarding excipient or a rate-limiting membrane. Implants can be of four basic types: solid compacts essentially free of other excipients; active agent dispersed in retarding excipients; pure compacts with or without excipients that are membrane-coated; and osmotic pumps.

If the solubility of a drug is optimal, the dissolution from a pure, solid compact could provide adequate pharmacological blood concentrations. Calsamiglia et al. (1992) administered 400 mg of pelleted, methionyl bSt every 14 days via subcutaneous implants and observed 17.5% milk production increase over controls. A patent application was filed on bSt compressed as powders and administered to bovine to increase milk production or improve feed efficiency and average daily weight gain (Azain et al., 1990).

Dispersing bSt in release-retarding excipients is another way to develop implants. The retarding excipients used included polyanhydride (Ron et al., 1989, 1992), polycaprolactone (Shalati and Viswanatha, 1988), polyesters (Sivaramakrishnan et al., 1989), cholesterol (Kent, 1984), and ethylcellulose (Janski and Yang, 1987, 1988). Although these investigators show sustained-release was achieved in vitro, none of these implants were administered to animals. Often the in vitro release initially shows a square-root-of-time matrix release, with a relatively rapid flattening of the profile. The net result is adequate early bSt release but insignificant release at later times.

4. Microparticles

One method of increasing release from implants is to increase the exposed surface area. Investigators have attempted this through the used of microparticles containing bSt. Cady et al. (1989) have produced beeswax/glyceryl tristearate, glyceryl disearate (GDS), and glyceryl tristearate (GTS) microspheres. Other excipients in these microspheres include sodium benzoate, Pluronic F68, Tween 80, and carbonate buffer. Microspheres were dispersed in saline, soybean oil, or Miglyol 812 oil before being administered subcutaneously to dairy cows. Somatotropin plasma concentrations obtained depended on the vehicle composition, bSt microsphere load, and hydrophobicity. Milk production depended on dosage and dosing interval. In another study an unspecified type of microspheres produced by Cyanamid was injected as a 350-mg biweekly dose and compared to 10.3-mg bSt administered daily (Zhao et al., 1994). From early through late lactation, overall mean plasma bSt, plasma IGF-I, and milk bSt concentrations increased. Milk bSt was significantly lower ($P < .05$) for the daily injected treatment compared to the sustained-release treatment, while mean plasma bSt and IGF-1 concentrations were similar. Cyanamid has several composition and process patent applications on microspheres made from beeswax, GDS, and GTS (Steber et al., 1987, 1992; Steber, 1991).

Another substance used to produce bSt microparticles is the fatty-acid anhydrides (Maniar and Domb, 1992). Fatty acid anhydrides such as stearic, lauric, and palmitic anhydride are combined with stabilizing agents like polysorbate 80, sucrose and sodium sulfate, and bSt. These solutions are spray-dried or lyophilized and then sieved to produce approximately 100-μm microparticles (10- to 400-μm range). After dispersing in an oleaginous carrier vehicle, the microspheres administered subcutaneously to cows produced elevated bSt concentrations for 9 days.

An advantage of using microparticles compared to implants is that they may be dispersed in carrier vehicles which can be more easily delivered parenterally than implants. However, if in vivo delivery of the microparticles dispersed in the carrier vehicle results in inhibited bSt release caused by aggregated protein blocking the tortuous channels in the microparticles, or if aggregated bSt causes decreased intrinsic dissolution, then an alternative carrier vehicle or delivery technique would be required to protect the bSt from initial protein aggregation. Ideally, selection of the proper carrier vehicle and microparticle excipients to minimize protein aggregation should result in maximum total release of the loaded drug. Although a high percent of the loaded bSt may be released from the dosage form, the challenge is discovering the excipients and carrier vehicle to release the bSt at a constant rate for the desired week or longer.

5. Oleaginous Vehicles

Using an oleaginous vehicle allows for parenteral administration, protects the bSt from rapid exposure to the aqueous milieu (i.e., helps prevent protein aggregation), can create sufficient surface area for bSt dissolution to occur, can be made with a relatively simple manufacturing method, and helps maintain the bSt at the injection site for commercially acceptable time periods (e.g., usually 1 to 4 weeks). Several scientific groups working independently correctly ascertained that these attributes meant that a hydrophobic-based delivery system appeared to be commercially promising for bSt. In 1990 Lilly patented a formulation for increasing daily milk production for 28 days which contained 10% to 25% bSt suspended in a carrier that contained 8% to 20% of a wax and 80% to 92% of an oil (Ferguson et al., 1990). Earlier they had filed an European Patent application for bSt in an oil when in combination with a fatty-acid salt (e.g., calcium stearate) and an absorption regulating agent (e.g., dextrans)

(Bramely et al., 1988). An 18% bSt suspension dispersed in sesame oil (50%) with 8.5% dextran and 23.5% calcium stearate elevated bSt concentrations for 24 days when injected into sheep.

Monsanto patented an oil suspension in 1991 (Mitchell, 1991). They suggested using peanut or sesame oil with 10% to 50% weight of bSt. The oils could be used neat or gelled with aluminum monostearate, which also acts as an antihydration agent. Depending on the formulation, the concentration of bovine somatotropin in rats could be elevated for 14 days. Patents were also granted for the claimed hydration-retarding effect that aluminum monostearate provides, and the product exists as a unit dose of approximately 300 mg bSt (Mitchell, 1992, 1995a,b). In a major study sponsored by Monsanto (n = 241 cattle), zinc methionyl bSt at 37% w/v in an oil-based formulation was shown to increase 3.5% fat-corrected milk by 10.2% to 26.5% at doses ranging from 250 to 700 mg (Hartnell et al., 1991). A dose was administered every 2 weeks. Fat, protein, and lactose contents in milk were not changed. Intake of energy increased in the bSt-treated cows so that body weight gain, body condition score, and net energy balance did not differ among the treatment groups. Other Monsanto studies have concluded that bSt at commercially used doses does not affect health and reproduction of cattle (Cole et al., 1992; Eppard et al., 1991).

6. Nonoleaginous Vehicles

One of the oldest methods suggested to sustain and enhance the effect of bSt was reported in 1961 (Organon Laboratories Limited). Organon Laboratories Limited patented the preparation of bovine growth hormone complexed with zinc or other insoluble metals. When dispersed in water this complex is 50% more potent than noncomplexed growth hormone when a Tibia test is used.

A water-in-oil-in-water (W/O/W) emulsion containing bSt has been suggested as a sustained-release formulation (Tyle and Cady, 1990). Administration of 2 or 4 mg bSt/kg to wether lambs produced elevated St concentrations for the 22-day duration of the study.

Cady and co-workers (1986) developed a nonoleaginous formulation by producing a water solution complex between bSt and carbohydrate. A low-viscosity corn dextrin and bSt were mixed with a carbonate-buffered saline until a homogeneous paste was formed. The paste containing 175 mg was administered weekly to cows. After 3 weeks, an average 6% increase in milk per week was observed. Most likely these formulations, which contained water, were not developed further because long-term stability could not be obtained.

B. Porcine Somatotropin

Exogenously administered porcine somatotropin is capable of increasing weight gain, improving feed efficiency, and increasing carcass leanness in swine

(Machlin, 1992; Etherton et al., 1986, 1987; Evock et al., 1988; McNamara et al., 1991). Shown in Table 4 are the results from one study where porcine somatotropin (pSt) was administered daily (Etherton et al., 1987). Increased milk production in dairy cows can be economically feasible using formulations which provide 1-day duration of effect without modifying producers' normal herd operating procedures. Each cow is milked two or three times a day. In contrast, swine producers do not normally handle animals frequently, so formulation discovery has focused on providing longer durations. The possibility of daily administration to swine was not totally discarded, though. As is true for most other drugs, solutions are generally the first formulations that are evaluated.

1. Solutions

In a study by Etherton et al. (1987), solutions of pSt were administered daily. The solutions were freshly prepared daily, presumably because pSt degrades in solution. Meat quality from pigs (i.e., color, texture, marbling) improved when given daily injections of pSt (Christian and Miller, 1991). The suggested dose was 2 to 8 mg per pig per day; interestingly, they stated the carrier material

Table 4 Effect of Porcine Growth Hormone (pGH) on Growth Performance and Carcass Composition[a]

Item	pGH, µg/kg body weight				SE
	0	10	30	70	
Final weight, kg	74[e]	77[e,f]	76[e,f]	79[f]	0.9
Ave. daily gain, kg	0.90[e]	0.98[e,f]	0.95[e,f]	1.03[f]	0.02
Feed/gain[b]	2.9[e]	2.7[e,f]	2.6[e,f]	2.4[f]	0.09
Backfat, cm	2.4[e]	2.4[e]	2.2[e]	2.1[e]	0.1
Carcass length, cm	73[e]	75[e]	74[e]	75[e]	0.2
Loineye area, cm^2	22[e]	23[e]	25[f]	27[f]	0.7
Skinned ham, kg	5.8[e]	5.8[e]	6.1[f]	6.4[f]	0.1
Protein, %[c]	14.8[e]	14.9[e]	16.5[f]	16.7[f]	0.3
Lipid, %[c]	28.7[e]	28.7[e]	24.4[e,f]	21.6[e]	1.4
Dry matter, %[c+]	45[e]	45[e]	42[f]	40[f]	1.1
Adipose tissue, kg[d]	11.5[e,f]	12.1[e]	9.9[e,f]	9.1[f]	0.7
Muscle, kg[d]	26[e]	28[e,f]	29[e,f]	31[f]	0.8

Source: Modified and used with permission from Etherton et al. (1987).
[a]Pigs were treated with the noted doses of pGH daily for 35 days by IM injection; n = 12 per treatment.
[b]Kg of feed consumed/kg body weight gain.
[c]Values are on a percent of soft-tissue basis, n = 8.
[d]Values were estimated as described in Etherton et al. (1982).
[e,f]Values in a row without a common superscript differ (P < .05).

was not critical. This statement implies that the pSt is stable in the vehicle. The study was unclear in that they stated pSt was injected, yet listed solid carrier materials (dextrose, glucose, and mannitol). Presumably these stabilizers were used in combination with a liquid vehicle to provide a formulation for daily injection. Other studies describe use of carbonate buffers. Chung et al. (1985) used 1 mg pSt/ml concentration in 25 mM $NaHCO_3$, 25 mM Na_2CO_3, 0.154 NaCl buffer at pH 7.4. Wray-Cahen et al. (1991) produced a 2.5-mg/ml solution using the same bicarbonate buffer. McLaren et al. (1990) used 1.5 to 9 mg/ml solutions in a sterile bicarbonate buffer that were administered in the neck region of swine using a multiple-dose "pistol-grip" syringe.

2. *Oleaginous Vehicles*

Because pSt aggregation and degradation occur in aqueous solutions, any long-term storage in an aqueous medium is not feasible. One method of obviating this situation is to use a hydrophobic base as has been done for bSt. Surprisingly little has been published on using hydrophobic liquid carriers with pSt. Martin and Kramer (1986) have developed a zinc complex pSt dispersed in a gelled oil. An example formulation included a 5% aluminum monostearate gelled peanut oil which contained 8 mg/ml of pSt. When 40 mg of this formulation was injected subcutaneously into the neck of four pigs, elevated pSt concentrations were obtained for at least 9 days. Kim et al. (1991) used the oily vehicle tocopheryl with a delaying agent (e.g., aluminum monostearate, waxes, choline derivatives) to elevate pSt for at least 20 days in hypophysectomized rats.

3. *Implants*

Implant types tested with pSt include matrix, coated, and osmotic. The most efficient, constant, and longest pSt delivery durations have been obtained with Alza's osmotic pumps (Eckenhoff et al., 1989a,b, 1990, 1991). These osmotically driven systems (see Chapter 2 for description) are capable of delivering in vitro an average 2.5 mg pSt for 60 days (Fig. 5) (Eckenhoff et al., 1989b, 1990). The pSt was formulated at 30% in a gelatin carrier vehicle. The carrier vehicle included gelatin, glycerol, and L-histidine. This combination helped to structurally stabilize the pSt and provided a flowing formulation which could be expressed through the pump orifice (Fig. 5). Another study used 0, 1, or 2 pumps where each implant contained 100 mg pSt (Hacker et al., 1993). The delivery rate was estimated to be 2.2 to 2.4 mg/day per osmotic pump. No details were given on how the pSt was formulated prior to placement in the pumps. Efficiency was monitored using a number of parameters which included improved growth, feed:gain ratio, and carcass characteristics. In animals that were given the 200-mg dose the feed:gain ratio improved and last rib fat and leaf fat were reduced. The authors concluded that their work represented a significant step toward showing that the long-term delivery of pSt was successful,

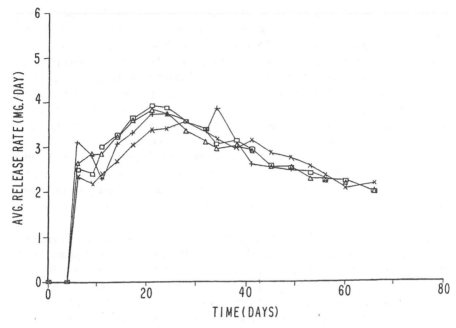

Figure 5 In vitro pSt release from four different osmotic pump implants. (Reprinted with permission from Eckenhoff et al., 1989b.)

especially compared to the laborious daily injection of pSt. A large amount of data on stabilizing pSt in solutions for ultimate use in osmotic pumps has been published (Azain et al., 1989a).

Azain et al., (1989b) concluded that a polyol (e.g., glycerol, tris(hydroxy-methyl)-aminomethane) and buffering (e.g., histidine hydrochloride, citrate) to a pH between 4.5 and 7 or the isoelectric point helped minimize dimer formation. These osmotic pumps, which delivered 2 mg/day, were subcutaneously implanted in the back of the swine's ear for a 6-week study (Azain et al., 1992). Their comparison study to daily administration showed that the implant was not as effective in improving feed efficiency and carcass quality. Another study used these same osmotic pumps to show that pigs treated with pSt had less leaf fat and less 10th-rib backfat than control pigs (Becker et al., 1992). Shorter-duration osmotic pumps have also been used (Azain et al., 1993). For example, Alzet miniosmotic pumps have been filled with pSt or bSt dissolved in 25 mM sodium bicarbonate buffer (pH 9.5) at various concentrations up to 33.3 mg/ml. The minipumps have effectively delivered pSt subcutaneously in rats for 2 to 4 weeks.

Implants that are less expensive to produce than the osmotic pumps consist of pSt dispersed in a retarding excipient. Kim et al. (1993) developed 20% pSt-loaded polyethylene glycol/paraffin wax implants. The molecular weight of the polyethylene varied from 1540 to 35,000 daltons. These pSt implants (50 mg, 7 mm in diameter, 1.4 mm in thickness) placed in hypophysectomized rats produced greater growth during 14 days than controls. Durations longer than 14 days were not reported. In an extremely large study (n = 2160 barrows and gilts), a pelleted implant containing 12 mg pSt mixed with copper sulfate and magnesium stearate was used (Knight et al., 1991). Doses of 12, 24, 36, or 48 mg/week were administered by implanting one, two, three, or four pellets subcutaneously at the base of the ear. The construction of the pellets has been patented (Azain et al., 1989b). The study showed improved feed:gain ratio and percentage of carcass protein with no negative effect on cooked lean pork palatability.

More elaborate pSt pelleted implants have been shown to produce an equivalent feed:gain ratio and reduced feed intake over 10 days when compared to 2 mg pSt injected daily (DePrince and Viswanatha, 1988). One-hundred-milligram pellets consisting of 25% zinc pSt and 75% porcine serum albumin (PSA) were compressed. Two of these pellets were inserted into siliconized tubing (ID 3.2 mm) with a Teflon plate in the middle. Thirty-five-micron microporous polyethylene disks covered the ends of the tubes. The pSt diffused from this tube with the PSA acting as a stabilizer.

Another method of stabilizing pSt from aggregation when using a solid implant has been reported (Janski and Drengler, 1989). They claim that sodium dodecyl sulfate (SDS), which substantially coats implants (40% or more of the SDS), will allow wetting and adequate in vivo release. They showed limited data supporting the claim using rat growth data.

The use of amino acids with metal-associated pSt has been suggested as a method to increase pSt solubility (Raman et al., 1993). The amino acid, preferably with a basic side chain which chelates the metal, was blended with powdered pSt. As an example, 20% zinc pSt, 20% arginine, and 60% sucrose pellets were compressed (101 mg, 7 mm in length, 4 mm in diameter). Three pellets were then inserted in a silicone tube and sealed at one end with glass beads. The other end contained a semipermeable membrane composed of L-leucine. Sustained in vitro release up to 14 days was shown.

Coated pellets of pSt have been used to sustain drug release. Steber et al. (1992) used pSt pellets compressed with a fat or wax (e.g., glyceryl trimyristate, glyceryl tristearate), sugar (e.g., sucrose, lactose), and buffer (e.g., monobasic and dibasic sodium phosphate), and coated with various types of poly-(acrylates). The pSt concentrations measured in pigs by radioimmunoassay remained elevated for 4 weeks when the pellets were administered in the ear.

Coated pSt pellets with polyvinyl alcohol (PVA) have effectively increased the growth rate in coho salmon over controls (Pitt et al., 1991, 1992; McClean et al., 1992). The pellets consisted of a 60:40 ratio of lyophilized pSt to chitosan with PVA coatings 25 to 75 μm thick.

4. Microparticles

Microparticle formulations release pSt in a fashion similar to larger solid implants but have the advantage of being delivered dispersed in a carrier vehicle. In addition to the bSt microparticle formulations which are often used with pSt, two more complex microparticulate formulations have been described. Raman et al. (1994) coated pSt onto nonpareil beads with loadings of 0.24% to 12%. A second coating, normally consisting of glycine, was placed on the beads to help stabilize solubilized pSt. A third coating containing a water-insoluble material to delay the pSt release was applied. This third layer can be a wax or polymer such as ethyl cellulose. For example, coating with a mixture of partially hydrogenated cottonseed oil, beeswax, and a surfactant created beads which provided 48-h in vitro release. In an attempt to sustain pSt release beyond 48 h, Sivaramakrishnan and Miller (1990) used a similar approach. They created zinc pSt/sucrose/arginine complex particles smaller than 250 μm. These particles were placed between a layer of carnauba wax which was then coated with a rupturable wax coating of beeswax, carnauba wax, and a surfactant. Weekly administration of these microparticulate implants to swine increased average weight, total weight gain, average daily gain, and decreased feed:gain ratio.

The pSt compatibility with the retarding microparticle excipient must be adequate. Wyse and co-workers (1989) have suggested that pSt incorporated at 18% in poly(glycolic) acid results in a pSt-polymer interaction. Like other sustained-release pSt and bSt formulations, incomplete in vitro release was observed (<30% of the pSt loaded). Using RP-HPLC they observed another peak at 9 min retention time compared to the native 10 min retention time. The degradation product was not identified.

C. Growth Hormone–Releasing Factor

Growth hormone-releasing factor (GRF) stimulates pituitary somatotropin release. Somatotropin, through the action of insulinlike growth factor 1, decreases lipogenesis and promotes protein accretion, resulting in enhanced growth in food-producing animals such as cattle and swine. The enhanced somatotropin release can also increase milk yield. These physiological responses to GRF suggest that potential indications for GRF include enhanced milk lactation in dairy cows (Dahl et al., 1990; LaPierre et al., 1988; Enright et al., 1988), improved carcass quality in swine (Pommier et al., 1990), and improved feed efficiency, growth rate, and carcass quality in cattle (Enright et al., 1993). The increased lactation observed in cows when 0, 1, 3 and 12 mg GRF was infused is shown in Figure 6.

Becuase bovine GRF is a 44-amino acid peptide, which is shorter than the 191-amino acid protein bSt, the physicochemical properties may be more amenable to formulation. A smaller molecule may also be less costly to produce. It is thought that GRF is more potent than somatotropin, thus requiring less drug to obtain the same effect. Furthermore, some development programs may have patent protection with GRF with the intention of further developing GRF. Except for the latter patent argument, only research (i.e., physicochemical and animal testing) will truly reveal whether a protein (e.g., somatotropin) or peptide (e.g., GRF) offers formulation advantages.

1. Solutions

The first formulations containing GRF, like bSt and pSt, were solutions. The solutions varied depending on the GRF analog being tested and the investigator. When examining the effect that human GRF $(1-44)NH_2$ or human GRF $(1-29)NH_2$ might have on growth hormone release in pigs and heifers, a phosphate buffer was utilized (Petitclerc et al., 1987). The peptide was first dissolved in

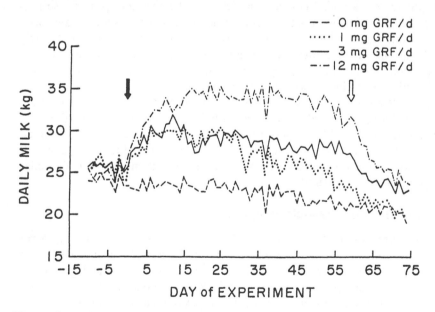

Figure 6   Daily milk yield of cows (6/treatment) continuously infused for 60 days with 0, 1, 3, and 12 mg of recombinant bovine somatotropin-releasing factor (rbGRF)/d. Beginning and end of rbGRF infusion indicated by the solid and open arrows, respectively. The SE of the difference among treatments was 1.6 kg/day. (Reprinted with permission from Dahl et al., 1990.)

0.1N HCl, then neutralized with 0.1N NaOH followed by dilution with 0.1M, pH7.4 phosphate buffer. This stock solution was diluted 2.5 to seven times with normal saline prior to injection. No GRF concentrations were stated. However, the bulk drug purity was mentioned as being 86% and 95% for the human GRF (1-44)NH_2 and human GRF (1-29)NH_2, respectively, with a peptide content of 93.4% and 99.0%.

This same solubilizing method has been used to study the effect of administering porcine GRF(1-29)NH_2 and/or thyrotropin-releasing factor on somatotropin, prolactin, triiodothyronine, and thyroxine release in growing pigs (Dubreuil et al., 1988). A modification to the solution was made when studying the effect which human GRF(1-29)NH_2 had on both milk production and growth hormone release in dairy cows (LaPierre et al., 1988). La Pierre et al. (1988) added gelatin to the diluted saline/buffered solution and subcutaneously administered 10 mg of human GRF(1-29)NH_2 daily. Average milk yield increased by 14.3% during the 9 days the drug was given. Hart et al. (1985) also used a phosphate buffer when delivering human pancreatic GRF (hpGRF 1-44, hpGRF 1-40, and hpGRF 1-29NH_2), rat hypothalamic GRF (rhGRF 1-29NH_2), and hpGRF (hpGRF1-10 with carboxyl-terminal groups of -NH_2, -OH and -OCH_3) to sheep. The GRF was dissolved in 0.1 M phosphate-buffered saline containing 0.1% bovine serum albumin and diluted with normal saline. When delivered IV, increases in GH were observed with all analogs administered except the 10-amino acid hpGRF.

Sterile water is another vehicle used to deliver GRF. To increase milk production in Holstein cows up to 23%, doses of bovine GRF 1-44 NH_2 from 0 to 50 mg/24 h were administered via IV infusion (Enright et al., 1988). Fresh solutions prepared daily with sterile water varied in concentrations from 0.2 to 3.9 mg/ml. Enright et al. (1993) also used sterile water as the diluent when examining the effect of GRF on growth, feed efficiency, carcass characteristics, and blood hormone concentrations in beef heifers. Daily SC injections for 86 day of 2.76 to 3.40 ml of [des NH_2Tyr^1, D-Ala^2, Ala^{15}]-GRF (1-29)-NH_2 were given. The 1 μg/kg/day dose yielded small but positive improvements in animal performance. Dahl et al. (1990) used sterile water as the diluent when IV infusing 0.08 to 0.9 mg/ml recombinant bovine GRF 1-45 homoserine lactone in dairy cows. In this study 1% bovine serum albumin dissolved in sterile water was used to coat the infusion catheters. Besides increasing milk production, doses of 3 and 12 mg increased serum somatotropin from 0.7 ng/ml (controls) to 8.2 and 10.3 ng/ml, when averaged across 1, 30, and 59 days.

Sterile saline as the solubilizing medium was used by Moseley et al. (1987) when showing that sustained GH elevations are achieved for at least 5 days when administered IV human GRF 1-44NH_2 to Holstein steers. A total of 20.8 ml/h was infused continuously for 21 days. A concentration of 7.2 ug/ml and dose of 3.6 mg/day was used. Etherton et al. (1986) used sterile saline to dissolve human GRF 1-44NH$_2$ before IV or IM administration to pigs. The saline had been adjusted to pH 3 prior to addition of the GRF which produced 1 mg/ml solutions. When delivered IM with human GRF 1-44NH$_2$ and porcine growth hormone, the effects produced by pGH were more pronounced than GRF. For example, growth rate was increased 10% by pGH but was not affected by GRF.

2. Oleaginous Vehicle

Use of a hydrophobic carrier to sustain the release of GRF has been attempted. Polyglycerol esters (e.g., diglycerol tetrastearate, hexaglycerol distearate) have been combined with oil (e.g., sesame) and optionally polysaccharides (e.g., dextran) to form a carrier vehicle for GRF (Brooks and Needham, 1994). The excipient ratio included up to 40% polyglycerol and 5% polysaccharides. By weight, 20% GRF was added to this vehicle to produce a light paste. Sheep were injected SC with 60 mg GRF in this hydrophobic vehicle. Increased somatotropin concentrations were observed for up to 17 days. Sustained release was claimed when a powder complex of GRF and gelatin was dispersed in an oil (Fujioka et al., 1986). The solids content was less than 2%, and the GRF loading was less than 0.5%. No in vitro or in vivo data were shown with the GRF formulation.

3. Lyophilized Powders

Besides glycine, other excipients have been utilized to stabilize GRF. Fujioka et al. (1985) have shown that glycine or human serum albumin can stabilize GRF solutions. Using these excipients and a buffering agent, lyophilized GRF formulations were produced that were physically and chemically (RP-HPLC) stable. However, the excipient to GRF ratio was 10- to 40-fold, which may produce unacceptable solution viscosity or volume for injection, or pellet size for implantation.

4. Implants

Several successful sustained-release implantable GRF formulations have been reported. Godfredson et al. (1990) implanted osmotic pumps containing a human growth hormone-releasing factor analogue (hGRFA) weekly into lambs. These Alza minipumps (model 2001) were filled with GRFA dissolved in a 1:1 ratio of dimethyl sulfoxide (DMSO) to water. The implants were placed SC in the axillary region of a forelimb. Delivering 0.7 µg/kg/h improved feed conversion and increased average daily weight gain. A similar Alza osmotic pump (model 2MLI) containing hGRF(1–29)NH$_2$ was used to show that circulating GH can be increased in steers and wethers (Wheaton et al., 1988). In the first part of this study hGRF(1–29)NH$_2$ was dissolved in water to give concentrations of 0.235, 2.35, and 23.5 mg/ml. The highest concentration reached the solubility limit of around 25 mg/ml. These implants, placed in the shoulder of calves, failed to produce increased growth hormone concentrations. Reformulation using a 1:1 ratio of DMSO to water allowed 2.35 and 23.5 mg/ml concentrations to be prepared. When these implants were placed in steers and wethers, increased circulating GH concentration was found. The authors concluded that "solubility and volume limitations render hGRF(1–29)NH$_2$ delivery via osmotic pumps problematical."

At least four implantable matrix GRF formulations have been reported. These implants contain retarding excipients, including cholesterol (Lenoard and Harman, 1991), collagen (Fujioka et al., 1989), polyesters (Mariette et al., 1993), and fatty-acid salts (Cady et al., 1988). The cholesterol/GRF implant was constructed by a melt and compression method (Lenoard and Harman, 1991). At a ratio of 30% GRF to 70% cholesterol acetate, pellets 2 mm in length and 3 mm in diameter were made. In buffered saline at 25°C an average release of 5.7 µg/mm of pellet per day was observed from day 6 to day 20. The release during the entire 20 days followed a square-root-of-time profile. Further evaluation of in vitro GRF release using a matrix implant has been reported (Mariette et al., 1993). These investigators used a modified version of 75/25 poly(lactide-co-glycolide) as the retarding excipient and GRF(1–29)NH$_2$. The GRF(1–29)NH$_2$ activity was 600 or 786 µg/mg and had a purity of 95%. Drug loads in the compressed matrices varied from 7.5% to 20%. Knowing the solubility, they found that when using distilled water in a flow-through apparatus the predicted release was obtained. The release rate was decreased using either saline or pH 7.4 phosphate buffer as the release medium. Their results show that the solubility characteristics of the protein/peptide are extremely important. The solubility and aggregating mechanism of the molecule must be studied under in vivo conditions to design an implant which will provide desired release profiles.

Fujioka et al. (1989) tried to modify the in vitro release of GRF implants by using acid compounds (e.g., citric acid, alanine, glycine) in an atelocollagen matrix. By creating a 15% loaded implant they were able to achieve constant release for at least 14 days in phosphate-buffered medium. Another method of altering release was to change the fatty-acid slats of the GRF (Cady et al., 1988). They used a synthetic GRF analog that was a hexapeptide and created C$_{10}$–C$_{20}$ fatty-acid salts. The 0.1875-inch diameter implants were then coated with Silastic or poly(lactide-co-glycolide). By leaving the ends uncoated, they achieved more appropriate (ie. greater) release rates and elicited a physiological response. Release rates varied from 0.6 to 10.9 mg/day depending on the fatty acid used. An average in vitro release of 3.1 mg/day and in vivo 3.2 mg/day during a 43-day study was observed with one particular implant.

5. Microparticles

Microparticles have been used to delivery GRF. Grangier et al. (1991) produced approximately 150-nm nanoparticles by lyophilizing GRF(1–29)NH$_2$ solutions with poly(alkylcyanoacrylate). In vitro release in Ringer's lactate medium and in Ringer's lactate medium containing esterase and rat plasma was studied. More rapid release was obtained in the Ringer's lactate medium containing esterase and rat plasma, with almost 100% release occurring over 8 h. Depending on the polymer, less than 10% GRF release was observed using the Ringer's lactate medium over the same time period. They concluded that release was due mainly to bioerosion rather than diffusion. The same researchers administered 10% loaded GRF(1–29)NH$_2$ poly(alkylcyanocryanate) nanoparticles subcutaneously to rats. After 24 h GRF was detected in the plasma. In contrast, when free GRF was used, a maximum plasma concentration was achieved at 2 min and no GRF was detected after 100 min. Transmission electron microscopy showed nanoparticles at the injection site 24 h after administeration. Whole-body autoradiography was used to determine the movement of the nanoparticles or biodegraded fragments for 24 h.

Polyester microspheres have also been used to deliver GRF. Craft et al. (1994) loaded 4-methylhippuroyl(1)porcine GHRH (2–76)–OH (GRH or somatogenin) at 10% into a 85/15 copolymer of poly(lactide-co-glycolide). Elevated serum growth hormone and decreased BUN concentrations were observed for 3 to 6 days when various doses were administered to barrows. Baseline values were observed on day 9. In research reported in the same patent by Craft and co-workers (1994), these same microparticles were administered in ovo with the claimed advantages of being easily administered, not requiring sterile techniques, and minimizing the risk of injury to the developing chicken embryo. The microspheres were injected into the air cell in the egg with release from the microspheres, delivering the active agent to the surrounding fluid and blood vessels.

In our laboratories we have observed sustained, elevated serum somatotropin concentrations when using aqueous, oil, and microsphere suspensions containing a previously discovered metabolic stable GRF analog (GRFA) (Ile^2, Ser^{8,28}, Ala^{15}, Leu^{27}, Hse^{30}bGRF(1-30)-NH-ethyl). Serum ST concentrations were measured periodically during a 35-day period after the formulations were administered to meal-fed Holstein steers. Areas under the serum ST daily curves (ST-AUC) formed the basis for determining the time at which daily ST-AUC returned to baseline for each treatment group (Fig. 7). Time to return to baseline was 21 days when delivering subcutaneously a 200- or 400-mg dose of a 20% oil suspension and 14 days for 200-mg dose at 10% load in poly(lactide-co-glycolide) microspheres. The longest duration, 35 days, was achieved with 200 mg dose of 18% aqueous nanosuspension. A zero-order release was observed with the 18% aqueous formulation as indicated by the ST-AUC not changing from day 7 through day 35. The observed constant release was hypothesized to occur because of slow dissolution from GRFA fibrils or crystals which were found in the aqueous formulation (Fig. 8).

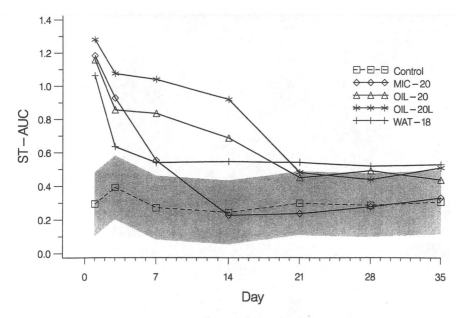

Figure 7 Daily ST-AUC for three growth hormone-releasing hormone analog factor formulations with 95% confidence interval constructed around daily control group means.

D. Other Proteins/Peptides Used in Animals

Two other proteins used in animals include lysostaphin and gonadotropin-releasing hormone (GnRH). Lysostaphin (rLYS) is a 246-amino acid recombinant mucolytic protein which has been evaluated as a treatment for bovine mastitis caused by *Staphylococcus aureus* (Oldham and Daley, 1991). Holstein-Friesian dairy cattle originally free of mastitis were inoculated with *S. aureus* in each quarter. The infected quarters were infused with 100 mg rLYS dissolved in 60 ml sterile phosphate-buffered saline (PBS), sodium cephapirin in sterile PBS or Cefa-Lak (cephapirin formulated in a peanut oil base). The disease cure rate was 20% when treated with rLYS. A statistically similar 29% was observed when the cephapirin was used while a higher cure rate of 57% was obtained when the commercial product Cefa-Lak was used. The authors hypothesize that a "properly formulated rLYS" might be more efficacious and would provide an alternative to antibiotic mastitis therapy.

The controlled release of GnRH has been studied to assist fish farmers in synchronizing spawning in seabream, Atlantic salmon, and coho salmon (Zohar et al. 1990). The dosage forms used included poly(lactide-co-glycolide) micro-spheres (MIC), poly(lactide-co-glycolide) rods (ROD) or hemispherical ethylene-

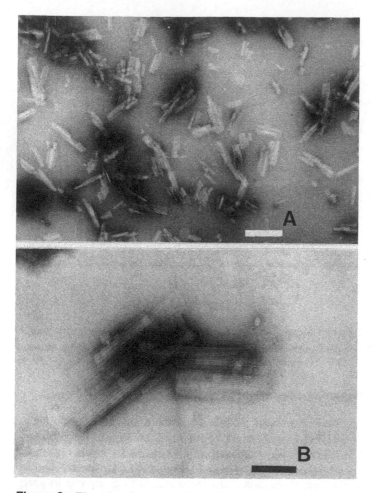

Figure 8 Electron micrograph of protein fibrils from a 3-month-old 180 mg/ml aqueous growth hormone-releasing factor analog suspension. An aliquot was diluted 1000-fold, adsorbed to Formvar-coated specimen grids, and negatively stained for 2 to 4 min with 0.1 M uranyl acetate, pH 4 (A) or 6% ammonium molybdate, pH 5.5 (B). Samples were air-dried and viewed using a JEOL, Inc. JEM-100EX transmission electron microscope operated at 60 kV. Bars equal 500 nm (A) and 100 nm (B).

vinyl acetate copolymer (HEMI) devices. The formulations contained from 25 to 200 μg GnRH or an analog. Administration of the HEMI (implanted SC) or MIC (intramuscularly injected) to seabream resulted in 80% of the females spawning for periods of 7 to 90 days (longer than 6 days is required for fish farming). Injecting a non-sustained release formulation resulted in only 25-30%

of the fish spawning for longer than 6 days. Similar results were obtained when the GnRH was administered to salmon.

Synchronizing estrus in sheep is another indication for use of GnRH. Williams et al. (1987) designed formulations that consist of GnRH mixed with water-soluble (e.g., lactose) or water-insoluble (e.g., calcium phosphate) excipients compressed into pellets which are subsequently coated with a water-permeable polymer (e.g., copolymer of ethyl acrylate and methyl methacrylate). Prolonged GnRH release was observed in vitro up to 266 days when varying the coating thickness up to 9%. Increased ovulation and estrus were observed in ewes implanted with the coated pellets compared to uncoated pellets. It was observed that 20% to 25% of the ewes ovulated followed by 75% to 100% estrus when 12 µg GnRH with a 1.25% to 4% water-permeable coating was administered, while 20% of the ewes treated with the uncoated pellets ovulated followed by only 5% breeding.

There are some interesting nontraditional indications for use of bST. These include stimulation of lactation in buffaloes (Ludri et al., 1989), enhanced growth in lambs (McLaughlin et al., 1993, 1994; Koppel et al., 1988; Rosemberg et al., 1989), increased lactation in ewes (Stelwagen et al., 1993), enhanced growth in coho salmon (*Oncorhynchus kisutch*) (Markert et al., 1977; Higgs et al., 1977), and chickens (Buonomo and Baile 1988) and even improved growth of chickens administered bSt in ovo (Dean et al., 1993). The formulations normally tested are the same as those used in cattle. Ludri et al. (1989) used somidobove suspended in sterile phosphate buffer at concentrations of 0.25% and 0.50% when assessing the effect in buffalo. A daily dose of either 25 or 50 mg (10 ml) was injected IM in the gluteal region. Milk production increased 16.8% and 29.5% compared with controls for the two doses, respectively. Lactation also increased (42%) in ewes given 0.1 mg bSt/kg IM daily (Stelwagen et al., 1993). The formulation used was not specified.

Improved feed efficiency and carcass composition were observed in lambs when bSt was administered as either a sustained-release oleaginous formulation or a daily aqueous solution (McLaughlin et al., 1993, 1994). The oleaginous formulation provided by Monsanto was not described. Doses of 25, 50, and 75 mg were administered every week while doses of 50, 100, and 150 mg were given every 2 weeks. A 1-ml syringe with a 16-gauge needle was used to administer the product SC in the neck. Daily doses of 4 and 8 mg of methionyl bovine somatotropin were given with the drug dissolved in 75 mM $NaHCO_3$. A 3-ml syringe with a 20-gauge needle was used to inject the solution SC in the dorsal area of the shoulder. These syringes, containing the solutions, were frozen at −20°C for up to 8 weeks prior to product administration. Rosemberg et al. (1989) studied growth rates in lambs administered bSt as a sterile sodium bicarbonate buffer with oxytetracycline. Daily SC injections of 2.6 ml containing 3.33 mg bSt were administered for 30 days. A sterile bicarbonate buffer was also used by Koppel et al. (1988) in studies of the effect of bovine growth hormone in suckling lambs.

Bovine growth hormone was used to increase the growth rate in coho salmon (Markert et al., 1977; Higgs et al., 1977). In these studies 10 μg bSt/g fish per week was administered by injecting 50 μl of a solution into the dorsal muscle. Enhanced growth and better food and protein conversion occurred in the fish that were given bSt.

Recombinant bovine somatotropin was given daily by interscapular subcutaneous administration to female broiler chickens (Buonomo and Baile, 1988). Daily doses of 0.5 or 2.5 mg/kg were given as a solution using a 26-gauge needle. The bSt was dissolved in 0.5 M sodium bicarbonate (pH 9.0) solution, loaded into 1-ml syringes, and stored at –20°C until used. The bSt was 95% pure as defined by RP-HPLC before use. Growth rates were increased 6.1% and 6.9% for the first week of treatment when using 0.5 and 2.5 mg/kg doses, respectively. During the second week of treatment difference in growth was noted (Fig. 9). Alterations in thyroid metabolism were observed when bSt was administered in ovo to fertile eggs (Dean et al., 1993). The pituitary and recombinant bSt were dissolved in pH 8.3 solution containing 0.03 M $NaHCO_3$ and 0.15 M NaCl. Two hundred and fifty (250) μg bSt in 100 μl carbonate solution was administered into the albumen on day 11 of embryogenesis. Pituitary, but not recombinant, bSt increased body weights and skeletal growth of

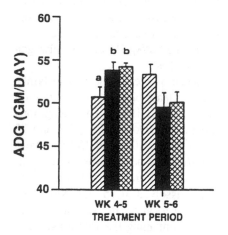

Figure 9 Average daily gain in chickens receiving daily injections of vehicle □, 0.5 mg/kg rbST ■ or 2.5 mg/kg rbSt ▨; n = 15 chicks. [a,b]Means with different superscripts are significantly different at $P < .05$. (Reprinted with permission of the publisher of Buonomo and Baile, 1988, by Elsevier Science Inc.)

male broilers at 3, 5, and 7 weeks after hatching. Decreased hatches with the bSt treated groups were possibly caused by hypothyroidism.

IV. FORMULATION DISCOVERY AND DEVELOPMENT

The discovery and development of a protein formulation for use in animals are extremely difficult. Like all good problem solving, a detailed description of the challenges should be identified. This description should include a profile of the planned product. This profile of product attributes will include delivery duration, dosing accuracy, packaging and product appearance, ease of preparation and administration, allowable product and packaging waste, and product storage conditions. The second step is a good review of the published formulation attempts with the molecule of interest and any closely related proteins/peptides. This benchmarking will help streamline product testing by knowing what other formulators have used as vehicles, excipients, packaging, syringe sizes, needle sizes, storage conditions, and elapsed time between manufacturing and administration. The third step is a literature review of the physicochemical characterization data collected for the molecule. In practice, defining the product profile occurs simultaneously with studying the formulation and physicochemical literature.

An example of the desired product profile of a bSt formulation for increased lactation in cows may state:

> Once-a-week administration to minimize labor; dosing accuracy in the mg range (i.e., want to accurately deliver 1 mg); packaging appearance consistent with companies' standard colors; preparation should take less than 10 sec to prepare for dosing; amounts less than 5 ml should be delivered; deliver subcutaneously in the neck with no observed swelling lasting more than 24 hours; no product waste; sterile product; price to animal producer less than $3.00; and 2-year room temperature storage.

The next steps include literature reviews, and hopefully this chapter provides an adequate review of veterinary protein/peptide formulations and molecule physicochemical characterization.

With the product profile, a list of formulations previously tested, and the physicochemical review, the main challenges can be described and undertaken. Continuing to use the bSt example, three main challenges are: discovering a stable dosage form that can be stored at room temperature; formulating sustained-release for 1 week; and providing product at a reasonable cost. Ways to approach these challenges are explored. A list of the major challenges existing with animal health proteins and suggestions on possible solutions are presented in Table 5. For example, if inadequate blood St concentrations are preventing 1-week increased milk production, a thorough investigation of the bSt at the injection site can be useful. In situ aggregation which creates insoluble mono-

Table 5 Major Challenges with Formulating Veterinary Proteins/Peptides and Suggestions on Possible Solutions

Challenges	Possible solutions
Rapid aqueous chemical degradation	Select different analog, store at lower temp., change formulation medium or use cosolvents, change pH, add excipients (e.g., sugars, amino acids, polyols, surfactants, other proteins)
Aggregation in solution	Assess bulk drug impurities (e.g., metals, excessive salt, solvents), select different analog, minimize ionic strength, vary buffer, vary pH, reduce agitation, add excipients (e.g., other proteins, amino acids, EDTA, surfactants, polyols, ions), vary counter ion using Hofmeister series
Hygroscopic	Use proper storage, formulate with excipients having propensity for water sorption, lyophilize, design formulation to protect from water (e.g., oil vehicle)
Adsorption	Add excipients (e.g., surfactants, other proteins, amino acids), conduct surface treatments, reduce molecule's hydrophobic sites/area, increase protein concentration, minimize agitation
Inadequate animal blood concentrations	Modify formulation; injection site evaluation including scintigraphy, visual examination, and formulation retrieval; PEG modification of protein; use penetration enhancers; change delivery route (e.g., subcutaneous, intramuscular); add enzyme inhibitor
Gelation	Reduce molecule's hydrophobic sites/area, uses excipients to prevent (e.g., carbohydrates, polymers), adjust pH, modify vehicle, eliminate electrolytes
Unacceptable powder (e.g., lyophilization) stability	Add stabilizers (e.g., amino acids, other proteins, polymers, carbohydrates), remove oxygen or add oxygen scavenger, prevent light exposure, change storage temperature, modify prelyophilization solution pH, increase glass transition temperature, optimize lyophilization cycle, assess cold denaturation, optimize excipient to protein ratio, study moisture content of stoppers
Drug and/or excipient tissue irritation and residue	Minimize injection volumes, select excipient based on low tissue irritation and rapid clearance

meric bSt would suggest modification of the formulation vehicle as one method to resolve the problem. If insurmountable hurdles are found, the desired product profile must be changed, as otherwise the protein/peptide cannot be formulated and ultimately marketed.

Once a suitable formulation is discovered, the development phase can begin. The necessary development activities for a protein/peptide formulation are generally the same as for a nonprotein entity and are described throughout this book. Some unique challenges that might be encountered include longer times for the validation and implementation of assays, larger quality variability between early batches, difficulty with determining the stability indicating assay, difficulty with defining active isoforms and impurities, and higher costs associated with formulation manufacturing scale-up of an expensive compound.

The path for the first protein/peptide approved for use in animals has been created. The political and socioeconomic issues are being fought. Just as these challenges continue, the search for ideal veterinary formulations will continue. The requirements for stable, economical, and once-only administration make protein/peptide formulation discovery and development for animals extremely challenging and exciting.

ACKNOWLEDGMENTS

I was graciously assisted by Drs. Leslie C. Eaton, Michael J. Hageman, and Alice C. Martino, who provided comments on the content of this chapter. I thank them for their help. I also appreciated receiving the scanning electron photomicrographs of GRF fibrils from Dr. Thomas J. Raub, and all the assistance received from Pharmacia and Upjohn Animal Health Discovery Research in conducting the GRF formulation research.

REFERENCES

Abdel-Meguid SS, Shieh HS, Smith WW, Dayringer HE, Violand BN, Bentle LA. (1987). Three-dimensional structure of a genetically engineered variant of porcine growth hormone. Proc Natl Acad Sci USA: Biochemistry 84(18):6434–6437.

Althen TG, Gerrits RJ. (1976). Metabolic clearance and secretion rates of porcine growth hormone in genetically lean and obese swine. Endocrinology 99(2):511–515.

Andrews P, Folley SJ. (1963). Molecular weights of bovine, ovine and porcine pituitary growth hormones estimated by gel filtration on sephadex. Proc Biochem Soc 87:P3.

Arbona JR, Marple DN, Russell RW, Rahe CH, Mulvaney DR, Sartin JL. (1988). Secretory patterns and metabolic clearance rate of porcine growth hormone in swine selected for growth. J Anim Sci 66(12):3068–3072.

Arendt VD. (1992). Composition and method for reducing vial breakage during lyophilization. U.S. Patent 5,169,834.

Asimov GJ, Krouze NK. (1937). The lactogenic preparations from the anterior pituitary and the increase of milk yield in cows. J Dairy Sci 20:289–306.

Azain MJ, Kasser TR, Sabacky MJ. (1989a). Composition for controlled release of polypeptides. Eur Patent Appl 89870204.8.

Azain MJ, Eigenberg KE, Kasser TR, Sabacky MJ. (1989b). Somatotropin prolonged release. U.S. Patent 4,863,736.

Azain MJ, Eigenberg KE, Kasser TR, Sabacky MJ. (1990). Method for prolonged release of somatotropin. Eur Patent Appl 90202047.8.

Azain MJ, Bullock KD, Kasser TR, Veenhuizen JJ. (1992). Relationship of mode of porcine somatotropin administration and dietary fat to the growth performance and carcass characteristics of finishing pigs. J Anim Sci 70:3086–3095.

Azain MJ, Kasser TR, Sabacky MJ, Baile CA. (1993). Comparison of the growth-promoting properties of daily versus continuous administration of somatotropin in female rats with intact pituitaries. J Anim Sci 71:384–392.

Bastiras S, Wallace JC. (1992). Equilibrium denaturation of recombinant porcine growth hormone. Biochemistry 31(38):9304–9309.

Bauman DE, Eppard PJ, DeGeeter MJ, Lanza GM. (1985). Responses of high-producing dairy cows to long-term treatment with pituitary somatotropin and recombinant somatotropin. J Dairy Sci 68:1352–1362.

Becker BA, Knight CD, Buonomo FC, Jesse GW, Hedrick HB, Baile CA. (1992). Effect of a hot environment on performance, carcass characteristics, and blood hormones and metabolites of pigs treated with porcine somatotropin. J Anim Sci 70:2732–2740.

Bell LN, Hageman MJ, Bauer JM. (1995). Impact of moisture on thermally induced denaturation and decomposition of lyophilized bovine somatotropin. Biopolymers 35(2):201–209.

Bewley TA, Li CH. (1972). Molecular weight and circular dichroism studies of bovine and ovine pituitary growth hormones. Biochemistry 11(5):927–931.

Bohlen P, Brazeau P, Bloch B, Ling N, Gaillard R, Guillemin R. (1983a). Human hypothalamic growth hormone-releasing factor (GRF): evidence for two forms identical to tumor derived GRF-44-NH$_2$ and GRF-40. Biochem Biophys Res Commun 114(3):930–936.

Bohlen P, Esch F, Brazeau P, Ling N, Guillemin R. (1983b). Isolation and characterization of the porcine hypothalamic growth hormone releasing factor. Biochem Biophys Res Commun 116(2):726–734.

Bongers J, Heimer EP, Lambros T, Pan Y-CE, Campbell RM, Felix AM. (1992). Degradation of aspartic acid and asparagine residues in human growth hormone-releasing factor. Int J Peptide Protein Res 39:364–374.

Bramley MR, Carter AB, Dunwell DW. (1988). Somatotropin formulations. Eur Patent Appl 88309997.0.

Brazeau P, Bohlen P, Esch F, Ling N, Wehrenberg WB, Guillemin R. (1984). Growth hormone-releasing factor from ovine and caprine hypothalamus: isolation, sequence analysis and total synthesis. Biochem Biophys Res Commun 125(2):606–614.

Brems DN, Plaisted SM, Havel HA, et al. (1985). Equilibrium denaturation of pituitary- and recombinant-derived bovine growth hormone. Biochemistry 24(26):7662–7668.

Brems DN, Plaisted SM, Kauffman EW, Havel HA. (1986). Characterization of an associated equilibrium folding intermediate of bovine growth hormone. Biochemistry 25(21):6539–6543.

Brems DN, Plaisted SM, Dougherty JJ, Holzman TF. (1987). The kinetics of bovine growth hormone folding are consistent with a framework model. J Biol Chem 262(6):2590–2596.

Brems DN, Plaisted SM, Havel HA, Tomich CS. (1988). Stabilization of an associated folding intermediate of bovine growth hormone by site-directed mutagenesis. Proc Natl Acad Sci USA: Biochemistry 85(10):3367–3371.

Brooks ND, Needham GF. (1994). Injectable extended release formulations and methods. U.S. Patent 5,352,662.

Buckwalter BL, Cady SM, Shieh HM, Chaudhuri AK, Johnson DF. (1992). Improvement in the solution stability of porcine somatotropin by chemical modification of cystein residues. J Agric Food Chem 40(2):356–362.

Buonomo FC, Baile CA. (1988). Recombinant bovine somatotropin stimulates short-term increases in growth rate and insulin-like growth factor-I (IGF-I) in chickens. Domest Anim Endocrinol 5(3):219–229.

Burger HG, Edelhoch H, Condliffe PG. (1966). The properties of bovine growth hormone. I. Behavior in acid solution. J Biol Chem 241(2):449–457.

Cady SM, Fishbein R, Schroder U, Eriksson H, Probasco B. (1986). Water dispersible and water soluble carbohydrate polymer compositions for parenteral administration. Eur Patent Appl 86102752.2.

Cady SM, Fishbein R, SanFilippo M. (1988). The development of controlled release implants for growth hormone releasing hexapeptide. Proc Int Symp Controlled Release Bioactive Materials 15:56–57.

Cady SM, Steber WD, Fishbein R. (1989). Development of a sustained release delivery system for bovine somatotropin. Proc Int Symp Controlled Release Bioactive Materials 16:22–23.

Calsamiglia S, Hongerholt DD, Crooker BA, Stern MD. (1992). Effect of fish meal and expeller-processed soybean meal fed to dairy cows receiving bovine somatotropin (Sometribove). J Dairy Sci 75:2454–2462.

Campbell RM, Lee Y, Rivier J, Heimer EP, Felix AM, Mowles TR. (1991). GRF analogs and fragments: correlation between receptor binding, activity and structure. Peptides 12:569–574.

Campbell RM, Bongers J, Felix AM. (1995). Rational design, synthesis, and biological evaluation of novel growth hormone releasing factor analogues. Biopolymers (Peptide Sci) 37:67–88.

Cardamone M, Puri NK, Sawyer WH, Capon RJ, Brandon MR. (1994). A spectroscopic and equilibrium binding analysis of cationic detergent-protein interactions using soluble and insoluble recombinant porcine growth hormone. Biochim Biophys Acta 1206:71–82.

Carlacci L, Chou KC, Maggiora GM. (1991). A heuristic approach to predicting the tertiary structure of bovine somatotropin. Biochemistry 30(18):4389–4398.

Cascone O, Biscoglio de Jimenez Bonino M, Santome JA. (1980). Oxidation of methionine residues in bovine growth hormone by chloramine-T. Int J Peptide Protein Res 16(4):299–305.

Chang JP, Tucker RC, Ghrist BF, Coleman MR. (1994). Non-denaturing assay for the determination of the potency of recombinant bovine somatotropin by high-performance size-exclusion chromatography. J Chromatogr A675(1–2):113–122.

Charman SA, McCrossin LE, Charman WN. (1993). Validation of a peptide map for recombinant porcine growth hormone and application to stability assessment. Pharm Res 10(10):1471–1477.

Chen T. (1992). Formulation concerns of protein drugs. Drug Dev Indust Pharm 18(11&12):1311–1354.

Christian LL, Miller LF. (1991). Porcine somatotropin to improve meat quality of pigs. U.S. Patent 5,015,626.

Chung CS, Etherton TD, Wiggins JP. (1985). Stimulation of swine growth by porcine growth hormone. J Anim Sci 66(1):118–130.

Clarke S, Stephenson RC, Lowenson JD. (1992). Lability of asparagine and aspartic acid residues in proteins and peptides: spontaneous deamidation and isomerization reactions. In: Ahern TJ, Manning MC, eds. Stability of Protein Pharmaceuticals: Part A: Chemical and Physical Pathways of Protein Degradation. New York: Plenum Press, pp. 1–29.

Clore GM, Martin SR, Gronenborn AM. (1986). Solution structure of human growth hormone releasing factor: combined use of circular dichroism and nuclear magnetic resonance spectroscopy. J Molec Biol 191:553–561.

Cole WJ, Eppard PJ, Boysen BG, et al. (1992). Response of dairy cows to high doses of a sustained-release bovine somatotropin administered during two lactations. 2. Health and reproduction. J Dairy Sci 75:111–123.

Coy DH, Murphy WA, Sueiras-Diaz J, Coy EJ, Lance VA. (1985). Structure-activity studies on the N-terminal region of growth hormone releasing factor. J Med Chem 28:181–185.

Craft LS, Ferguson TH, Heiman ML, Thompson WW. (1994). Methods for administering biological agents and microparticle compositions useful in these and other methods. Eur Patent Appl 94303655.8.

Dahl GE, Chapin LT, Zinn SA, Moseley WM, Schwartz TR, Tucker HA. (1990). Sixty-day infusions of somatotropin-releasing factor stimulate milk production in dairy cows. J Dairy Sci 73:2444–2452.

Davio SR, Hageman MJ. (1993). Characterization and formulation consideration for recombinantly derived bovine somatotropin. In: Wang YJ, Pearlman R, eds. Stability and Characterization of Protein and Peptide Drugs: Case Histories. New York: Plenum Press, pp. 59–89.

Dean CE, Hargis BM, Burke WH, Hargis PS. (1993). Alterations in thyroid metabolism are associated with improved posthatch growth of chickens administered bovine growth hormone in ovo. Growth Devel Ageing 57:57–72.

Delfino JM, Fernandez HN, Santome JA. (1986). Carboxylate groups in bovine somatotropin involved in growth promoting activity. Int J Peptide Protein Res 28:307–314.

Dellacha JM, Enero MA, Faiferman I. (1966). Molecular weight of bovine growth hormone. Experientia 22(1):16–17.

Dellacha JM, Santome JA, Paladini AC. (1968). Physicochemical and structural studies of bovine growth hormone. Ann NY Acad Sci 148(2):313–327.

DePrince RB, Viswanatha R. (1988). Stabilized porcine growth hormone. U.S. Patent 4,765,980.

Dougherty JJ, Snyder LM, Sinclair RL, Robins RH. (1990). High-performance tryptic mapping of recombinant bovine somatotropin. Anal Biochem 190(1):7–20.

Dubreuil P, Pelletier G, Petitclerc D, et al. (1988). Influence of growth-hormone-releasing factor and(or) thyrotropin-releasing factor on somatotropin, prolactin, triiodothyronine and thyroxine release in growing pigs. Can J Anim Sci 68:699–709.

Dubreuil P, Petitclerc D, Pelletier G, et al. (1990). Effect of dose and frequency of administration of a potent analog of human growth hormone releasing factor on hormone secretion and growth in pigs. J Anim Sci 68:1254–1268.

Eckenhoff JB, Magruder JA, Cortese R, Peery JR, Wright JC. (1989a). Device compromising means for protecting and dispensing fluid sensitive medicament. Eur Patent Appl 89302699.7.

Eckenhoff JB, Magruder JA, Peery JR, Wright JC. (1989b). Device compromising means for protecting and dispensing fluid sensitive medicament. U.S. Patent 4,855,141.

Eckenhoff JB, Magruder JA, Cortese R, Peery JR, Wright JC. (1990). Method for delivering somatotropin to an animal. U.S. Patent 4,959,218.

Eckenhoff JB, Magruder JA, Cortese R, Peery JR, Wright JC. (1991). Device compromising liner for protecting fluid sensitive medicament. U.S. Patent 4,996,060.

Edelhoch H, Burger HG. (1966). The properties of bovine growth hormone. II. Effects of urea. J Biol Chem 241(2):458–463.

Enright WJ, Chapin LT, Moseley WM, Tucker HA. (1988). Effects of infusions of various doses of bovine growth hormone-releasing factor on growth hormone and lactation in Holstein cows. Dairy Sci 71(1):99–108.

Enright WJ, Prendiville DJ, Spicer LJ, et al. (1993). Effects of growth hormone-releasing factor and (or) thyrotropin-releasing hormone on growth, feed efficiency, carcass characteristics, and blood hormones and metabolites in beef heifers. J Anim Sci 71:2395–2405.

Eppard PJ, Hudson S, Cole WJ, et al. (1991). Response of dairy cows to high doses of a sustained-release bovine somatotropin administered during two lactations: 1. Production response. J Dairy Sci 74:3807–3821.

Eppard PJ, White TC, Birmingham BK, et al. (1993). Pharmacokinetic and galactopoietic response to recombinant variants of bovine growth hormone. J Endocrinol 139(3):441–450.

Esch F, Bohlen P, Ling N, Brazeau P, Guillemin R. (1983). Isolation and characterization of the bovine hypothalamic growth hormone releasing factor, Biochemical and Biophysical Research Communications 117(3):772–779.

Etherton TD, Walker OA. (1982). Characterization of insulin binding to isolated swine adipocytes. Endocrinology 110:1720–1724.

Etherton TD, Wiggins JP, Chung CS, Evock CM, Rebhun JF, Walton PE. (1986). Stimulation of pig growth performance by porcine growth hormone and growth hormone-releasing factor. J Anim Sci 63:1389–1399.

Etherton TD, Wiggins JP, Evock CM, et al. (1987). Stimulation of pig growth performance by porcine growth hormone: determination of the dose-response relationship. J Anim Sci 64:433–443.

Evock CM, Etherton TD, Chung CS, Ivy RE. (1988). Pituitary porcine growth hormone (pGH) and a recombinant pGH analog stimulate pig growth performance in a similar manner. J Anim Sci 66:1928–1941.

Felix AM, Heimer EP, Mowles TF, et al. (1986). Synthesis and biological activity of novel growth hormone releasing factor analogs. In: Theodoropoulus D, ed. Peptides Berlin: Walter de Gruyter and Co., pp. 481–485.

Felix AM, Heimer DP, Wang C-T, et al. (1988). Synthesis, biological activity and conformational analysis of cyclic GRF analogs. Int J Peptide Protein Res 32:441–454.

Ferguson TH, Thompson WW, Moore DL, Rodewald JM. (1988). Chemical and physical characterization of somidobove, a recombinantly produced bovine somatotropin. Proc Int Symp Controlled Release Bioactive Materials 15:55a–55b.

Ferguson TH, Harrison RG, Moore L. (1990). Injectable sustained release formulation. U.S. Patent 4,977,140.

Fischer M, Lebens MR, Chaleff DT. (1991). Somatotropins with alterations in the alpha-helix 1 region and combinations with other mutations. Aust Patent Appl AU-A-88300/91.

Friedman AR, Ichhpurani AK, Brown DM, et al. (1991). Degradation of growth hormone releasing factor analogs in neutral aqueous solution is related to deamidation of asparagine residues. Int J Peptide Protein Res 37:14–20.

Friedman AR, Ichhpurani AK, Moseley WM, et al. (1992). Growth hormone-releasing factor analogs with hydrophobic residues at position 19. Effects on growth hormone releasing activity in vitro and in vivo, stability in blood plasma in vitro, and secondary structure. J Med Chem 35:3928–3933.

Frohman LA, Downs TR, Williams TC, Heimer EP, Pan Y-CE, Felix AM. (1986). Rapid enzymatic degradation of growth hormone releasing hormone by plasma in vitro and in vivo to a biologically inactive product cleaved at the NH_2 terminus. J Clin Invest 78:906–913.

Frohman MA, Downs TR, Chomczynski P, Frohman LA. (1989a). Cloning and characterization of mouse growth hormone releasing hormone (GRH) complementary DNA: increased GRF messenger RNA levels in the growth hormone deficient lit/lit mouse. Molec Endocrinol 3(10):1529–1536.

Frohman LA, Downs TR, Heimer EP, Felix AM. (1989b). Depeptidylpeptidase IV and trypsin-like enzymatic degradation of human growth releasing hormone in plasma. J Clin Invest 83:1533–1540.

Fronk TJ, Peel CJ, Bauman DE, Gorewit RC. (1983). Comparison of different patterns of exogenous growth hormone administration on milk production in Holstein cows. J Anim Sci 57(3):699–705.

Fujioka K, Sato S, Takada Y. (1985). Stable growth hormone releasing factor preparation. Eur Patent Appl 85309462.1.

Fujioka K, Sato S, Takada Y. (1986). Sustained-release preparation. Aust Patent Appl AU-A-55983/86.

Fujioka K, Sato S, Tamura N, Takada Y, Sasaki Y, Maeda M. (1989). Improved controlled release formulation. Eur Patent Appl 89101428.4.

Grangier JL, Puygrenier M, Gautier JC, Couvreur P. (1991). Nanoparticles as carriers for growth hormone releasing factor. J Controlled Release 15:3–13.

Glaser CB, Li CH. (1974). Reaction of bovine growth hormone with hydrogen peroxide. Biochemistry 13(5):1044–1047.

Godfredson JA, Wheaton JE, Crooker BA, Wong EA, Campbell RM, Mowles TF. (1990). Growth performance and carcass composition of lambs infused for 28 days with a growth hormone releasing factor analogue. J Anim Sci 68:3624–3632.

Gooley PR, Plaisted SM, Brems DN, Mackenzie NE. (1988). Determination of local conformational stability in fragment 96-133 of bovine growth hormone by high-resolution H NMR spectroscopy. Biochemistry 27(2):802–809.

Graf L, Li CH. (1974). On the primary structure of pituitary bovine growth hormone. Biochem Biophys Res Commun 56(1):168–175.

Graf L, Li CH, Bewley TA. (1975). Selective reduction and alkylation of the COOH-terminal disulfide bridge in bovine growth hormone. Int J Peptide Protein Res 7:467–473.

Grangier JL, Puygrenier M, Gautier JC, Couvreur P. (1991). Nanoparticles as carriers for growth hormone releasing factor. J Controlled Release 15:3–13.

Guillemin R, Brazeau P, Bohlen P, Esch F, Ling N, Wehrenberg WB. (1982). Growth hormone-releasing factor from a human pancreatic tumor that caused acromegaly. Science 218:585–587.

Hacker RR, Deschutter A, Adeola O, Kasser TR. (1993). Evaluation of long-term somatotropin implants in finishing pigs. J Anim Sci 71:564–570.

Hageman MJ. (1988). The role of moisture in protein stability. Drug Devel Indust Pharm 14(14):2047–2070.

Hageman MJ, Bauer JM, Possert PL, Darrington RT. (1992). Preformulation studies oriented toward sustained delivery of recombinant somatotropins. J Agric Food Chem 40:348–355.

Hamilton EJ, Burleigh BD. (1987). Stabilization of growth promoting hormones. Eur Patent Appl 87307406.6.

Hanson MA, Rouan SKE. (1992). Introduction to formulation of protein and pharmaceuticals. In: Ahern TJ, Manning MC, eds. Stability of Protein Pharmaceuticals. Part B. In Vivo Pathways of Degradation and Strategies for Protein Stabilization. New York: Plenum Press, pp. 209–233.

Hara K, Hsu-Chen CJ, Sonenberg M. (1978). Recombination of the biologically active peptides from a tryptic digest of bovine growth hormone. Biochemistry 17(3):550–556.

Haro LS, Collier RJ, Talamantes FJ. (1984). Homologous somatotropin radioreceptor assay utilizing recombinant bovine growth hormone. Molec Cell Endocrinol 38:109–116.

Hart IC, Chadwick PME, Coert A, James S, Simmonds AD. (1985). Effect of different growth hormone-releasing factors on the concentrations of growth hormone, insulin, and metabolites in the plasma of sheep maintained in positive and negative energy balance. J Endocrinol 105:113–119.

Hartman PA, Stodola JD, Harbour GC, Hoogerheide JG. (1986). Reversed-phase high-performance liquid chromatography peptide mapping of bovine somatotropin. J Chromatogr 360(2):385–395.

Hartnell GF, Franson SE, Bauman DE, et al. (1991). Evaluation of sometribove in a prolonged-release system in lactating dairy cows—production responses. J Dairy Sci 74:2645–2663.

Havel HA, Kauffman EW, Plaisted SM, Brems DN. (1986). Reversible self-association of bovine growth hormone during equilibrium unfolding. Biochemistry 25(21):6533–6538.

Havel HA, Chao RS, Haskell RJ, Thamann TJ. (1989). Investigations of protein structure with optical spectroscopy: bovine growth hormone. Anal Chem 61(7):642–650.

Higgs DA, Fagerlund UHM, McBride JR, Dye HM, Donaldson EM. (1977). Influence of combinations of bovine growth hormone, 17α-methyltestosterone, and L-thyroxine on growth of yearling coho salmon (*Oncorhynchus kisutch*). Can Zool 55:1048–1056.

Hodate K, Johke T, Ohashi S. (1986). Response of growth hormone release to human growth hormone releasing factor and its analogs in the bovine. Endocrinol Jpn 33(4):519–525.

Holladay LA, Hammonds RG, Puett D. (1974). Growth hormone conformation and conformation equilibria. Biochemistry 13(8):1653–1661.

Holzman TF, Dougherty JJ, Brems DN, MacKenzie NE. (1990). pH-induced conformational states of bovine growth hormone. Biochemistry 29(5):1255–1261.

Hu OY, Chang WC, Yang TS. (1995). Pharmacokinetic properties of recombinant porcine growth hormone in pigs. Pharmacology 50(1):63–68.

Janski AM, Yang R-D. (1987). Controlled release implants for administration of recombinant growth hormones. Eur Patent Appl 86305431.8.

Janski AM, Yang R-D. (1988). Cylindrical implants for the controlled release of growth hormones. U.S. Patent 4,786,501.

Janski AM, Drengler SM. (1988). Method for stabilising proteins. Eur Patent Appl 88302317.8.

Jenny BF, Grimes LW, Pardue FE, Rock DW, Patterson DL. (1992). Lactational response of Jersey cows to bovine somatotropin administered daily or in a sustained-release formulation. J Dairy Sci 75:3402–3407.

Kent JS. (1984). Cholesterol matrix delivery system for sustained release of macromolecules. U.S. Patent 4,452,775.

Kim NJ, Rhee BG, Cho HS. (1991). A composition durably releasing bioactive polypeptide. Aust Patent Appl 70937/91.

Kim NJ, Cho HS, Song MS, Choi YJ, Rhee BG. (1993). Release-controlled implantable somatotropin composition. Aust Patent Appl 52718/93.

Knight CD, Kasser TR, Swenson GH, et al. (1991). The performance and carcass composition responses of finishing swine to a range of porcine somatotropin doses in a 1-week delivery system. J Anim Sci 69:4678–4689.

Koppel J, Kuchar S, Rynikova A, Mozes S, Noskovic P, Boda K. (1988). Effects of bovine growth hormone in suckling lambs. Exp Clin Endocrinol 91(2):223–226.

Kosen PA. (1992). Disulfide bonds in proteins. In: Ahern TJ, Manning MC, eds. Stability of Protein Pharmaceuticals. Part A. Chemical and Physical Pathways of Protein Degradation. New York: Plenum Press, pp. 31–67.

Kubiak TM, Kelly CR, Krabill LF. (1989). In vitro metabolic degradation of a bovine growth hormone-releasing factor analog Leu27-bGRF(1–29)NH2 in bovine and porcine plasma. Correlation with plasma dipeptidylpeptidase activity. Drug Metab Dispos 17(4):393–397.

Kubiak TM, Friedman AR, Martin RA, et al. (1992a). High in vivo bioactivities of position 2/Ala[15]-substituted analogs of bovine growth hormone releasing factor (bGRF) with improved metabolic stability. In: Smith JA, Rivier JE, eds. Peptides, Chemistry and Biology. Leiden: ESCOM Scientific Publishers, pp. 85–87.

Kubiak TM, Martin RA, Hillman RM, et al. (1992b). Implications of improved metabolic stability of peptides on their performance in vivo: resistance to DPP-IV mediated cleavage of GRF analogs greatly enhances their potency in vivo. In: Smith JA, Rivier JE, eds. Peptides, Chemistry and Biology. Leiden: ESCOM Scientific Publishers, pp. 23–25.

Kubiak TM, Friedman AR, Martin RA, et al. (1993). Position 2 and position 2/Ala[15]-substituted analogs of bovine growth hormone releasing factor (bGRF) with enhanced metabolic stability and improved in vivo bioactivity. J Med Chem 36:888–897.

Lance VA, Murphy WA, Sueiras-Diaz J, Coy DH. (1984). Super-active analogs of growth hormone releasing factor (1–29)-amide. Biochem Biophys Res Commun 119(1):265–272.

Langley KE, Lai PH, Wypych J, Everett RR, et al. (1987a). Recombinant-DNA-derived bovine growth hormone from *Escherichia coli*. 2. Biochemical, biophysical, immunological and biological comparison with the pituitary hormone. Eur J Biochem 163(2):323–330.

Langley KE, Berg TF, Strickland TW, Fenton DM, Boone TC, Wypych J. (1987b). Recombinant-DNA-derived bovine growth from *Escherichia coli*. 1. Demonstration that the hormone is expressed in reduced form, and isolation of the hormone in oxidized, native form. Eur J Biochem 163(2):313–321.

LaPierre H, Pelletier G, Petitclerc D, et al. (1988). Effect of human growth hormone-releasing factor (1–29) NH$_2$ on growth hormone release and milk production in dairy cows. J Dairy Sci 71:92–98.

Lehrman SR, Tuls JL, Havel HA, Haskell RJ, Putnam SD, Tomich CS. (1991). Site-directed mutagenesis to probe protein folding: evidence that the formation and aggregation of a bovine growth hormone folding intermediate are dissociable. Biochemistry 30(23):5777–5784.

Leonard RJ, Harman SM. (1991). Partially fused peptide pellet. U.S. Patent 5,039,660.

Leung FC, Jones B, Steelman SL, Rosenblum CI, Kopchick JJ. (1986). Purification and physiochemical properties of a recombinant bovine growth hormone produced by cultured murine fibroblasts. Endocrinology 119(4):1489–1496.

Lewis UJ, Cheever EV, Hopkins WC. (1970). Kinetic study of the deamidation of growth hormone and prolactin. Biochim Biophys Acta 214:498–508.

Liberti JP, Alfano J, Sonenberg M. (1969). Chemical and biological characterizaton of methylated bovine growth hormone. Biochim Biophys Acta 181(1):176–183.

Ling N, Esch R, Bohlen P, Brazeau P, Wehrenberg WB, Guillemin R. (1984a). Isolation, primary structure, and synthesis of human hypothalamic somatocrinin: growth hormone-releasing factor. Proc Natl Acad Sci USA: Biochemistry 81:4302–4306.

Ling N, Baird A, Wehrenberg WB, Ueno N, Munegumi T, Brazeau P. (1984b). Synthesis and in vitro bioactivity of C-terminal deleted analogs of human growth hormone releasing factor. Biochem Biophys Res Commun 123(2):854–861.

Ludri RS, Upadhyay RC, Singh M, Guneratne JRM, Basson RP. (1989). Milk production in lactating buffalo receiving recombinantly produced bovine somatotropin. J Dairy Sci 72:2283–2287.

Machlin LJ. (1972). Effect of porcine growth hormone on growth and carcass composition of the pig. J Anim Sci 35(4):794–800.

Machlin LJ. (1973). Effect of growth hormone on milk production and feed utilization in dairy cows. J Dairy Sci 56:575–580.

MacKenzie NE, Plaisted SM, Brems DN. (1989). Proton nuclear magnetic resonance study of the histidine residues of pituitary bovine growth hormone. Biochim Biophys Acta 994(2):166–171.

Maniar M, Domb AJ. (1992). Controlled release microparticulate delivery system for proteins. Int Patent Appl PCT/US92/01351.

Manning MC, Patel K, Borchardt RT. (1989). Stability of protein pharmaceuticals. Pharm Res 6(11):903–918.

Mao B. (1990). Molecular topology and dimerization of recombinant bovine somatotropin. Biopolymers 30(5-6):645–647.

Mariette B, Coudane J, Vert M, Gautier J-C, Moneton P. (1993). Release of the GRF29NH$_2$ analog of human GRF44NH$_2$ from a PLA/GA matrix. J Controlled Release 24:237–246.

Markert JR, Higgs DA, Dye HD, MacQuarrie DW. (1977). Influence of bovine growth hormone on growth rate, appetite, and food conversion of yearling coho salmon (*Oncorhynchus kisutch*) fed two diets of different composition. Can J Zool 55:74–83.

Martin JL, Kramer JF. (1986). Prolonged release of growth promoting hormones. Aust Patent Appl 61092/86.

Martin RA, Cleary DL, Guido DM, Zurcher-Neely HA, Kubiak TM. (1993). Dipeptidyl peptidase IV(DPP-IV) from pig kidney cleaves analogs of bovine growth hormone-releasing factor (bGRF) modified at position 2 with Ser, Thr or Val. Extended DPP-IV substrate specificity? Biochim Biophys Acta 1164(3):252–260.

McLaren DG, Bechtel PJ, Grebner GL, et al. (1990). Dose response in growth of pigs injected daily with porcine somatotropin from 57 to 103 kilograms. J Anim Sci 68:640–651.

McLaughlin CL, Byatt JC, Hedrick HB, et al. (1993). Performance, clinical chemistry, carcass responses of finishing lambs to recombinant bovine somatotropin and bovine placental lactogen. J Anim Sci 71:3307–3318.

McLaughlin CL, Hedrick HB, Veenhuizen HH, et al. (1994). Performance, clinical chemistry and carcass responses of finishing lambs to formulated sometribove (methionyl bovine somatotropin). J Anim Sci 72:2544–2551.

McLean E, Teskeredzic E, Donaldson EM, et al. (1992). Accelerated growth of coho salmon *Oncorhynchus kisutch* following sustained release of recombinant porcine somatotropin. Aquaculture 103:377–387.

McNamara JP, Brekke CJ, Jones RW, Dalrymple RH. (1991). Recombinant porcine somatotropin alters performance and carcass characteristics fed alternative feedstuffs. J Anim Sci 69:2273–2281.

Mentlein R, Gallwitz B, Schmidt WE. (1993). Dipeptidyl-peptidase IV hydrolyses gastric inhibitory polypeptide, glucagon-like peptide -1(7-36)amide, peptide histidine methionine and is responsible for their degradation in human serum. Eur J Biochem 214:829–835.

Mitchell JW. (1991). Prolonged release of biologically active somatotropin. U.S. Patent 5,013,713.

Mitchell JW. (1992). Methods of using prolonged release somatotropin compositions. U.S. Patent 5,086,041.

Mitchell JW. (1995a). Prolonged release of biologically active somatotropin. U.S. Patent 5,411,951.

Mitchell JW. (1995b). Prolonged release of biologically active somatotropin. U.S. Patent 5,474,980.

Moseley WM, Huisman J, VanWeerden EJ. (1987). Serum growth hormone and nitrogen metabolism responses in young bull calves infused with growth hormone-releasing factor for 20 days. Domest Anim Endocrinol 4(1):51–59.

Moseley WM, Paulissen JB, Goodwin MC, Alaniz GR, Claflin WH. (1992). Recombinant bovine somatotropin improves growth performance in finishing beef steers. J Anim Sci 70:412–425.

Oeswein JQ, Shire SJ. (1991). Physical biochemistry of protein drugs. In: Lee VHL, ed. Peptide and Protein Drug Delivery. New York: Marcel Dekker, pp. 167–202.

Oldham ER, Daley MJ. (1991). Lysostaphin: use of a recombinant bactericidal enzyme as a mastitis therapeutic. J Dairy Sci 74:4175–4182.

Oppezzo OJ, Fernandez HN. (1991). Contact area of bovine somatotropin dimer: involvement of tyrosine 142. Int J Peptide Protein Res 37(4):277–282.

Organon Laboratories Limited. (1961). UK Patent Specification 885,798.

Parker DB, Coe IR, Dixon GH, Sherwood NM. (1993). Two salmon neuropeptides encoded by one brain cDNA are structurally related to members of the glucagon superfamily. Eur J Biochem 215:439–448.

Pearlman R, Nguyen TH. (1991). Analysis of protein drugs. In: Lee VHL, ed. Peptide Protein Drug Delivery. New York: Marcel Dekker, pp. 247–301.

Peel CJ, Bauman DE, Gorewit RC, Sniffen CJ. (1981). Effect of exogenous growth hormone on lactational performance in high yielding dairy cows. J Nutr 111:1662–1671.

Petitclerc D, Pelletier G, Lapierre H, et al. (1987). Dose response of two synthetic human growth hormone-releasing factors on growth hormone release in heifers and pigs. J Anim Sci 65:996–1005.

Pitt CG, Cha Y, Donaldson EM, McLean E. (1991). Polyvinyl alcohol coated pellet of growth hormone. Int Patent Appl PCT/US91/08129.

Pitt CG, Cha Y, Donaldson EM, McLean E. (1992). Polyvinyl alcohol coated pellet of growth hormone. Int Patent Appl 92/07556.

Pommier SA, Dubreuil P, Pelletier G, Gaudreau P, Mowles TF, Brazeau P. (1990). Effect of a potent analog of human growth hormone-releasing factor on carcass composition and quality of crossbred market pigs. J Anim Sci 68:1291–1298.

Puri NK. (1991). Refolding of recombinant porcine growth hormone in a reducing environment limits in vitro aggregate formation. Fed Eur Biochem Soc 292(1,2):187–190.

Puri NK, Cardamone M. (1992). A relationship between the starting secondary structure of recombinant porcine growth hormone solubilised from inclusion bodies and the yield of native (monomeric) protein after in vitro refolding. Fed Eur Biochem Soc 305(3):177–180.

Raman SN, DePrince RB, Blum A. (1993). Delayed release for transition metal/protein complexes. Int Patent Appl PCT/US93/00274.

Raman SN, Gray MW, Smith RL. (1994). Compositions and processes for the sustained release of drugs. U.S. Patent 5,328,697.

Randall CS, Malefyt TR, Sternson LA. (1991). Approaches to the analysis of peptides. In: Lee VHL, ed. Peptide and Protein Drug Delivery. New York: Marcel Dekker, pp. 203–246.

Reynaert R, Franchimont P. (1974). Radioimmunoassay of bovine growth hormone. Ann Endocrinol 35(2):139–148.

Rivier J, Spiess J, Thorner M, Vale W. (1982). Characterization of a growth hormone-releasing factor from a human pancreatic islet tumor. Nature 300:276–278.

Ron E, Turek T, Mathiowitz E, Chasin M, Langer R. (1989). Release of polypeptides from poly(anhydride) implants. Proc Int Symp Controlled Release Bioactive Materials 16:338–339.

Ron E, Chasin M, Turek T, Langer R. (1992). Polyanhydride bioerodible controlled release implants for administration of stabilized growth hormone. U.S. Patent 5,122,367.

Rosemberg E, Thonney ML, Butler WR. (1989). The effects of bovine growth hormone and thyroxine on growth rate and carcass measurements in lambs. J Anim Sci 67:3300–3312.

Santome JA, Dellacha JM, Paladini AC, et al. (1973). Primary structure of bovine growth hormone. Eur J Biochem 37(1):164–170.

Secchi C, Biondi PA, Berrini A, Simonic T, Ronchi S. (1988). A biotin-avidin sandwich enzyme-linked immunosorbent assay of growth hormone in bovine plasma. J Immunol Methods 110(1):123–128.

Seeburg PH, Sias S, Adelman J, et al. (1983). Efficient bacterial expression of bovine and porcine growth hormones. DNA 2(1):37–45.

Shalati MD, Viswanatha R. (1988). Sustained release implant and method for preparing same. U.S. Patent 4,761,289.

Sivaramakrishnan KN, Rahn SL, Moore BM, O'Neil J. (1989). Sustained release of bovine somatotropin from implants. Proc Int Symp Controlled Release Bioactive Materials 16:14–15.

Sivaramakrishnan KN, Miller LF. (1990). Controlled release delivery device for macromolecular proteins. Int Patent Appl PCT/US90/01340.

Sonenberg M, Beychok S. (1971). Circular dichroism studies of biologically active growth hormone preparations. Biochim Biophys Acta 229:88–101.

Stanisiewski EP, McAllister JF, Ash KA, Taylor VN, Kratzer DD, Lauderdale JW. (1994). Production performance of dairy cattle administered recombinantly derived bovine somatotropin (USAN, Somavubove) daily: a dose range study. Domest Anim Endocrinol 11(3):239–260.

Steber W. (1991). Stable compositions for parenteral administration and method of making same. Eur Patent Appl 91100650.0.

Steber W, Fishbein R, Cady SM. (1987). Compositions for parenteral administration and their use. Eur Patent Appl 87111217.3.

Steber WD, Cady SM, Johnson DF, Rice T. (1992). Implant compositions containing a biologically active protein, peptide or polypeptide. Eur Patent Appl 92107278.1.

Stelwagen K, Grieve DG, Walton JS, Ball JL, McBride BW. (1993). Effect of prepartum bovine somatotropin in primigravid ewes on mammogenesis, milk production and hormone concentrations. J Dairy Sci 76:992–1001.

Stevenson CL, Donlan ME, Friedman AR, Borchardt RT. (1993). Solution conformation of Leu^{27} hGRF(1-32)NH$_2$ and its deamidation products by 2D NMR. Int J Peptide Protein Res 42:24–32.

Stodola JD, Walker JS, Dame PW, Eaton LC. (1986). High-performance size-exclusion chromatography of bovine somatotropin. J Chromatogr 357(3):423–428.

Su C-M, Jensen LR, Heimer EP, Felix AM, Pan Y-CE, Mowles TF. (1991). In vitro stability of growth hormone releasing factor (GR) analogs in porcine plasma. Horm Metab Res 23:15–21.

Suhr ST, Rahal JO, Mayo KE. (1989). Mouse growth hormone releasing hormone: precursor structure and expression in brain and placenta. Mol Endocrinol 3(11):1693–1670.

Toutain PL, Schams D, Laurentie MP, Thomason TD. (1993). Pharmacokinetics of a recombinant bovine growth hormone and pituitary bovine growth hormone in lactating dairy cows. J Anim Sci 71(5):1219–1225.

Tsuji K. (1993). Evaluation of sodium dodecyl sulfate non-acrylamide, polymer gel filled capillary electrophoresis for molecular size separation of recombinant bovine somatotropin. J Chromatogr A 652(1):139–147.

Tyle P, Cady SM. (1990). Sustained release multiple emulsions for bovine somatotropin delivery. Proc Int Symp Controlled Release Bioactive Materials 17:49–50.

Vaughan JM, Rivier J, Spiess J, et al. (1992). Isolation and characterization of hypothalamic growth hormone releasing factor from common carp, *Cyprinus carpio*. Neuroendocrinology 56:539–549.

Violand BN, Takano M, Curran DF, Bentle LA. (1989). A novel concatenated dimer of recombinant bovine somatotropin. J Protein Chem 8(5):619–628.

Violand BN, Schlittler MR, Toren PC, Siegel NR. (1990). Formation of isoaspartate 99 in bovine and porcine somatotropins. J Protein Chem 9(1):109–117.

Wallis M. (1973). The primary structure of bovine growth hormone. FEBS Lett 35(1):11–14.

Wang Y-CJ, Hanson MA. (1988). Parenteral formulations of proteins and peptides: Stability and stabilizers. J Parenteral Sci Technol 42(suppl):S4–S26.

Wheaton JE, Al-Raheem SN, Godfredson JA, et al. (1988). Use of osmotic pumps for subcutaneous infusion of growth hormone-releasing factors in steers and wethers. J Anim Sci 66:2876–2885.

Williams AH, Staples LD, Thiel WJ, Oppenheim RC, Clarke LJ. (1987). Biocompatible implants. Int Patent Appl PCT/AU87/00139.

Wingfield PT, Graber P, Buell G, Rose K, Simona MG, Burleigh BD. (1987a). Preparation and characterizaton of bovine growth hormones produced in recombinant *Escherichia coli*. Biochem J 243:829–839.

Wingfield PT, Graber P, Rose K, Simona MG, Hughes GJ. (1987b). Chromatofocusing of N-terminally processing forms of proteins. Isolation and characterization of two forms of interleukin-1 beta and of bovine growth hormone. J Chromatogr 387:291–300.

Wolfenstein-Todel C, Santome JA. (1983). Modification of arginines in bovine growth hormone. Int J Peptide Protein Res 22(5):611–616.

Wood DC, Salsgiver WJ, Kasser TR, et al. (1989). Purification and characterization of pituitary bovine somatotropin. J Biol Chem 264(25):14741–14747.

Wray-Cahen D, Ross DA, Bauman DE, Boyd RD. (1991). Metabolic effects of porcine somatotropin: nitrogen and energy balance and characterization of the temporal pattern of blood metabolites and hormones. J Anim Sci 69:1503–1514.

Wyse JW, Takahashi Y, DeLuca PP. (1989). Instability of porcine somatotropin in polyglycolic acid microspheres. Proc Int Symp Controlled Release Bioactive Materials 16:334–335.

Yamasaki N, Kikutani M, Sonenberg M. (1970). Peptides of a biologically active tryptic diges of bovine growth hormone. Biochemistry 9(5):1107–1114.

Young FG. (1947). Experimental stimulation (galactopoiesis) of lactation. Br Med Bull 5(1104):155–160.

Zhao X, McBride BW, Trouten-Radford LM, Golfman L, Burton JJ. (1994). Somatotropin and insuin-like growth factor-I concentrations in plasma and milk after daily or sustained-release exogenous somatotropin administrations. Domest Anim Endocrinol 11(2):209–216.

Zohar Y, Pagelson G, Gothilf Y, et al. (1990). Controlled release of gonadotropin releasing hormones for the manipulation of spawning in farmed fish. Proc Int Symp Controlled Release Bioactive Material 17:51–52.

4

Formulation of Vaccines

RUSSELL BEY
College of Veterinary Medicine, University of Minnesota, St. Paul, Minnesota

RANDY SIMONSON
Bayer Animal Health, Worthington, Minnesota

NATHALIE GARCON
SKB Biologicals, Rixensart, Belgium

I. INTRODUCTION

Frequently, the treatment or prevention of a disease occurs before scientists understand the infectious disease process or how the preventive measure works. Such was the case with the early, mid-18th-century attempts to immunize humans against smallpox. Animals, on the other hand, were often immunized by passing live virus from a diseased animal to an apparently healthy animal, as was done to immunize against rinderpest (1). Pasteur's work on the attenuation of the chicken cholera virus in the latter half of the 1870s and his work with anthrax and rabies virus were the first major advances after Jenner's work with smallpox. Pasteur developed a strategy, in the latter part of the 19th century, to attenuate live virulent organisms that could provide immunity against virulent challenge. Most of the work of these early investigators was done in animals, which set the stage for what ultimately has become routine vaccination of farm livestock (1,2).

In 1872 Louis Pasteur adopted the term "vaccination" in recognition of Edward Jenner's (1749–1823) work with cowpox (1,2). Thus, a vaccine (*L. vaccinus*, relating to cowpox virus), was defined as the live virus inoculated into the skin of a human as prophylaxis against smallpox (1–3). This definition was readily extended to include any substance intended to invoke immunity in an animal, making that animal refractory to infection.

Veterinary vaccines today may include inactivated (nonviable) bacteria, viruses or parasites, attenuated live microorganisms, attenuated live organisms carrying genes of a different agent (subunit antigens of different organisms), or soluble substances produced during bacterial growth. Experimental vaccines include the use of a synthetic analog of the disease-causing agent or DNA from a specific pathogen.

It is apparent from this very brief history that the antigenic components (antigens) of animal vaccines have changed considerably over the last 150 years, as has their manufacture. Similarly, the technology to formulate these vaccines has also advanced. Methods to produce safe, efficacious, consistent, and stable products have been, and are continually, improving. It is the purpose of this chapter to review some of these formulations and hopefully provide a reference point for those seeking information in this area.

II. BACKGROUND

Regulatory agencies throughout the world vary in scope and practice with regard to veterinary biologics (4–6). The veterinary vaccine industry is dynamic. Technologies are evolving that are leading to safer, more effective, and more cost-efficient vaccines. As such, the terms used in the biologics industry is changing or are used differently in various parts of the world. In Europe, a killed bacterial is referred to as a bacterial vaccine (5), while in the United States, a killed bacterial product is referred to as a bacterin (6) and the term "vaccine" is reserved for an immunizing agent that contains live components. In addition to understanding basic definitions and some general regulatory concerns regarding biologicals, it is important to understand some general development strategies to better appreciate how vaccines are formulated.

A. Basic Definitions

1. *Biological Product*

Refers to "all viruses, bacteria, sera, toxins, and analogous products of natural or synthetic origin, live microorganisms, killed microorganisms, and antigen immunizing components of microorganisms intended for use in diagnosis, treatment, or prevention of diseases in animals" (6).

2. *Toxoid*

A sterile, soluble, nontoxic material prepared from growing bacteria, separating the bacterial cells from the culture fluids and rendering the soluble liquid nontoxic. Toxoids may also be prepared by chemical synthesis or using recombinant DNA technologies.

3. Modified Live Vaccine

A vaccine containing a live microorganism that has been rendered avirulent or incapable of causing disease when given to the target or host animal.

4. Vector Vaccine

A live bacterium or virus determined to be nonpathogenic, by genetic manipulation or other means, and has encoded and expressed genes from another microorganism.

5. Killed or Inactivated Vaccine

This refers to a whole viral or whole bacterial or whole parasitic vaccine in which the active component has been rendered nonviable.

6. Subunit Vaccine

A fragment of the microorganism that has been shown to be immunogenic and can be used in place of the whole microorganism.

B. Regulatory Considerations

Regulatory agencies throughout the world vary in scope and practice in regard to governing the production of veterinary biologicals. In the United States, veterinary biologicals are regulated by the United States Department of Agriculture (USDA) as provided by the March 4, 1913, Act of Congress (37 stat. 832-833, 21 U.S.C. 151-158). There are three main branches of the USDA which are directly involved in the licensing of veterinary biologics: licensing, located in Riverdale, Maryland, which is responsible for licensing all new veterinary biologics; Veterinary Biologics Field Office (VFBO, Ames, Iowa), which has responsibility for inspecting manufacturing facilities for compliance and investigating incidence reports involving the use of veterinary biologics; and the National Veterinary Services Laboratory (NVSL, Ames, Iowa), which has responsibility for confirmatory testing, pre- and postlicensing as well as diagnostic services to the veterinary biologics industry (6,7).

In Europe, veterinary biologics were controlled at the country and/or national level. However, in early 1993, the Committee for Veterinary Medicinal Products (CVMP) was formed and given authority over all member states of the European Union (EU), which were required to comply with the new centralized directives (4–8).

In addition to the administrative differences noted above, these agencies may require different product formulations. For example, the USDA has standards which allow certain antibiotics to be used as preservatives (6) while this practice is prohibited by the CVMP (4–8). In the United States multivalent (component) vaccines are common, while in Europe they are rare.

At the present time various organizations, Animal Health Institute (AHI), and Fédération Européenne de la Santé Animale (FEDESA) are working with regulatory agencies throughout the world to help consolidate and unify standardization guidelines for biologics.

C. Basic Biologic Development Strategies

The basic requirements for developing, registering, producing, and selling veterinary vaccines are proof of purity, potency, efficacy, and safety. Purity tests must be conducted to ensure that the final product is free of extraneous microorganisms or material. Potency refers to the relative strength (i.e., degree of immunity induced) of a biological product. Efficacy must be demonstrated in the target species for which protection or treatment with the product is intended, while safety refers to target species and may include other species and/or the environment (6–8).

In addition to the above requirements, animal vaccines need to be cost-effective. This is especially true for food-producing animals where market prices fluctuate and modern record-keeping systems allow the producer (farmer/rancher) to closely monitor production costs. The economics of the market has, in the past, provided little incentive for veterinary vaccine companies to develop veterinary vaccines comparable to the highly purified, precisely defined vaccines that are required for human use. The technology to produce more defined and pure vaccines has been, for the most part, impractical or lacking. Approval for the first USDA-licensed vaccine involving recombinant DNA technology was granted in 1982. This vaccine was prepared by cloning the genes that encode for the *E. coli* adhesive antigens (pili) involved in neonatal enteritis into a laboratory stain of *E. coli* which enhanced production of the cloned proteins—in this case, pili. In 1986, Dougan et al. (9) cloned the gene encoding the production of the K88 pili antigen from *E. coli* into *S. typhimurium*, demonstrating that heterologous antigens can be expressed in a potential vector vaccine. In the future, we can expect to see a newer generation of veterinary vaccines that will have the potential to totally change how vaccines are formulated and administered to animals.

III. VIRUS VACCINES

A. Types

Vaccines that provide protection against diseases caused by viruses are among the most successful immunoprophylactic agents, and their application has resulted in the control of many diseases in animals. There are at present four types of virus vaccines approved for use in immunizing animals. The conventional virus vaccines, modified live vaccine (MLV), and inactivated or killed vaccines make up the majority of these products sold and used worldwide. These products are available for virtually all species of animals. The fourth type of virus vaccine is a recombinant vector vaccine.

The inactivated or killed virus vaccines are used for viruses that are reliably inactivated while preserving the optimal antigen conformational structure and are able to induce adequate protective immunity in the host. Inactivated virus vaccines are subject to failure through either improper inactivation or poor immunogenicity. To circumvent the immunogenicity problems, animals must be immunized more frequently to achieve adequate levels of protection as is the case with inactivated equine influenza vaccines. These vaccines are often administered every 4 to 6 months for several years to competition horses before adequate protective immunity is achieved. In addition, large amounts of material must be generated for production purposes, which may be difficult. These vaccines are predominantly used in breeding or gestating mammals (pigs, horses, and cattle) because of concern for the potential of adverse reactions associated with modified live vaccines. There is some support for this concern as some modified live bovine viral diarrhea (BVD) vaccines were found to be contaminated with a virulent noncytopathogenic strain of BVD that went undetected in vaccines. When administered to susceptible cattle, these contaminant viruses resulted in substantial economic loss to the producer, causing substantial embarrassment to the veterinarian and biologics manufacturer. In addition, some modified live infectious bovine rhinotracheitis (IBR) vaccines were known to cause fetal infections which often resulted in early embryonic death or abortion (10). These vaccines tended to induce development of humoral immunity with little cell-mediated immunity, which was often necessary for adequate protection to occur (11).

The situation in poultry is similar to that of other species (12). The majority of chick and poult vaccines used are modified live, while killed immunogens are used in breeding birds. However, modified live vaccines are also used. A primary concern in vaccinating breeding birds is the effect on egg laying. Thus, most vaccines are given prior to setting the birds.

Modified live virus vaccines are among the most successful immunizing agents developed and available for use in animals. Although their have been some problems, these have occurred in a small number of cases which attracted considerable attention and adverse publicity. However, this has also fostered the development of better and safer attenuated vaccines. These vaccines have the advantage of being able to stimulate long-term immunity (years) not achievable with killed virus vaccines (11). However, this is not the case with all attenuated viral vaccines, as evidenced by the equine herpes virus vaccines which provide short-term protection in immunized animals. Because the virus is replicating within the host, the vaccine is often able to stimulate both cell-mediated and humoral immunity (elevated levels of neutralizing antibody), which results in better protection of the immunized host. This is probably due to attenuated viral epitopes which structurally closely resemble that of the wild-type virulent virus. The primary disadvantage is reversion of the attenuated virus to the virulent state, resulting in disease as mentioned previously or was the case with hepatitis virus-inducing ocular damage in immunized dogs. It is for this reason that killed vaccines are often preferred for use in pregnant animals.

The third type of viral vaccine is one that uses the subunit approach. This approach is dependent on identifying which specific viral proteins (antigens) are required for the induction of a protective immune response. The simplest and most basic form of subunit vaccine is one in which the infectious agent has been solubilized or separated into its component parts. These met with variable results which depended on the agent being used for the vaccine. With the advent of recombinant DNA technologies (cloning and sequencing), it has been possible to determine the precise amino acid residues of the protective protein. This technology also meant that foreign genes could be inserted into an expression vector and into a host cell that acted as "production factories" for the inserted protein encoded by the genes. For a subunit vaccine approach to be practical for routine use, the virus has to be readily propagated in vitro. This has not always been possible, which limited the use of this approach. Some feline leukemia virus (FeLV) vaccines may be considered "subunit." The cultivation method in transformed cells made this technique feasible (13). The use of subunit vaccines is advantageous when the particular virus does not lend itself to safe attenuation.

The fourth type of vaccine available today is a recombinant vector vaccine. The most common viral vector is the vaccinia virus. Several veterinary vaccines (e.g., rabies, hog cholera, bovine leukemia, feline leukemia, avian bronchitis, Newcastle disease) have been investigated using this approach. However, in 1994 Syntrovet was issued a license for a vector vaccine using fowlpox virus which contains the HN gene and the F gene for Newcastle disease virus. In this case the parent virus did not require further genetic manipulation or attenuation prior to use (14). The rationale for using a recombinant live virus vaccine of this type is that it will immunize the host for fowlpox and Newcastle disease simultaneously. Using fowlpox virus as a parent organism does not have to be limited to poultry, as recombinant virus capable of expressing rabies glycoprotein was inoculated into mice, dogs, and cats with the resulting immune response sufficient to protect the immunized host against rabies challenge.

Other potential viral vectors that may soon be available are the adenoviruses or herpesvirus for use as parent organisms. With these vector systems, attenuation of the host virus by selectively deleting identified genes, will be required (14–16). First-generation products using the potential vector minus inserts are currently on the market.

B. Manufacturing Methods

Several methods are used to propagate viruses intended for use as vaccines. Many viruses that were to be used as biologicals started by being cultivated in embryonated eggs (e.g., rabies, infectious bursal disease, Newcastle disease, infectious bronchitis, and influenza). This approach was highly labor-intensive, requiring extensive testing of the eggs to ensure that they were free from extraneous contaminating agents. Vaccines produced in this way are generally quite efficacious, but may provide the basis of adverse reactions due to the presence of egg proteins.

To circumvent this problem and reduce labor costs, tissue cell culture replaced egg cultivation methods. With this procedure, tissue cells, frequently from the kidney (bovine, swine, canine, or chicken), are cultured in vitro. During the first few in vitro passages they are considered primary cell lines. Some veterinary vaccines are still produced using primary cells. (The use of primary cell lines is what contributed to the vaccine problem associated with the BVD vaccines mentioned earlier.) Today, most virus-containing products are produced using established cell lines. These cell lines are extensively tested for adventitious agents (i.e., mycoplasma or viruses), and karyology testing is performed to ensure that cell transformation has not occurred. After the cells have been extensively tested, a "master cell" is constructed by freezing a large amount of the cells for use in virus vaccine cultivation (6,7). This is a better way to ensure purity, safety, and continued efficacy for the vaccine.

Plastic or glass roller bottles are still used for cell culture and virus vaccine cultivation. This is a proven method but is very labor-intensive in that a production lot generally requires 50 or more bottles of vaccine. The potential for introducing a contaminant during the "pooling" of these materials is another disadvantage that can be costly, since the entire lot would be rejected.

Bioreactors are increasingly being used for virus propagation in the veterinary biologics industry. Cells cultivated in the bioreactors are of two types— those requiring attachment to a solid surface, and those that do not. These machines have been demonstrated to be more cost-effective by reducing labor costs, decreasing potential for contamination, and increasing control over the cell growing cycle, thus providing a more consistent, efficacious product. Following propagation by one of the above methods, the virus is further processed depending on the type of vaccine being produced and the nature of the respective virus being cultivated.

A modified live vaccine generally requires that the virus be stabilized (to reduce loss of viability) in some way prior to desiccation (lyophilization). (Stabilizers will be discussed in a later section.) If the intended product is an inactivated vaccine, then an inactivation step, with a chemical agent, is performed on the pooled materials. (Inactivating and preserving agents are discussed later.) At this time the virus is quantified, stabilized, and stored, either frozen or refrigerated, prior to formulating the final product and dispensing it into sterile vials. The length of time the agents are stored is variable.

Either killed or modified live vaccines may require concentration to ensure adequate potency. This is generally done by centrifugation or filtration. Another method that may be used to concentrate viruses is the addition of a metallic salt, which attaches to any proteins present, causing them to settle out with the metallic salt. These metallic salts often act as adjuvants (discussed later in this section) which enhance the immune response to the antigens present. For example, when aluminum hydroxide (AlOH) is added to a solution, proteins, including virus particles, bind noncovalently to the AlOH and upon standing settle out of solution. The clear supernate can then be removed.

C. Formulation Strategies

Formulation of a particular vaccine is directly related to the studies done to determine safety, efficacy, and potency. The amount of vaccine virus needed to immunize an animal is determined by host animal efficacy studies. Generally, killed virus vaccines are more stable than modified live vaccines, and additional virus is not necessary. However, in practice it is necessary to ensure potency, so killed virus vaccines are formulated with more antigen than minimum protective dose studies indicate. Modified live vaccines are formulated by adjusting the virus titer, determined by plaque-forming units (PFU) or tissue culture infectious dose 50% (TCID$_{50}$), to the predetermined, by host efficacy studies, amount. This will vary depending on the virus but will range from 3 $\log_{10}$ to 6 $\log_{10}$ per dose. As a general rule enough virus is added to allow for a potent product through expiration dating and error in the tests being run. The amount of extra virus added ranges from 0.7 $\log_{10}$ to 1.2 $\log_{10}$.

Safety is also a concern in any formulation so all virus vaccines, in their final formulation, are tested at two to 10 times the recommended dose in the most susceptible animal known. If any overt clinical signs are noted or adverse reactions occur the vaccine is considered unsafe and is not offered for sale.

IV. BACTERIAL VACCINES

A. Types

Vaccines for the prevention of disease caused by bacterial infectious agents may take one of five forms: inactivated toxin (toxoid), chemically inactivated whole organisms (bacterin), attenuated live organism (modified live), recombinant produced and/or purified cellular proteins (subunit), or live recombinant organisms. Of these, the toxoid and chemically inactivated whole organisms make up the majority of vaccines licensed for use in animals.

Toxoids are prepared from bacteria-produced toxins, which are inactivated and used to prevent or control diseases caused by these toxins. Toxoids are used to control infections caused by members of the following genera; *Pasteurella* (pneumonia in cattle, pigs, and sheep); *Staphylococcus* (mastitis); *Bordetella bronchiseptica* (atrophic rhinitis in pigs); and *Clostridia* sp., which are responsible for causing such diseases as tetanus in horses; enterotoxemia of lambs; malignant edema of horses, cattle, sheep, and swine; black disease in sheep and sudden death in cattle and pigs. Tenanus toxoid with adjuvant is a highly potent antigen capable of inducing immunity against the tetanospasm. However, at least two doses, 4 to 6 weeks apart, are required to induce solid immunity in the horse. This immunity must be reinforced at yearly intervals to ensure that protective immunity is maintained. The use of toxoids to control disease in companion animals (dogs and cats) is relatively rare.

Inactivated whole bacterial cell suspensions are probably the most common type of vaccine used in domestic animals. They are generally administered once or twice annually. These vaccines have historically been extremely effective and safe to use in young or pregnant animals. Vaccines prepared in this manner proved highly effective in protecting against respective diseases. Such was the case with early *Leptospira* vaccines that were prepared from one serovar and provided protection against only that offending serovar. These vaccines provide only short-term immunity with highly specific efficacy. However, whole-cell vaccines prepared from gram-negative organisms (i.e., *E. coli* and *Salmonella*) occasionally caused abortion, disseminated intravascular coagulation, or fever in some animals due to the presence of endotoxin. Using whole-cell inactivated vaccines tended to induce primarily circulating antibody, failing to provide immunity against facultative intracellular or obligate intracellular pathogens, which required the induction of cell-mediated immunity (17). In addition, these vaccines often failed, particularly with the gram-negative enteric pathogens— e.g., *E. coli*—and those bacteria responsible for causing pinkeye in cattle (*Moraxella bovis*). These failed because not enough (quality and quantity) of the appropriate antigen could be administered to induce adequate protective immunity.

Modified live bacterial vaccines (e.g., *Brucella abortus*, *Erysipelothrix rhusiopathiae*, and *Pasteurella multocida*) have been developed in an attempt to alleviate some of the short-term protection problems encountered with killed whole-cell vaccines. These vaccines were able to induce cellular immunity or secretory immunity in the host species, which was necessary to prevent infection (18). However, they have not gained in popularity (except *Brucella*) due to the fear that they may induce disease in some animals which may have been immunocompromised at the time of immunization. Choosing the proper delivery system and maintaining adequate viable cell numbers required for the induction of protection or other difficulties must be overcome. In addition, these vaccines often induced a transient illness in the immunized host, which was very undesirable.

In an effort to circumvent the problems associated with whole-cell vaccines (failure to protect the immunized animal or adverse reactions), research efforts were launched at determining exactly which antigens were directly associated with the induction of adequate protective immunity. One of the first products that directly benefited from the identification of the antigens responsible for adhesion of the bacterium to the intestinal villi was the *E. coli* vaccines. If adhesion could be prevented or blocked by antibodies, then, in theory, the animal would be refractory to infection. Early work demonstrated that animals immunized with experimental vaccines prepared from purified *E. coli* K88 antigens were protected from experimental challenge (18). Since this first commercial success, several other, similar vaccines have been produced and gained wide acceptance among veterinary practitioners—e.g., *Morxella bovis* (pinkeye); *E. coli* K99, 987P, F41 (neonatal scours); and *Bacteroides nodosus* (foot rot in sheep). To facilitate less laborious means of antigen purification and decrease production costs, veterinary biologics manufacturers turned to recombinant DNA technology to help solve the problems. Specific gene sequences encoding the production of specifically desired immunogenic proteins were cloned into bacteria or yeast, and these surrogate organisms produced large quantities of the desired antigen. This approach has gained wide acceptance and has proved a valuable means of producing specific antigens required for adequate immunization of the host species.

More recently this approach has been coupled with the use of live avirulent bacteria which propagate in the host while simultaneously expressing the cloned antigen in a noninfectious manner (17). This allows continuous exposure of the host's immune system to the specific antigen. These vaccine vectors, which are attenuated pathogens expressing foreign antigens, when injected into an animal, elicit a protective immune response. Vectors are generally derived from either pathogenic bacteria or viruses, and the protective antigens are cloned from the pathogen.

Gene-deleted avirulent vaccines (*Salmonella*) are being developed for use in domestic animals. In these organisms, specific genes for virulence (i.e., toxic gene) are deleted or a specific metabolic pathway is altered by *aro* A mutation so the organism requires some precursor compound for growth (20). This compound is not present in mammalian tissues which restrict bacterial replication in vivo. This approach has proven to be a highly safe and effective way of developing a modified live avirulent vaccine. Another vaccine that uses viable organisms in which virulence genes have been deleted is the Sterne vaccine against anthrax. This vaccine consists of anthrax spores which lack genes necessary for capsular material which is required for virulence. The inoculated spores germinate in the inoculated animals living long enough to produce the toxin which stimulates antibody production against factors II and III. Antibody against factor II prevents factor III from binding to the target cell.

B. Manufacturing Methods

Prior to the production of any vaccine, a "master seed" is prepared (6,7). In this a quantity (5 to 10 ml) is aliquoted into ampules, and a cryoprotective agent is added and frozen. This master seed is extensively tested for adventitious agents. If any are found, the master seed is discarded and another is prepared. It is this master seed that provides the basis for all production lots of vaccine.

The most common method of vaccine production is the cultivation of large quantities of the desired organism in a bioreactor (fermentor). This can be accomplished in several ways depending on the organism and culture conditions (pH, aeration, agitation, etc.) necessary for the organism to express the appropriate protective antigens. This can range from growth in 50-L carboys to several hundred-liter bioreactors where all culture conditions can be continuously monitored. The latter has replaced the carboy as several hundred thousand doses or more of vaccine can be generated in one run of the bioreactor. However, the carboy method is still used for the production of small batches of antigen (vaccine). With the bioreactors, continuous cultivation processes are feasible for some bacteria, (*Pasteurella*, *Salmonella*, and *E. coli*). Oftentimes, if culture conditions are maintained optimally, adequate antigen concentration with adequate vaccine potency can be achieved without concentration. After the appropriate concentration of cells per milliliter is attained, a chemical inactivating agent such as formaldehyde (0.2% to 0.5% residual), merthiolate (thimerosal (1:10,000 to 1:15,000), β-propriolactone, or binary ethylenimine (BEI) is added. Should concentration be necessary, it can be accomplished by centrifugation, filtration, or precipitation after inactivation and a portion of the supernate aseptically removed. Another method used to concentrate bacteria is the addition of AlOH. When added to a bacterial culture, proteins including bacteria bind noncovalently to the AlOH and, upon standing (generally at 4°C), settle out of solution. The clear supernate can then be removed.

Bacterial concentration can be measured by several means. The method of measuring is often dependent on the bacterium being cultivated. However, the most common means is turbidimetric analysis. Using this technique a relationship can be established between cell number (which is determined by standard plate count or Petroff-Hausser counting chamber values) and optical density value determined by a spectrophotometer or colorimeter. If absorbance readings are matched with plate counts or counted cell number of the same culture, this correlation can be used in future estimations of bacterial numbers by measuring liquid culture turbidity.

C. Formulation Strategies

Bacterial vaccine formulation is directly related to efficacy, potency, and safety studies. The quantity of antigen needed to induce an adequate (protective) response is determined by host animal efficacy studies. Vaccine doses can range from 10^5 to 10^{10} cells per dose. This depends on the immunogenic agent. More often than not, with killed bacterial vaccines a large quantity is needed to provide adequate concentrations of protective antigen to achieve satisfactory immunity. Generally, excess antigen is added to the minimal formulation (that concentration of antigen that is able to confer immunity) to ensure adequate potency past product expiration dating and to provide a "safety cushion" to compensate for improper dose administration. Should a bacterium be a poor antigen or possess low amounts of the required antigen, an adjuvant may be necessary to ensure adequate potency of a particular product (below).

With some subcellular vaccines, oftentimes one or more antigenic components are combined, as was done with the *E. coli* vaccines. In this situation, as more became known about the *E. coli* agent, other pili types were added to the original bacterin formulation, making the vaccine more efficacious against a wide range of *E. coli* pathogens. These types of vaccines are still very desirable as the host is immunized with only the essential protective components of the bacterium, thus reducing the possibility of adverse reactions caused by other bacterial cell components.

A major concern when formulating any vaccine is safety. To ensure that the product is safe, samples from each vaccine candidate (serial) of the formulation being considered for sale are administered at two to 10 times the recommended dose in the most susceptible animal species known and the host animal. If any overt clinical signs, such as vasculitis syndromes associated with immune complex disease (especially in those animals with high levels of antibody), edema, pyrexia, stiffness, lethargy, or restlessness, are observed the vaccine serial is considered to be unsafe. If reactions occur with a vaccine that was found to be safe, it is often due to improper administration (to frequent administration or administration to subclinically ill animals).

In general, the vaccine formulation ultimately decided upon must be able to induce an adequate host response, be it cellular, secretory, or humoral immunity, in a safe and efficacious manner. It must not only be effective in the controlled clinical trial but be effective in field efficacy studies.

The future for vaccine improvement is unlimited. Existing vaccines and new vaccination and vaccine technology should quantitatively and qualitatively improve the immune response. Increases in duration of immunity and lower costs with more effective and economical delivery systems will also be attained. New vaccines will parallel new control technologies and strategies striving toward disease eradication where possible. Many of the new developments will be achieved through the incorporation of new adjuvants, recombinant DNA technologies, and new vaccine delivery systems.

V. PARASITIC VACCINES

A. Types

In the area of parasitology, definition of the appropriate antigens is extremely difficult, in part by the complexity and variability of the organisms and the multistage development process. Currently, there are a limited number of parasitic vaccines available for use in domestic animals (i.e., *Eimeria* spp., *Toxoplasma* spp., *Babesia* spp., *Theileria* spp.).

B. Manufacturing Methods

It is well known that animals that have been infected with many types of parasites develop some immunity. The degree of immunity developed seems to vary with the infecting parasite. Initial attempts to immunize animals were made with attenuated or irradiated parasite organisms. Although this has been successful in experimental situations, it has rarely led to the production of a commercial vaccine. The exceptions include cattle babesiosis, chicken coccidiosis, and *Babesia canis* in dogs. These vaccines were at least effective in stimulating immunity of some type in the host. However, they had problems with liability, short shelf-life, batch inconsistency, the induction of limited immunity, and high incidence of undesirable side effects, all of which led to limited production and use.

The trend in vaccine manufacture is toward subunit vaccines where possible. The quality and quantity of antigen administered can be carefully controlled under these conditions. Molecular biology has made the possibility of vaccine production in the very near future more feasible once the protective antigens are identified. Whether these vaccines will be as effective as those for bacteria and viruses remains to be determined.

VI. STABILIZERS

A. Purpose

The primary purpose of stabilizers in vaccines is to aid in maintaining proper immunogenic quality by preserving the natural configuration of the charge and conformational structure of the antigen. They are also used to prevent the loss of viability or titer of live vaccines which assures the maintenance of adequate potency during shipping and storage.

B. Types and Formulation Strategies

Several types of stabilizers are available for use in preserving vaccine stability under a variety of conditions. While the type and quantity of stabilizer used in a vaccine is often proprietary, agents such as lactose or other saccharides, skim milk, or serum may be added to liquid vaccines or, prior to lyophilization, to improve stability during storage. Other measures such as maintenance of a cold chain is sufficient to prevent product degradation. Precautions must be taken to ensure that any added chemical is compatible with all other components of the vaccine formulation and does not elicit adverse reactions in the host. Vaccine reactivity is an important consideration for companion animals and food-producing animals, where carcass damage reduces meat value.

VII. PRESERVATIVES

A. Purpose

When biologicals are manufactured, a series of hurdles are placed in the pathway of a contaminant microorganism. It must overcome these hurdles to propagate and render a product unusable. Biologic preparations are manufactured in a clean environment, the type of which may vary by country but which generally follows the good manufacturing practice (GMP) recommendations. To enable multidose biologic preparations to cope adequately with contaminants that have gained access during repeated withdrawal of doses from the container, the presence of a potent antimicrobial preservative is essential. The first use of preservatives in vaccines occurred approximately 150 years ago when glycerin was used as an inactivating agent and to prevent bacterial growth in vaccinia virus (2). The development of this agent made available a ready supply of stable vaccine that had reasonably consistent potency (2). The preservative selected must remain largely undissociated at the pH of the formulation and from other agents present. Consequently, the preservation of biologics involves the formulation, GMP, and maintenance of appropriate conditions.

Plastic containers, rubber or plastic caps, or liners may absorb the preservative, thereby decreasing the quantity available for antimicrobial activity. In almost every biological product there are factors for inhibiting microbial growth (pH, osmotic effects, or the presence of toxic molecules). How these vaccines are formulated will determine the microbial growth or killing rate.

Preservatives are not allowed, under GMP specifications, for use in modified live vaccines. Vaccines not produced under GMP specifications may contain an antifungal or antimicrobial agent, but those agents cannot adversely affect the viability of the vaccine agent. However, cryoprotective agents, used to prevent or reduce the loss of viability on storage or lyophilization, may have some preservative capability for a limited time. But as such, little or no preservatives are used in modified live bacterial vaccines.

B. Types

There are a variety of chemical molecules exhibiting antimicrobial properties that are incorporated in vaccines to be used parenterally. Among the most common are thimerosal, benzyl alcohol, phenol, phonoxyethanol, phenyl mercuric acetate, organic acids, EDTA, and formaldehyde. Antimicrobial agents such as fungistatic agents (amphotericin B or nystatin) and antibiotics (tetracycline, penicillin, streptomycin, polymyxin B, neomycin, or gentamicin) may also be used as preservatives in some biologicals with the caveat that only one antifungal agent and one antibiotic shall be used in a biological product (6).

C. Actions

The preservative selected must be able to prevent the growth of even the hardiest of organisms such as *Pseudomonas* spp., *Proteus* spp., *Aspergillus niger*, or *Candida albicans*. In addition, they should possess good antimicrobial activity against a variety of other organisms. Generally these agents adversely affect some part of the microbial cell, such as the cell wall, membrane, enzymes, or nucleic acid.

VIII. INACTIVATING AGENTS

A. Purpose

Live vaccines have several advantages over killed vaccines. However, killed vaccines do have certain advantages that contrast with the shortcomings of live vaccines. By definition, killed vaccines means the bacteria present cannot multiply or disseminate to cause disease they are intended to prevent. The purpose of inactivating agents is to kill, as gently as possible, the virus or bacteria while preserving critical integrity of the antigens necessary for the induction of a protective immune response.

B. Types

There are a number of inactivating agents, some of which are used as preservatives—phenol, thimerosal, and formaldehyde. Some newer agents, such as binary ethylenimine (BEI), have gained in popularity as biological manufacturers have tried to reduce the degree and incidence of adverse reactions at the injection sites.

C. Actions

Most inactivating agents act relatively quickly and efficiently. They alter or inactivate essential enzyme systems through protein inactivation, or alter nucleic acids in an irreversible manner. Whatever the agent, the critical antigen on the cell must not be substantially altered such that it becomes significantly different from the native configuration. It is these antigens that are responsible for stimulating protective immunity. For example, formaldehyde, which acts on amino and amide groups in proteins and on non-hydrogen-bonded amino groups in purine and pyrimidine bases of nucleic acids, forms crosslinks and confers structural rigidity. Other agents that crosslink nucleic acid chains are also suitable for killing organisms because they leave the surface protein antigens unchanged.

IX. ADJUVANTS

A. Purpose

As vaccines become more refined, their natural adjuvanticiy to enhance the immune response is lost. As a result, external adjuvant agents must be added to ensure an adequate and appropriate immune response. The function of these external adjuvants it to augment the immune response to a particular antigen or group of antigens. However, selection of the immunogen is always of paramount importance.

B. Types

There are a wide variety of adjuvants available. Some of the most commonly used are the metallic salt adjuvants such as aluminum hydroxide and alum, which have been used for more than 50 years. Oil adjuvants, such as those containing mineral oil and, more recently, poly(oxyethylene) sorbitan mono-oleate and trio-oleate (Tween 80 and 85) or sorbitan monolaurate (Tween 20), have also been used to increase the immune response to antigens that would otherwise be relatively weak immunogens. Surface-active agents such as saponin have also been used. Isolated from bacteria, muramyl dipeptide (N-acetymuramyl-L-alanyl-D-isoglutamine) and endotoxin (lipid A) have demonstrated the ability to stimulate humoral antibody response to various antigens in vaccines (21). Levamisole and isoprinosine have been shown to stimulate T-helper cell and delayed-type hypersensitivity reactions (22). Carriers (immunogenic substances that, when coupled to a hapten) render that molecule more immunogenic have been used—i.e., cholera toxin B. The ISCOM (immunostimulating complex) is an adjuvant in which Quil A and cell proteins are formed into micelles. All these agents have been shown to markedly enhance the immune response to an antigen. However, which one to use is solely dependent on the type of immune response desired, either T-cell or B-cell, and the type of antigen. A relatively new group of adjuvants is the cytokines, which are protein factors produced by cells that are able to produce multiple effects on many immune cells (i.e., upregulation of T-helper, cytotoxic, or NK cells) or may be able to target specific immune cells and produce the desired and enhanced immune response (23).

C. Actions

The exact mechanism of action of current adjuvants is largely unknown. It is thought that the antigen with adjuvant is deposited at the injection site for a prolonged period of time. The resulting granuloma formation may contain immune cells that are able to produce antibody (24). In addition, this depot allows for slow or prolonged release of antigen (25). For an effective immune response to occur, host's immune cells must have been challenged and the antigen must have been processed and presented in the proper manner by the macrophage to the appropriate B and T lymphocytes. However, some vaccines—e.g., killed strangles vaccine for use in horses—contain antigen that is capable of inducing local reactions, which may be undesirable and must be monitored closely. The use of an adjuvant with such a vaccine may be undesirable, as the adjuvant may enhance the local reaction.

For subunit or specific epitope vaccines, there is a requirement for a more target-specific approach to generation of the desired specific immune response. Thus, the use of adjuvants by the biological manufacturers will increase dramatically in the future development of vaccines.

It has been demonstrated in mice that the size of the particle used is critical for its penetration within the mucosal immunocompetent system. Microspheres < 5 μm in diameter are not retained in the Peyer's patches but are found in the spleen and the lymph nodes, where they induce systemic immunity, while those with the size range 5 to 10 μm remain in the Peyer's patches and induce mucosal immunity. It may then be possible to exploit different trafficking of mucosally delivered microspheres to induce systemic and mucosal immune responses together or separately.

Various polymers have been developed in an attempt to allow for both resistance to the gastric low pH and appropriate delivery to the Peyer's patches. They are based on condensed polyaminoacids that form microspheres resistant at a low pH, which then degrade at neutral pH. polylactic coglycolide polymer microspheres have also been tested by direct intraduodenal administration with some success. In those cases however, the stomach and its aggressive environment were bypassed and the amount of antigen required to induce an immune response was several logs higher than what was necessary through a parenteral injection.

Storage of those systems can lead to loss of immonogenicity. New methods are being explored such as deuterium or addition of sugars as cryoprotectors. Achieving sterility of the product may also provide difficult, conventional methods being deleterious to the antigen as well as the carrier material. Using aseptic process is a possible alternative but expensive.

Those systems may nonetheless, have several advantages such as some protection against proteolysis as well as copresentation of the antigen and an immunostimulant capable of enhancing the immune response. Several drawbacks, however, will have to be overcome before those systems will be widely used in vaccines. Indeed, their preparation requires large amounts of antigen due to the low encapsulation yield and the losses that occur in the digestive tract as well as the low targeting of the Peyer's patches. It is also known that production of such systems on a large scale is difficult as far as uniformity and residual solvent are concerned.

Finally, It is very important to make sure that those products are not overengineered; indeed, one must thus keep in mind that the cost of those systems may at this time outweigh their potential benefit and make them still expensive delivery systems for cheap vaccines.

X. CONTROLLED-RELEASE TECHNOLOGIES

A new type of vaccine delivery system has emerged in the past years that would allow for slow release of the vaccine if given parenterally or for oral delivery of vaccines if given orally. Indeed, it is now well recognized that delivery systems other than live vectors will play an important role in the development of new strategies for improved vaccines. Among those strategies, microspheres have been developed to allow for both slow release of the antigen after parenteral administration and mucosal delivery. In addition, due to their lyophilized storage condition, they could avoid the necessity of a cold chain.

A. Parenteral Delivery Systems

A vaccine that would allow in one single injection, the administration of the multiple injection, necessary to the full course of immunization will be the ideal situation. This could theoretically be realized by using microspheres capable of releasing the antigen at various time points, representing the various times of immunization. Those microspheres could be of various sizes or biodegradation kinetics, thus allowing for different degradation and antigen liberation times.

Different polymers have been tested in animal models based on their low or zero toxicity, biodegradability, and ease of preparation of the microspheres with no deleterious effect on the antigen. The most commonly used is probably the polylactic coglycolide polymer (PLPG), but others are being investigated such as polyphosphazine and polyanhydrides. It is possible (in theory at least) to engineer release rates, loading, surface, and bulk distribution. By varying the ratio of each compound of PLPG, it is possible to modify the time at which the active principle will be released.

Some interesting data have been obtained by one group in which a systemic response after one shot of two different microsphere populations gave results comparable to two separate shots of the antigen.

B. Mucosal Delivery Systems

Another attractive area of improvement for vaccines is the ability to deliver mucosally the various vaccines intended primarily for pathogens for which the portal of entry is at the mucosal level (respiratory tract or oral tract mainly).

It has been demonstrated in mice that the size of the particle used is critical for its penetration within the mucosal immune competent system. Microspheres < 5µm in diameter are not retained in the Peyer's patches but found in the spleen and the lymph nodes where they induce systemic immunity, while those with the size range 5-10µm remain in the Peyer's patches and induce mucosal immunity. It may then be possible to exploit different trafficking of mucosally-delivered microspheres to induce systemic and mucosal immune response together or separately.

Various polymers have been developed in an attempt to both allow for resistance to the gastric low pH, and appropriate delivery to the Peyer's patches. They are based on condensed polyaminoacids that form microspheres resistant at a low pH, which then degrade at neutral pH. PLPG microspheres have also been tested by direct intra duodenal administration with some success. In those cases, however, the stomach and its aggressive environment was bypassed and the amount of antigen required to induce an immune response was several logs higher than what was necessary through a parenteral injection.

C. Stability

Storage of those systems can lead to loss of immunogenicity. New methods are being explored, such as deuterium or addition of sugars as cyroprotectors.

Achieving sterility of the product may also prove difficult, conventional methods being deleterious to the antigen as well as the carrier material. Using aseptic process is a possible alternative but expensive.

Those systems, nonetheless, may have several advantages, such as some protection against proteolysis, as well as copresentation of the antigen and an immunostimulant capable of enhancing the immune response. Several drawback however will have to be overcome, before those systems will be widely used in vaccines. Indeed, their preparation requires large amounts of antigen due to, first, the low encapsulation yield, and second, the losses that occur in the digestive tract as well as the low targeting of the Peyer's patches. It is also know that production of such systems on a large scale is difficult as far as uniformity and residual solvent.

Finally, it is very important to make sure that those products are not overly engineered. Indeed, one must thus keep in mind that the cost of those systems may at this time outweigh their potential benefit, while remaining expensive for delivery systems of cheap vaccines.

REFERENCES

1. Dunlop RH, Williams DJ, eds. Veterinary Medicine: An Illustrated History. St. Louis; Mosby-Year Book, 1995.
2. Plotkin SA, Mortimer EA, eds. Vaccines. 2nd ed. Philadelphia: W.B. Saunders, 1994.
3. Stedman's Medical Dictionary, 26th ed. Baltimore: Williams and Wilkins, 1995.
4. Sauer F. Regulatory framework of immunologicals in the European Union. Biologicals 1994; 22:307–312.
5. Vaccina Ab USUM Veterinarium. E.U. Pharmacopoeia, 1995.
6. Code of Federal Regulations: Animals and Animal Regulations 9, Parts 1–199, 1995.
7. Espeseth DA, Greenberg JB. Licensing and regulation in the USA. In: Peters AR, ed. Vaccine for Veterinary Applications. Oxford: Butterworth-Heinmann, 1993.
8. Lee A. Registration and regulation in the E.C. In: Peters AR, ed. Vaccines for Veterinary Applications. Oxford: Butterworth-Heinmann, 1993.
9. Dougan G, Selwood R, Maskell D, et al. In vivo properties of a cloned K88 adherence antigen determinant. Infect Immun 1986; 52:344–347.
10. Merck Veterinary Manual. 7th ed. Rahway, NJ: Merck & Co., 1991.
11. Powell F, N. New York: Plenum Press, 1995.
12. Glisson JR, Kleven, SH. Poultry vaccines. In: Peters AR, ed. Vaccines for Veterinary Applications. Oxford: Butterworth-Heinmann, 1993.
13. Olsen RG, Lewis MG, In: Olsen RG, ed. Feline Leukemia, R.G. Boca Raton: CRC Press, 1981.
14. McMIllen JK, Cochran MD, Junker DE, et al. In: Brown F, ed. Recombinant Vectors in Vaccine Development, New York, SkargerAG, 1994.
15. Lubeck M, Bhalat M, Davis A, et al. Novel delivery systems for oral vaccines. In: O'Hagen D, ed. Boca Raton; CRC Press, 1994.
16. Francis M. Subunit vaccines and vectors. In: Peters AR, ed. Vaccines for Veterinary Applications. Oxford: Butterworth-Heinmann, 1993.
17. Kurstak E, ed. Modern Vaccinology. New York: Plenum Publishing Corp. 1994.
18. Walker PD. Bacterial vaccines: old and new. Vet medicinal. Vaccine 1992; 10:977–990.
20. Hodgson ALM. Bacterial vaccine vectors. In: Wood PR, Willadesen P, Vercoe JE, eds. Vaccines in Agriculture: Immunological Applications to Animal Health. Melbourne, Australia; CSIRO, 1994.

22. Hadden JW. T-cell adjuvants. Int J Immunopharm 1994; 16:703–710.
23. Taylor CE. Cytokines as adjuvants for vaccines: antigen-specific responses differ from polyclonal responses. Infect Immun 1995; 63:3241-3244.
24. Alexander J, Brewer JM. Adjuvants and their modes of action. Livestock Prod 1995; 42:153–162.
25. East IJ, Kerlin RL, Altman K, Waltson DL. Adjuvants for new veterinary vaccines. In: Pandy R, Hoglund S, Prasad G, eds. Veterinary Vaccines. New York: Springer-Verlag, 1993.

5

Administration Devices and Techniques

DAVID W. COOK
Syrvet Inc., Waukee, Iowa

I. INTRODUCTION

The fact that animals do not knowingly dose themselves, coupled with the diversity of domesticated species, creates a variety of widely different methods and devices for the efficient administration of health and productivity preparations. Husbandry methods, and the financial or emotional profit reasons for dosing, also influence the design of medication delivery systems. These factors constantly challenge scientists and the marketers of animal health preparations to create formulations that are acceptable to animals and easy for their attendants to administer. Many dosage forms require a specific device for their efficient administration. This chapter offers guidance to those involved in the development of specific medication delivery devices and reviews systems currently in use.

II. GENERAL DESIGN INFLUENCES

It is helpful, when reviewing the animal care products industry, to separate it into companion and food animal sectors because their general characteristics affect the design of medication systems. Companion animals are treated individually; food animals are frequently dosed as entire production units. Individual animals can be affordably treated using disposable equipment. The food animal production industry demands lower cost per animal treatment; reusable equipment, which may be more expensive, is usually preferred, particularly if it is supplied with the medication. Companion animals have a high emotional value; food animals are a commercial enterprise.

These factors have concentrated design efforts on the food animal sector, which demands reusable, high-speed, multiple-dose, cost-efficient delivery systems. The emphasis throughout this chapter will therefore be on food animal medication delivery systems.

During the last few decades veterinarian remedies have largely been replaced by packaged formulations. Continued intensification of livestock production has also emphasized the need to contain labor costs. These shifts have brought about a general reduction in dose volumes and demand for greater speed and accuracy of dosing. The maxim "form follows function" is most clearly demonstrated by the design and appearance of nearly all agricultural machinery; animal dosing equipment is no exception.

The popular devices in use today are the result of countless user experiences and subsequent manufacturer modifications. Designing cost-efficient and practical delivery devices for novel or new animal preparations is probably best approached by considering traditional and trusted designs coupled to a careful examination of the intended function. The elements contributing to the definition of a delivery systems function are described in the next section.

III. DEVELOPMENT OF DELIVERY SYSTEMS

Because the effectiveness of a medication or drug preparation is affected by the efficiency of its administration, developers should consider the following functional influences.

A. Passive and Active Systems

An animal medication delivery system could be considered to be in one of two categories:

1. Passive Systems

Medication is received as a consequence of the animal's usual, healthy, and voluntary behavior. Examples include incorporation into feed or drinking water, and application of a topical preparation by animal actions or movements. Passive systems are generally low on labor cost, but less accurate, in terms of dosage, because animal behavior is not always consistent or predictable.

2. Active Systems

Direct application of medication to individual animals. These are more labor-intensive than a passive system, but more accurate dosage is likely to be achieved.

Passive delivery of a product does not generally require the use of custom-designed dispensing equipment since simplicity and low cost of administration are its main benefits. With the exception of the metered incorporation of a therapeutic agent in livestock drinking water, a simple measuring device is often all that is needed.

For many active dosing situations widely available standard equipment, such as one of the many variants of the hand-operated piston and cylinder syringe system, is adequate. Under some active dosing circumstances power is applied to the system to reduce operator fatigue (Sec. IX, Powered Systems).

B. Dosing Factors

The choice of medication formulation and its delivery system is influenced by the following:

Purpose: prophylaxis, therapy, or performance enhancement
Dose volume: permitted by regulation; commonly accepted practice; user expectation; animal tolerance
Route and body site of administration
Frequency of dosing
Number of animals likely to be dosed per session
Accessibility of the animal(s) to be dosed
Availability of helpful or necessary animal restraints
Age and/or size of the animals to be dosed
Skill and cost of personnel likely to be dosing the animals

It should be noted that most countries use the metric system of measurement. In the U.S., metric, avoirdupois weights (ounces and pounds) and apothecaries' fluid measures (fluid ounces and pints) are all used. Dosage recommendations may be given in any of the three measurement systems; the delivery system equipment may be similarly calibrated. The U.S. trend seems to be toward the metric system for fluid measurements.

C. Cost of Dosing

The cost of a delivery system influences its design. If a unique or specific system is essential for efficient administration, the medication manufacturer should consider the preferences, and the value to the user, of the following:

Reusable vs. disposable equipment
Power assistance
Automation
Comparative novelty
User perceptions

If the medication can be administered using standard equipment not supplied by the medication manufacturer or distributor, then the purchase of such equipment is a direct and added cost to the user.

D. Essential Collaboration

In the case of new product development, early collaboration between delivery system engineers, likely users, and medication marketers is essential to avoid costly reformulation or a less than completely practical delivery system. The medication formulation and the design of a delivery device should be coordinated from an early stage.

As the work of bringing new compounds from research through development to market progresses, major characteristics of the compound and its likely formulation emerge. It is possible at such times to make decisions about the various physical forms the product will take, and to predict dosage and administration choices.

In cases where administration equipment is essential for use of the product it is also necessary to select appropriate containers and delivery systems. Failure to do so will be costly to the project; also less than ideal later compromises could curtail sales potential.

E. Regulatory Affairs

By the time a submission of medication data to regulatory authorities is made, all the major features of a product will have been determined. It is crucial therefore to know, before registration submission, and certainly before marketing, that suitable administration equipment exists or can be devised. This is necessary even if the equipment is not part of the submission.

When it is anticipated that a medication will mostly be used for the individual (active) dosing of many animals in sequence, the medication container becomes a key consideration to be included in the medication registration submission. At the time of submission the container should, ideally, be compatible with the delivery system. If it is not, compromises in its design might have to be made; this may result in a less than satisfactory system.

The combination of delivery device and container is the most tangible connection between the medication and its user. If the entire system does not perform satisfactorily, the medication will be administered either improperly or perhaps not at all.

F. User Expectations

The development of formulations, dosages, and the delivery system is, ideally, a collaborative effort led by the medication manufacturer. To meet as many of the user's administration expectations as possible, the limitations imposed by the laws of physics, chemistry, and economics on engineering a system must be considered. These affect hand-used and powered system design in the following ways:

1. Comfortable

The maximum volume a human hand can repeatedly power through a piston and cylinder syringe design is between 40 and 75 ml. The smaller the volume the better. The exact amount will be determined by the potency of the medication and:

> The strength of the hand. Design should be influenced, for example, by whether mostly male or female operators will be using the equipment.
>
> The viscosity of the medication. Aqueous solutions require less effort to dispense than thick suspensions or oily emulsions.
>
> The diameter of the dispensing orifice. The smaller the bore, the greater the pressure needed to dispense the medication.
>
> The temperature of the medication at dispensing time. As temperatures drop, medication viscosity often increases; this raises the hand power needed to dispense the medication.

It should be noted that in some situations, such as when dosing cattle with a topical pour-on, it may be more convenient and less tiring to dispense, for example, a 60-ml dose as 2 × 30-ml doses.

2. Manageable

The main influences are:

> The total weight of the filled instrument. A 100-ml glass bottle of medication, directly attached to an automatic refill syringe, is heavier than when used with a plastic or 50-ml bottle. A dial-a-dose semiautomatic instrument would also be lighter.
>
> The spread movement of the handles and their actuation. For example:
> - For injecting a "pistol"-style syringe (Fig. 1) is most often preferred because the fixed rear handle provides for firm, easily directed needle insertion; the trigger action of the fingers can then more easily control the speed of the injection.
> - For oral dosing, or topical applications, where speed may be more important than directional accuracy, a fixed front handle against which the palm pushes a movable rear handle is more suitable (Fig. 2).

3. Fast

Large-scale livestock processing is labor-intensive. Producers therefore prefer medications and delivery systems that fill, dispense, and refill quickly. Springless hydraulic valves in the system improve speed. The other factors influencing speed are formulation viscosity, dose volume, and delivery orifice diameter.

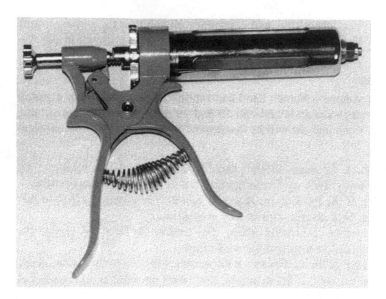

Figure 1 A "pistol"-style syringe, most often preferred for injecting because the fixed rear handle provides for firm, easily directed needle insertion; the trigger action of the fingers can then more easily control the speed of the injection. (Courtesy Syrvet Inc., Waukee, Iowa.)

4. Convenient The type of animals to be dosed and their location will influence the design of the medication and delivery system. Consider the three main dosing situations:

- Fixed dosing stations and restrained animals. Examples—breeding poultry; cattle feedlots. Large containers of medication connected by flexible tube to automatic refill dispensers are feasible, fast, and popular.

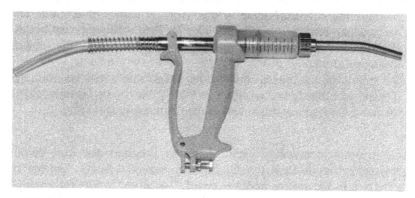

Figure 2 A fixed front handle against which the palm pushes a movable rear handle. (Courtesy Syrvet Inc., Waukee, Iowa.)

- Mobile dosing stations and restrained animals. Example: Animals such as dairy cows at milking can most conveniently be injected, for example, by a self-contained mobile system such as with a prefilled repeater syringe or the operator carrying the drug reservoir connected by tube to an automatic refill syringe.
- Mobile dosing stations and animals. Example: each member of an unweaned litter of pigs must be captured, restrained, and individually treated.

5. *Simple*

Delivering an accurate dose is a high priority for compliance with product registration and, increasingly, producer group quality standards. Medications should be formulated for simple dose calculation to reduce misunderstandings and mistakes.

Helpful concepts are:

Keep ratios simple: example—1.0 fl oz/100 lb bw.

Avoid measurement mixes: example—0.7 fl oz/1.0 kg bw.

Strive for consistent dose increments particularly under circumstances where dosing by animal weight covers a wide spectrum of animal sizes: Example—5 10 15 20 25 ml.

System engineers can build dose adjustment mechanisms of the infinitely variable or fixed increment type. The minimum and maximum dose volumes for each species to be treated must first, however, be determined.

6. *Trouble Free*

Users expect reusable administration equipment to faultlessly delivery many thousands of doses without interruption and with a minimum of maintenance. Formulation development should include steps that check its compatibility with components of the equipment. Engineers can test and make adjustments for effects such as:

Slow delivery caused by the reaction of the compound or carriers, emulsifiers and penetrating agents, with delivery system components.

Formulation breakdown caused by passing the product, under pressure, through small valve orifices.

Suspensions that block valve systems.

Poor lubrication resulting from reaction with formulation ingredients.

Temperature variations caused by storage, climate, or equipment friction.

7. *Economical*

When medication manufacturers provide the administration system with their product, users should benefit from the speed, simplicity, and labor-saving attributes of the equipment.

G. Critical Decisions

As soon as possible, after deciding to begin work on a new compound or formulation, developers should begin to consider, from the perspective of the eventual user, the practicality of its administration. In summary, the key issues to be addressed, assuming the target animals and their circumstances at the time of dosing are known, are these:

1. No matter whether a passive or an active delivery system is involved, the cost, suitability, and future availability of standard equipment should be determined.
2. If a novel delivery system is necessary—
   • Can standard equipment be suitably modified?
   • Should the system be reusable or disposable?
   • Is power assistance needed?
3. What cost of delivery system can the product and/or its user sustain?
4. Is it simple for users to quickly and correctly calculate dosing volumes and regimens and translate them to use by simple adjustments of the delivery system?
5. Is the formulation such that the dose volume can easily and speedily be delivered to the dose site within the limitations of user ergonomics and engineering feasibility?
6. If the medication container is to be connected to the delivery device system, can this be done directly and simply?

IV. PASSIVE SYSTEMS

A. Drinking-Water Metering Devices

Medication of large numbers of animals or birds can be accomplished by adding regulated amounts of the soluble drug to their drinking water. Antibiotics, vaccines, anthelmintics, electrolytes, disinfectants, and antibloat surfactants, for example, are administered in this fashion. The toxic-to-effective dose ratio must be high to ensure safety to animals or birds that may consume higher than normal volumes of the medicated drinking water.

To determine the effective concentration of medicament required in the drinking water of various farm animals, average daily water consumption must be measured. Tables 1, 2, and 3 give daily estimates for different birds and animals based on age, size, and number to be treated (1).

The water medication metering devices generally fall into two categories, inline devices and trough devices.

1. In-line devices, often known as medicators, can only be installed in situations where piped drinking water is permanently available to the animals or

Table 1 Daily Water Consumption for Turkeys, Broilers, and Pullets (gals/1000 birds)

Age (weeks)	Large turkeys	Small turkeys	Chicken broilers	Chicken pullets[a]
1	10	7	6	4.3
2	20	15	10	9
3	30	25	17	16
4	40	35	23	20
5	50	40	29	22
6	60	45	36	25
7	75	50	38	29
8	95	55	45	32
9	110	65	47	34
10	125	70	50	38
11	140	75	—	40
12	150	80	—	43
13	160	95	—	45
14	165	105	—	47
15	165[b]	120	—	49
16	170[b]	130	—	50
17	165[b]	—	—	50
18	165[b]	—	—	50
19	165[b]	—	—	50
20	165[b]	—	—	50

[a]Average of light and heavy breeds.
[b]Consumption will vary from 15 weeks to maturity from 140 to 190 gal/day depending on the weather.

birds to be treated. Most devices rely on water pressure to proportion the mixing and the rate of delivery, which can range from 2 gal/h to 20 gal/min.

2. Trough. Addition of medication to drinking water can be as simple as adding the dose to the estimated amount of water animals will drink from the trough during a fixed time, to more sophisticated systems such as the type developed by Phillips (2). The device floats when used with medicaments heavier than water (pluronics) or sinks to the bottom of the trough when used with medicaments lighter than water (alcohol ethoxylates). When used with the me-

Table 2 Water Consumption Formula for Cattle and Swine

Cool season	Warm season
1 gal/100 lb body weight	1-1/2 gal/100 lb body weight

Table 3 Daily Free-Will Water Consumption of Swine
on Full Rations (Full-Fed)[a]

Weight of pig (lb)	Consumption (gal)	
	Spring pigs	Fall pigs
25	0.55	0.45
50	0.69	0.56
100	1.12	0.94
150	1.25	1.12
200	1.12	1.20
300	0.75	0.75
400	0.65	0.65

Source: Ref. 1, © 1978 E. R. Squibb & Sons, Inc.
[a]During the winter gestation period 1 gal is required, while lactating sows require 4 to 6 gal daily. Temperature, seasons, and other environmental factors affect water consumption.

dicaments that are heavier than water, a tube extends from the base of the floating container to the bottom of the trough. An inlet tube also extends through the base of the container to a level above the surface of the liquid additive. Since the liquid additive is denser than water, it flows through the outlet tube to the bottom of the trough. At the same time, water enters the inlet tube and settles on the surface of the liquid additive. The liquid additive is stored in the bottom of the trough until active drinking begins. As the drinking rate increases, the incoming water picks up the additive from the bottom, thus depleting the reservoir and allowing more medicament to be delivered. Factors affecting rate of delivery are the density of the medicament, the head of liquid over the outlet, nonturbulent addition of water to the device through the inlet tube, and pickup and mixing with incoming trough water when drinking starts. Since a variety of potential problems are associated with this system, each medicament would have to be specially tailored to the system.

B. Feed

The most popular way to administer medications, growth promoters, flavors, and nutritional supplements to livestock and companion animals is through their feed. The preparation of nearly all supplemented feed is done by the manufacturer or feed mill. Feed forms include blocks, pellets, flowable solids, and liquids. Before mixing any drug into animal feed, further mixing of medicated animal feed or repackaging an animal feed, FDA approval is required in the U.S.

No special equipment, beyond standard feed systems, is required to deliver treated feed.

C. Movement

Some administration systems rely on the natural, voluntary movement of healthy animals to dispense medication.

1. Dust Bags

Cattle dust bags allow topical application of insecticidal powders to control horn flies, face flies, and lice (3). The dust bags apply the insecticidal powders to cattle as they brush against or walk underneath them. They (fig. 3) usually have an inner porous storage and dusting bag with lower and side parts for discharging powder. An outer skirt of flexible, weather-resistant material protects the inner bag and usually extends below it. The bags allow free-choice application or forced use depending on where they are hung. Free-choice application is effected when the cattle can voluntarily work it. With forced use, the bags are hung in doorways, lanes, gateways, etc., so that the cattle have to brush against them when passing.

A face fly fighter, based on a similar design, is attached directly over a salt or mineral feeder. When the animal feeds, direct contact with the bag is effected. Liquid insecticide in oil can also be self-administered (4). Cattle can rub against a suspended tank with apertures in the wall and surrounded by a fabric layer. The apertures communicate with tubes extending into the liquid in the tank and have cups at their ends. Rotation of the tank transfers liquid through the tube to the fabric layer for distribution to a depending apron.

2. Cattle Ear Tags

Flexible plastic ear tags have long been used to identify individual animals. The ear is pierced by a plierslike tagger and the tag is kept in place by a fitted but-

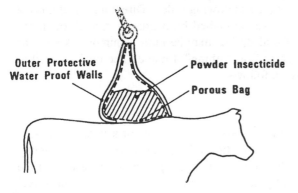

Figure 3 Configuration and common use of powder dust bag.

ton or self-locking design. The natural grooming habit of the animal spreads insecticide, which may be incorporated into the plastic of the tag or separately attached in a pouch, around the areas of the body afflicted with ectoparasites and flies.

3. Cattle and Hog Rubs

A length—usually about 10 ft—of a polyester and acrylic blend rope, with a diameter of 3 to 7 in., is soaked with a solution of insecticide in diesel fuel along its length. The rope is placed where cattle pass beneath or hogs rub. The rope is recharged by hand as and when necessary.

4. Automatic Oilers

A 3-in.-diameter rope is constantly saturated with oily insecticide by its connection to a 5-gal reservoir. A measured volume of oil is automatically released only when the animal is in contact with the rope; oil flow is in proportion to the duration of contact.

5. Flea and Tick Collars

Dog and cat flea and tick collars slowly release pesticide at a concentration lethal to the pest but innocuous to warm-blooded animals. Collars are usually made from plasticized solid thermoplastic resin and usually secured around the neck by a self-locking catch. No special application equipment is necessary, although the circumference of the collar may need to be reduced with scissors depending on the size of the animal.

V. ACTIVE SYSTEMS—ORAL DEVICES

By placing a solid (bolus/tablet/capsule), liquid (suspension/solution), paste, or powder medication in the animal's mouth, the preparation may be introduced into the gastrointestinal tract by the swallowing reflex. Direct introduction of a liquid into the stomach can be accomplished by passing a tube through the esophagus. These methods of oral dosing may be more commonly known as "balling" a solid, and "drenching" or "tubing" a liquid. The devices used to perform these procedures are as follows:

A. Balling Guns

Balling guns are relatively simple. A barrel, holding one or several boluses, is fitted with a flexible constriction at its tip and a plunger which, when gently depressed by hand, dislodges the blous through its tip into the esophagus of the animal. (5–14). The number of boluses delivered is determined by the travel distance of the plunger in the barrel.

The basic barrel/plunger system for delivering solid dosage forms has been developed into a wide variety of designs (Fig. 4). Guns for use with specific animals types: flexible barrels and plungers; single or multiple dosing; universal (for different size boluses); and small-animal tablet dispensers.

When dosing with a solid dosage form, immobilize the animal's head and pass the barrel of the gun or dispenser along the midline of the roof of the mouth over the base of the tongue. The tongue may have to be pulled out if it obstructs the tube. When animal swallows, the barrel is pushed gently into the esophagus, and the plunger depressed to expel the bolus or tablet. Care and patience should be exercised to avoid damage to delicate tissues.

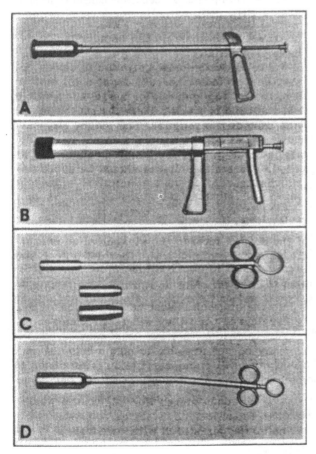

Figure 4 Balling guns. (A) 1-oz pistol grip. (B) Multidose, 1 to 5, 240-grain boluses. (C) triple-head sheep balling gun. (D) Curved equine balling gun. (Courtesy Ideal Instruments, Chicago, Illinois.)

Special forceps are available for dosing sheep with capsules.

B. Esophageal Delivery Devices

Devices used for esophageal administration of medication can be classified into two categories, syringes and tubes.

1. *Syringes*

Two types are available. Regular injection syringes can be adapted for direct stomach delivery by attaching a dose tube to the barrel outlet. Flexible tubes from 12 in. or more in length and 3/16 to 3/8 in. in external diameter are available. These usually have rounded or ball-tip ends to reduce the danger of trauma.

The second type is essentially a pipe. The barrel is passed via the esophagus to the stomach. This dose syringe usually has a pear-shaped distal end and is generally used only on cattle with a body weight >450 lbs (200 kg).

2. *Tubes*

Syringes, injection (stomach) pumps, and funnels may be used to deliver liquid medication through an esophageal tube inserted into the stomach through a speculum in the mouth or via a nostril. The stomach tube should always be passed completely into the stomach. If it is passed only to the thoracic inlet, fluid may return up along the tube when there is a partial obstruction of the tube opening by the esophageal wall; this fluid may be inhaled.

When inserting the tube, care should be taken to ensure that it is well lubricated for the full length that will pass through—for example, the nostril. When the medication has been administered, the inside of the tube should be rinsed with water using the stomach pump or dose syringe. Most veterinarians then blow the contents of the tube and hold a thumb over the end of the tube or kink it in such a manner that the tube cannot produce a siphon action when it is removed from the esophagus. The technique of the stomach tube passage in equines has been fully described by Adams (15). Direct administration of a liquid to the stomach has some advantages:

1.   The total dose enters the stomach. Most horses will resist a balling gun to a greater degree than a nasogastric tube, and drenching usually results in a loss of some of the fluid administered. Many horses are able to hold a bolus in the pharynx until irritation causes coughing, which will expel the bolus.

2.   Irritating substances that might otherwise cause esophageal or pharyngeal lesions can be administered.

3.   Feed, water, or medication can be delivered to animals that are unable or refuse to eat.

Esophageal tubes vary in length, internal and external diameter, and composition. They vary from catheter size for feeding small puppies to 1-1/2 in. O.D. for washing out the rumen. the most common size for equine use is the 1/2 to 5/8 in. O.D. variety. These tubes may be made of silicone rubber, polyvinyl chloride (PVC), cloth-reinforced red rubber, etc., and must be flexible but strong enough to resist kinking. Silicone rubber tubing maintains a constant degree of flexibility over a wide range of temperatures and, because of its nonwetting surface, tube adherence to tissues or fluids is greatly reduced.

Whenever there is a danger of transmitting infectious disease, stomach tubes should be sterilized before use.

C. Drenchers

The oral administration of liquid medication to livestock (drenching) is facilitated by a variety of equipment based on the syringe concept. The main types, made of plastic or chromium-plated brass with synthetic or rubber valve components, are the following.

1. Nonautomatic

Simple calibrated syringes: medication is drawn into the dosing barrel either by a rubber bulb or, more commonly, a plunger, the plunger may be fitted with a dose setting ring which rotates around a threaded plunger rod (Fig. 5). The simple syringe can be fitted with one of a variety of drenching nozzles. The capacity of this type of syringe may be from 2 to 200 ml. All are filled by immersing the nozzle end in the liquid and withdrawing the plunger from the nozzle end to the desired volume shown by the barrel calibration mark. A different version is shown in Figure 6.

2. Automatic

The barrel of this type of drenching syringe refills automatically after each dose has been dispensed through the oral dosing nozzle. The drencher draws doses from an upright or inverted (gravity feed—with air inlet valve) reservoir of medication, to which it is connected by flexible tubing. A typical example is shown in Figure 7.

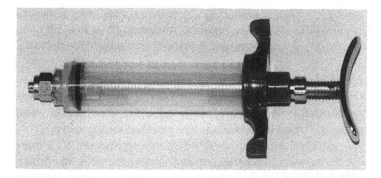

Figure 5 Plastic syringe with dose adjustment nut. (Courtesy Syrvet Inc., Waukee, Iowa)

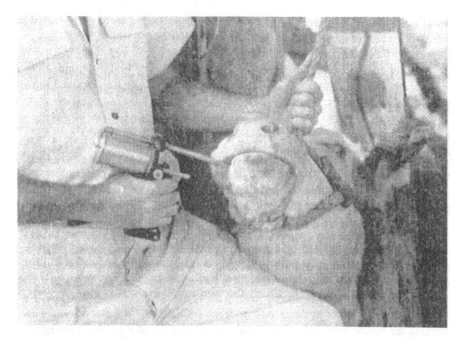

Figure 6 4- or 5-oz universal drencher. (Courtesy N. J. Phillips Pty. Ltd., Somersby, NSW, Australia.)

Various configurations and designs of the basic components have been developed to meet particular requirements:

a. Barrel (fill chamber).   Capacity may range from 5 to 150 ml. The chamber may be of glass or plastic and is etched with volumetric measurements to assist in setting the dose to be delivered.

b. Handle Action.   Movement of the plunger, to fill the barrel and discharge doses, may be by "palm-push" (Fig. 7), where a rear lever is pushed against a fixed front handle, or by the opposite "finger-pull" action (Fig. 8). Both types of movement are countersprung to return the plunger for its barrel refill action.

c. Valve Design.   Automatic dispensing and refill requires two double-action valve sets. One open set, on the medication side, allows the vacuum created by plunger withdrawal to fill the barrel while the other, closed set prevents expulsion during filling. These valve actions are reversed during delivery.

Two designs of valve are common: 1. A coiled wire spring and "T"-shape valve. The circular cross piece of the T acts as an aperture gate while the spring, which surrounds the vertical of the T, keeps the valve closed. 2. The springless

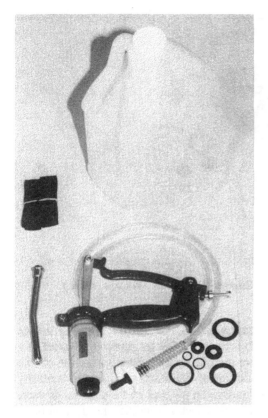

Figure 7   Supervet 50-ml drencher. (Courtesy Syrvet Inc., Waukee, Iowa.)

circular flap valve opens when positive pressure is applied to the medication by the plunger during delivery, while a second flap valve in the plunger remains closed. Their actions are reversed when the strong return spring, usually fitted to the handle of the instrument drencher, refills the barrel by withdrawing the plunger. Both types of valve are generally made of chemically resistant but pliable rubber compounds, as are any "O" rings and washers used in the design.

Generally the "T" spring type of valve allows viscous formulations to pass through more easily because a larger orifice is available, whereas the flap valve opens and closes more quickly and securely to offer higher-speed delivery.

   d.   Dose Adjustment.   Two designs of dose volume selection are common:

   1.   The infinitely variable type requires the user to align (by moving the handle against its counterspring) the middle of an "O" ring or mark on the

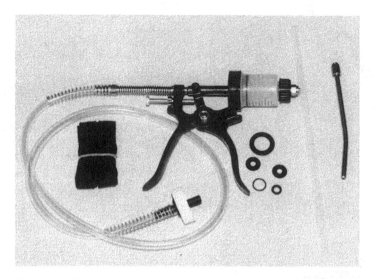

Figure 8 Supervet automatic doser. (Courtesy Syrvet Inc., Waukee, Iowa.)

plunger with the chosen dose mark on the barrel. The plunger "O" ring is then set precisely at the mark by a thumb set screw. The set screw is situated at a point which prevents the plunger from moving under pressure from the handle counterspring; this is often at the fulcrum of the handles or, on-low capacity models, through the barrel mount to impinge on the back of the plunger. In most designs the thumb set screw is fitted with a lock nut so that the dose volume can be secured against inaccuracy caused by accidental loosening of the screw. Responsibility for the accuracy of the dose delivered rests with the user. It can be checked by dispensing a known number of doses into a measuring cylinder, calculating the division result, and adjusting the dose setting mechanism accordingly. The procedure should be repeated until the desired accuracy is achieved.

2. The fixed-dose selection design limits the travel distance of the plunger by axially staggered fill volume stops engineered into the plunger push rod. The volume is selected by a ring indicating volume choices; this is usually mounted at the rear of the barrel. The ring has apertures which allow all stops to pass—except that of the selected dose volume.

Fixed-dose volumes may be in equal increments; for example, five 5-ml stops would allow the user to select any dose volume from 5, 10, 15, 20, or 25 ml. Unequal increment designs have been produced to deliver specific medications with particular dose volume to animal weight ratios. Fixed-dose designs depend, for accuracy, on 1. the engineering tolerances used in manufacture, and 2. the resistance to wear, and consequent inaccuracy, of the selection stops.

When dose volume is determined by animal body weight, a proportion of animals, unless they all weigh the same, will be more or less under- or over-dosed. If the dose volume must be changed for each animal, and accuracy is not a prime consideration but speed is paramount, the fixed-dose design may be preferred over the infinitely variable type.

e.   Nozzles.   Automatic drencher nozzles are usually made of metal to resist wear caused by the animals teeth. Nozzle tips should be free of sharp edges to prevent damage to the mouth. Many are fitted with a rounded tip containing an antidrip valve to reduce medication waste. Nozzles may be "fixed" directly to the drencher (Fig. 6) or indirectly by a flexible tube—the so-called floating nozzle (Fig. 9).

The length and shape—straight, curved, or hooked—of the nozzle pipe is determined by the species of animal to be dosed and the method of restraint. Sheep and pigs are usually restrained directly by the operator, whereas cattle are most frequently held in a chute fitted with a means (headgate) of keeping the animal's head available for operator restraint.

The choice between a fixed or floating nozzle is made by the operator, depending on the dosing circumstances. The floating hook nozzle is preferred almost exclusively by cattle workers, who use one hand to hook it into the animals mouth and the other to operate the dose delivery handles on the drencher.

f.   Hook Drencher.   The floating hook technique of dosing has been developed into a one-handed method by incorporating the automatic delivery and refill mechanism into the hook (Fig. 10). By pulling back on the pipe handle, after the nozzle has been "hooked" into the corner of the animal's mouth, the dose is dispensed from the built-in barrel/plunger. Releasing the hook from the animals mouth allows a return spring to refill the drencher.

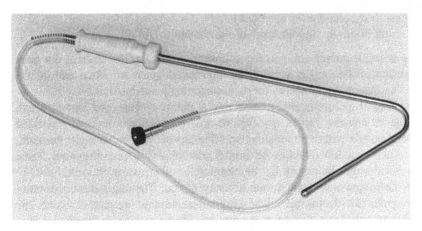

Figure 9   Supervet floating hook nozzle. (Courtesy Syrvet Inc., Waukee, Iowa.)

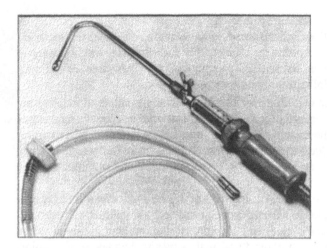

Figure 10   Supervet hook drencher. (Courtesy Syrvet Inc., Waukee, Iowa.)

D. Paste Dosers

Pastes or gels, presented in a prefilled syringe, cartridge, squeeze tube, or bottle, are popular with formulators of oral medications because insoluble compounds can be suspended in the paste or gel. From the users perspective their attractions are:

Easily portable; requires no back pack, straps, or connecting tubes
Ready to use
Cannot be spilled
Animals find paste harder to reject
Use rarely requires the purchase of equipment

Paste or gel is administered to the animal by depositing the dose on the posterior portion of the tongue through the interdental space on either side of the mouth. The dosing apparatus is directed posteriorly and medially.

Most paste or gel products are filled into a plastic syringe-type barrel which may range in capacity from 5 to 500 ml. The barrel has a smoothly molded oral dosing nozzle with a removable cap. The nozzle may be straight, hooked, or shaped in a way best suited to the animal and the dosing circumstances. Paste is extruded through the nozzle by the application of pressure on a plunger.

Paste cartridges may contain one or many doses. The single-dose presentation is popular for dosing individual companion animals because it avoids the possibility of infectious disease transmission. Horse owners in particular like avoiding the "tubing" method and the waste, by rejection, of drench.

Dispensing many doses from a prefilled cartridge requires a means of moving its plunger, in measured volume increments, along the length of the barrel. Three design concepts have been applied over the years:–

1. Syringe

A prefilled syringe barrel with capped nozzle fitted with a calibrated and threaded plunger rod. The rod is fitted with a rotating dial ring. The dose is selected by aligning the dial with the appropriate mark on the rod. Delivery is made, after removing the nozzle cap, by depessing the plunger rod with the thumb until its movement is halted by the dial ring. The commercial name for this system is Dial-A-Dose by Plas-Pak Industries Inc.

2. Caulking Gun

A standard household gun accepts a prefilled cartridge with capped nozzle and plunger. The barrel of the cartridge is marked with dose volumes. Measured doses are dispensed according to the position of the plunger against the barrel marks. The cartridge plunger is moved to express doses by means of the gun's ratchet rod and trigger mechanism. This system is most suitable for low-cost products, such as dietary supplements and probiotics, where accuracy is not critical. Some manufacturers of such products recommend, or offer, a gun modified to accept their particular design of cartridge.

3. Threaded Paste Cartridge

For drugs of high potency formulated at maximum usable paste concentration in low-dose volume, a sophisticated gun is required to deliver each dose accurately and repeatedly. The prefilled cartridge, with a capped, molded nozzle, is threaded at its outer back end. Turning the cartridge into the threaded barrel of a doser measures the volume to be dispensed. The doser trigger handle then pushes a rod against the cartridge plunger the distance required to dispense the exact dose volume. This system has largely been abandoned in favor of the syringe design—probably due to the expense of supplying dosing guns to users.

E. Powder Guns

Drenching with a dry powder was a preferred technique for a while. Following administration powder adhered to the inside of the mouth, was involuntarily mixed to saliva and swallowed. Rejection of the dose was less frequent than with a liquid drench–particularly by horses.

Powder administration by a gun (Fig. 11) involved cocking the spring-loaded plunger by pulling it back to click into position. The barrel was then filled with the required amount of powder. The animal's upper lip, at the corner of the mouth, was lifted to expose the slippery mucosa. The gun barrel was then

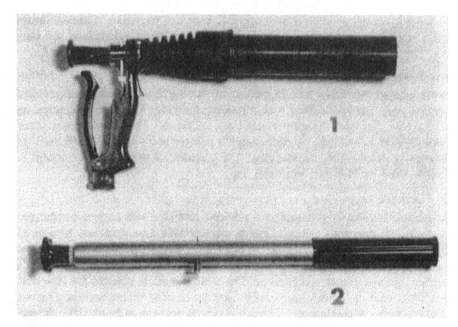

Figure 11 Examples of powder drench guns. (1) horse drencher (N. J. Phillips Pty. Ltd., Somersby, NSW, Australia); (2) Merck AgriVet thibenzole powder drench gun (Merck Research Laboratories, Rahway, New Jersey).

worked about 8 cm into the mouth, the trigger pulled, and the powder rapidly delivered with a slight rush of air. Aspiration of the powder into the animal's lungs was not reported to be a problem.

VI. ACTIVE SYSTEMS—TOPICAL APPLICATORS

A topical preparation is applied directly to the exterior of a portion of the body through which treatment is effected. The external surfaces of the body include not only hair, fur, and hide but also the skin and membranes of orifices.

A topical preparation may act directly or indirectly:

1. Direct action topicals treat or prevent the surface cause of a condition. Examples: a repellent to prevent flies from irritating, and reducing the rate of weight gain of beef cattle; a pessary to prevent postpartum infections of the uterus.

2. Indirect action topicals penetrate the surface to treat or prevent a condition. Examples: transdermal cattle anthelmintics; intranasal vaccination of pigs. Topical product applicators therefore apply material to an external body surface or are designed to deposit medication into an orifice.

A. Sprayers, Spray Races, and Dip Baths

Control of ectoparasites on cattle, poultry, sheep, goats, and swine can be accomplished by either spraying or dipping the animal. Animal premises are also sprayed to control insect pests. Spraying is done with a hand wand fitted to either a knapsack sprayer or bucket stirrup pump. In large-scale operations a power pump may provide a coarse, driving spray fitted to a race through which the animals are driven (16, 17). To dip animals the bath must be long, wide, and deep enough to ensure complete immersion without injury. Because of the nature of the process, the insecticide dip must not be inactivated by the body materials deposited by the many animals that pass through the bath, must maintain its stability throughout a range of temperatures and concentrations, and must be safe to the animals while toxic to ectoparasites.

In the U.S. the advent of high-potency ectoparasiticides, the introduction of endectocides, and heightened awareness of the potential environmental dangers associated with dipping have almost eliminated dipping as a livestock treatment method. Many dog owners, breeders, and grooming parlors use dips and sprays for ectoparasite control and cosmetic purposes. Such preparations are often combined with a shampoo or grooming aid. No special equipment is needed to apply these preparations.

B. Teat Dips

An important and integral part of mastitis prevention and milk quality control involves dipping the teats in an antiseptic liquid; such medications prevent dermal irritation and infection of the teat canal. Treatments of this kind are particularly necessary when milking machines are used on dairy cattle. Treatment is applied by immersing the lower extremity of each teat in a cup of disinfectant.

Dipper cups are often connected to a flexible squeeze fill reservoir of disinfectant. Various cup designs incorporate:

Nonsiphon action to prevent reservoir contamination
A flexible liner to ensure even coating of the teat
An angled neck to make dipping of the rear quarter easier
Soft splash-proof lips to reduce trauma and reduce waste

Some dairy operatives prefer to use a teat sprayer. A household, hand-held, finger-pump sprayer of 16 oz capacity is modified so that the spray head angle is at 45°—for easier upward application. Usually the spray mechanism is fitted with a stainless-steel valve.

To ensure a more complete coverage of the teat than might be achieved with a "straight" spray, another design surrounds the teat with a circular spray head. The antiseptic is hand-pumped through a series of misting holes on the inside of the head.

C. Pour and Spot-On Applicators

Formulations effecting direct, or systemic, activity by application to the skin or hide, offer the advantages of ease and speed of administration. Preparations of this kind are:

1. poured evenly along the animal's backline using a measuring cup, or applied with an automatic refill gun of the type shown in Figure 2. Such applicators offer adjustable doses of between 5 and 75 ml with each pump of the handle. Such equipment, under most circumstances, is more accurate and less wasteful than the cup pour-on method. Pour-on applicators also reduce operator exposure to medications; this is particularly helpful when a transdermal product is being applied. An applicator can be fitted with a delivery nozzle which dispenses product in a particular way. Among the common designs are nozzles which enable the instrument to project medication 6 to 10 ft, deliver it in a fan pattern, or apply multiple streams.

2. Applied to one spot in concentrated form. For the control of cattle grubs, for example, a squeeze bottle applicator with integral measuring cup (Fig. 12) is a popular design for dosing small groups of animals. The dosing cup is filled with the required dose by squeezing medication from the bottle reservoir; the dose is then spotted onto the backline of the animal.

For the control of fleas on dogs, the spot application of a concentrated organophosphate solution is fast and convenient. An individual-animal dose is filled into a small disposable, collapsible tube which is then sealed with a safety cap. Dose volumes range from 0.34 to 2.18 ml, depending on the weight of the animal. The dose is applied by firmly squeezing out the entire contents of the tube onto one spot on the skin.

D. Ointment and Paste Application

The necessarily bland nature of topically applied ointments does not require any applicator other than the human hand and fingers. Caustic paste is, however, used to dehorn animals. To avoid operator injury, paste is applied directly from a prefilled squeeze bottle to the horn button. In the case of calves, this procedure is usually carried out before the animal is 2 months old.

E. Intramammary Applicators

The mammary gland can be treated by inserting medication through the teat canal using a specially shaped infusion cannula. The cannula is attached to a syringe to deliver the dose.

Antibiotic formulations are supplied in prefilled, plastic, disposable syringes fitted with infusion cannulas. Some brands have cannulas with snap caps to facilitate partial insertion into the teat canal. Each syringe contains the amount of product needed to treat one quarter (teat) of the bovine mammary gland.

Figure 12   Squeeze-A-Spot applicator. (Courtesy Bayer Corporation, Shawnee, Kansas.)

F. Intravaginal

Vaginal drugs are primarily used to control estrus during the sexually active season. Synchronization of estrus in ewes, for example, increases productivity and contributes to more efficient timing of lamb production. A variety of vaginal drug dispensers have been described (18–21) together with the methods for delivery (22–25).

One device, which has proven successful in ewes, is a removable sponge impregnated with an ovulation control compound, such as flurogestone acetate or medroxyprogesterone acetate. The sponge is administered by an operator restraining the standing ewe while a second coats the sponge with an antiseptic cream and loads it into the applicator illustrated in Figure 13.

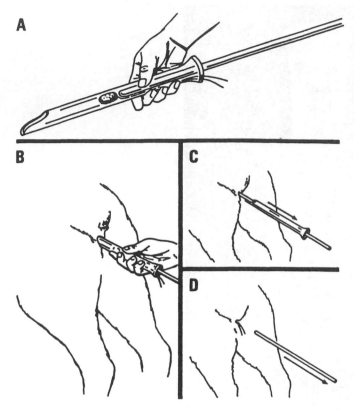

Figure 13 Application of vaginal sponge to ewes. (A) The sponge inserted into the wide end of the applicator and pushed down half its length with the plunger. (B) Applicator inserted into the vagina as far as possible. In maiden ewes the sponge may be positioned with the finger. (C) Applicator removed over the plunger. (D) Plunger removed leaving the nylon lines of the sponge protruding to aid its later removal. (Courtesy The Pharmacia and Upjohn Company, Kalamazoo, Michigan.)

G. Intranasal

Intranasal administration of liquid vaccines to one or a number of animals, sequentially or simultaneously, can be accomplished by using one of the following devices:

1. Automatic, multidose or single-dose syringe fitted with a special plastic nasal tip in place of an injection needle. Such tips vary in size depending on the species of animal to be dosed. The shape is somewhat conical, for easy positioning in the nasal cavity, and smooth to prevent damage to the mucous membranes.

2. A single-dose calibrated dropper. The dose is drawn and expelled by a rubber bulb.

3. An automatic spray device for day-old chicks (see Sect. IX B)
4. Day-old chick spray boxes (see section IX C)

H. Aerosol Dispensers

Advantages include:
1. Medication is delivered directly to the affected area in the desired form such as a spray, stream, quick break, or stable form.
2. Medication is applied with little mechanical irritation to sensitive areas.
3. A dose can be delivered without contaminating the remaining material. Pharmaceutical aerosols and the devices used to dispense them have been reviewed by Sciarra (26). Most animal products dispensed by this method are wound dressings and compounds for the control of flies, fleas, ticks, and lice.

There is a trend away from propellant-base aerosols due to environmental interactions and consequent government restrictions. Pump sprays are therefore replacing the aerosol formulations. The comparative advantages and disadvantages of propellant and pump aerosols have been reviewed (27).

VII. ACTIVE SYSTEMS—PARENTERAL

The introduction of a substance by injection into the body is most commonly accomplished by a syringe which dispenses it through an hypodermic needle, the outlet tip of which is placed at the deposit site. The physical characteristics of the medication to be injected, the injection site, and the reasons and circumstances of its use, all influence the design of the needle and syringe.

A. Injection Needles

The most common routes of injection for liquids are subcutaneous (SC), intramuscular (IM), and intravenous (IV). Operators have their own preferences of needle length, guage, and bevel. The species and age of the animal(s), the viscosity of the material to be injected, and the amount of restraint required influence the choice. The standard guide is given below (in. × SWG):

Animal	SC	IM	IV
Pig	1/2 × 20	1 × 20	1 × 20
Hog	3/4 × 18	1 × 20	2 × 18
Sow	1 × 18	1 1/2 × 18	4 × 18
Sheep	3/4 × 18	1 × 18	1 × 18
Calf	3/4 × 18	1 × 18	1 × 18
Yearling	3/4 × 16	1 × 16	1 × 16
Cow	1 × 16	1 1/2 × 16	1 1/2 × 16
Colt	3/4 × 20	1 × 20	1 × 18
Horse	1 × 20	1 1/2 × 20	1 1/2 × 18
Dog and cat	3/4 × 22	1 × 22	—

Source: Jeffers. West Plains, Missouri.

Injection needles for the common routes are available in reusable or disposable form. The most common needle hub, connecting the needle shaft to the syringe, is the Luer type, of either slip or lock design. An injection needle may be connected directly to the syringe or by a length of flexible tube. The latter arrangement, or "floating needle" (Fig. 14) is preferred by some operators.

B. Syringe Systems

The withdrawal of a plunger from a barrel creates negative pressure within the barrel; the partial vacuum draws liquid medication into the barrel for volumetric measurement and transportation. All syringe barrels carry volume marks so that users can either check or measure filled and dispensed volumes. It is advisable to randomly check the calibration marks on disposable syringes because they could be misplaced on the barrel.

Some syringe models have centered or eccentrically placed needle mounts.

Doses are dispensed by applying pressure to move the plunger along the barrel. Pressure on the plunger can be by hand or power (see Sect. IX).

Even though oral and topical medications are dispensed by the syringe principle, the mechanism is commonly considered to be primarily an injection, or parenteral, delivery system. It should be noted that veterinary authorities do not recommend injection of more than 10 ml into any one tissue site.

A syringe injection system should be sterile at the time of use. Systems that are not supplied sterile, or are not prefilled with sterile medication by the manufacturer, must therefore be able to withstand the sterilization processes available to users. Most commonly these are boiling water or autoclave (followed

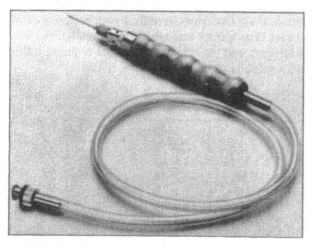

Figure 14 Supervet floating needle handle. (Courtesy Syrvet Inc., Waukee, Iowa.)

by a cooling period), or a chemical treatment. Users should check with the syringe maker to determine if any of these sterilization methods is contraindicated. If chemical sterilization is contemplated, the manufacturer of the medication should be asked if the method is compatible with its product.

With the exception of single-use (prefilled, multichamber, and disposable) injection systems, all sterilizable syringes may be used more than once. Such uses may range from injecting, on separate occasions, a variety of medications and dose volumes into small numbers of different animal species, to immunizing, at one session, many hundreds of animals with the same vaccine and dose volume. The variety of such factors and circumstances has given rise to a multitude of syringe designs, each one seeking to make the process of inoculation as simple and convenient as possible. Examples of currently available syringe types include the following:

1. Single Use

Single-use systems dispense one dose of medication and are then discarded. The most common is the single-dose liquid companion animal vaccine filled, by the vaccine manufacturer, into a ready-to-use sterile plastic syringe. The syringe may be fitted with an injection needle before it is sealed into a protective container for distribution.

Because of the inherent instability of some medication components in solution or suspension, they must be maintained in the dry state. Several designs of prefilled syringe keep the dry and liquid components apart, in separate chambers, until reconstitution at the time of injection. Component mixing is accomplished by puncturing the separating diaphram (28–44). For example, the two-chambered system (Fig. 15) uses a glass or plastic syringe barrel to which is added an interior trip step; a rubber separator plug divides the syringe into two chambers.

In use, the plunger is moved forward, causing the disk to trip and permit shake mixing. When complete, the reconstituted medication is ready for injection by completion of the plunger movement.

2. Single Dose

Single-dose syringes require each dose to be separately filled and dispensed. Most popular models are fitted with a threaded plunger rod and nut so that the required dose can be set and more easily repeated. Barrel capacity may be from 10 to 50 ml.

3. Simple Multiple Dose (see Fig. 5)

These syringes are most convenient for the injection of a small number of animals, each of which may require a different dose volume. Barrel capacities vary from 1 to 250 ml. Users select the capacity that can contain the total volume of medication needed to treat all the animals to be dosed in one session. Thus

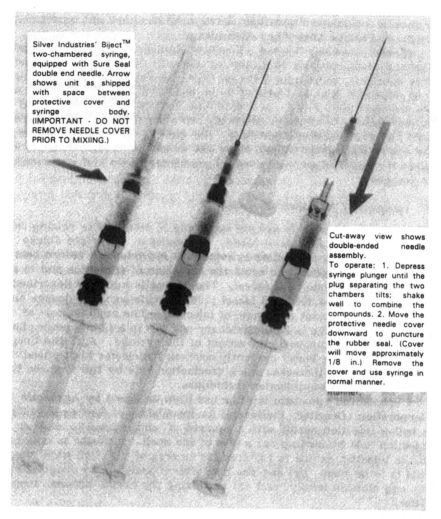

Silver Industries' Biject™ two-chambered syringe, equipped with Sure Seal double end needle. Arrow shows unit as shipped with space between protective cover and syringe body. (IMPORTANT - DO NOT REMOVE NEEDLE COVER PRIOR TO MIXIING.)

Cut-away view shows double-ended needle assembly.
To operate: 1. Depress syringe plunger until the plug separating the two chambers tilts; shake well to combine the compounds. 2. Move the protective needle cover downward to puncture the rubber seal. (Cover will move approximately 1/8 in.) Remove the cover and use syringe in normal manner.

Figure 15   Two-chambered syringe. (Courtesy Silver Industries, Norwich, Connecticut.)

one filling of a 1-ml syringe could dispense 5 × 0.2 ml doses, or a 60-ml capacity syringe could be filled with 50 ml to dose 25 animals with 2.5 ml each. If dosage is determined by body weight, and the animals to be dosed vary in this regard, the operator will need to change the volume of each injection.

4.   Repeater

Repeater syringes are filled the same way as single-use and simple multiple-dose syringes. The repeated delivery of each of a fixed dose volume is effected by a ratcheted plunger rod and handle mechanism which advances the plunger. Selection of dose volume is made by rotating a ring to alter the number of ratchet teeth advanced by the handle.

A 50-ml capacity pistol-grip repeater syringe, capable of dispensing 1, 2, 3, 4, or 5 ml doses, is shown in Figure 1. Users can fill a repeater syringe with the amount of medication required for the dosing session; also, any one of the fixed dose volumes can be "dialed" as each animal is presented for injection.

5. Automatic Refill

These syringes for the injection of liquid medications into a single tissue site differ in the following ways from the designs so far described:

The syringe is connected to the medication container

The medication reservoir is inverted, and held above the syringe, to provide gravity flow

The barrel automatically refills after each dose has been dispensed

Maximum dose delivery is 10 ml

A vacuum is created in the container as medication is withdrawn by the syringe. If the container is not collapsible, air must replace the withdrawn liquid for the syringe to function efficiently. Most automatic systems are therefore provided with an air inlet valve so that rigid containers can be accommodated.

Two syringe designs stem from the way in which the medication reservoir or container is connected to the syringe. It may be attached directly onto a vial holder built into the syringe, as illustrated in Figure 16. In this "bottle attached" design, and rubber-capped vial of up to 100 ml fill can be attached to the holder near the nose of the instrument. Attachment pierces the rubber vial cap with an outlet needle. Air enters the vial, as product is withdrawn, through an inlet valve in the holder. The utility of this design is limited because containers larger than about 125 ml can make the whole system too heavy for comfortable prolonged use. Also, the bulk of the syringe, with bottle attached, makes the system less maneuverable in confined spaces, such as between the bars of a cage or restraining chute. It is, however, conveniently portable for dosing small numbers of hand-held or loosely confined animals.

The alternative reservoir connection method is a length of flexible tube. One end is fitted to the syringe, either near the nose or at the rear; the other is fitted with a combination vial cap puncture spike and air inlet valve. Figure 17 illustrates the fill tube connected to the syringe through the rear handle. The inverted reservoir in this case is a plastic bottle of vaccine carried in a back pack, or hung by a strap around the neck of the operator.

Automatic syringes fill and refill using two pairs of inlet/outlet valves. Two types are described in section V.C.2.c. Dose selection and adjustment are usually

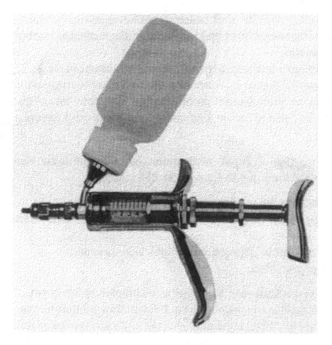

Figure 16 5-ml automatic top bottle syringe. (Courtesy N. J. Phillips Pty. Ltd., Somersby, Australia.)

of the infinitely variable plunger rod screw type. Automatic syringes may be sterilized by boiling in water and are therefore reusable.

Plastic, disposable, nonsterile designs, either dispensing a single fixed-volume dose or fitted with an infinitely variable dose adjustment mechanism, are available. Figures 18 and 19 illustrate automatic refill, single fixed, and adjustable dose models. Both have rear fill tube connections.

Less expensive disposable automatic syringes are designed to be discarded after no more than 1 day's continuous use. Sterilization, usually by chemical means only, is possible; approval of the medication manufacturer should be obtained before such a method is used.

Tube-connected automatic syringes allow operators to use much larger-volume containers than are possible with bottle-attached designs. This makes them popular in situations where large numbers of individually restrained animals, such as at a cattle feedlot processing chute, are each injected with the same volume of the same medication. Having fewer large-reservoir connections reduces total processing time. In such situations the medication reservoir is often suspended near the injection work area.

Figure 17 5-ml rear-fill automatic vaccinator in use of the mass vaccination of Australian Merino sheep. (Courtesy N. J. Phillips Pty. Ltd., Somersby, Australia.)

A tube-connected syringe, used in conjunction with a medication reservoir back pack, offers the user high numbers of portable doses.

6. One-Handed

When the animals to be injected cannot be closely restrained and/or the operator prefers a "slap technique," an instrument that automatically injects the dose when the animal is "hit" may be preferred. Figure 20 shows how such a syringe is held for use. The manufacturer's instructions state "with a firm thrust hit the animal at the injection site and then withdraw the instrument. The thrusting motion penetrates the hide and forces the plunger forward, dispensing medication." Figure 20 shows the construction of the one-handed injector.

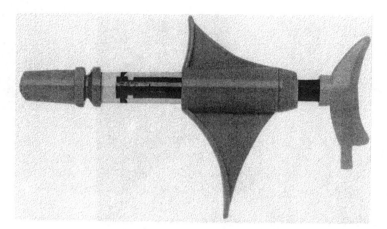

Figure 18 0.5-ml disponsable/reusable fixed dose syringe. (Courtesy N. J. Phillips Pty. Ltd., Somersby, Australia.)

The one-handed injector in Figure 20B dispenses medication either from a prefilled disposable syringe or through a flexible tube conneced to a remote reservoir.

7. Injector Pole

When a dangerous large wild, zoo or domestic animal, must be injected, safety is a concern. A syringe mounted on a 5ft pole would enable the operator to more easily avoid injury. Such an instrument must secure the injection needle at the site, automatically deliver the measured dose, and incorporate a mechanism for retrieving the syringe and needle. Figure 21 shows a system, capable of delivering up to 30mL of inoculum, which provides the operator with a maximum 25 ft safety zone.

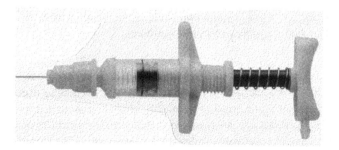

Figure 19 5-ml disposable/reusable variable dose syringe. (Courtesy N. J. Phillips Pty. Ltd., Somersby, Australia.)

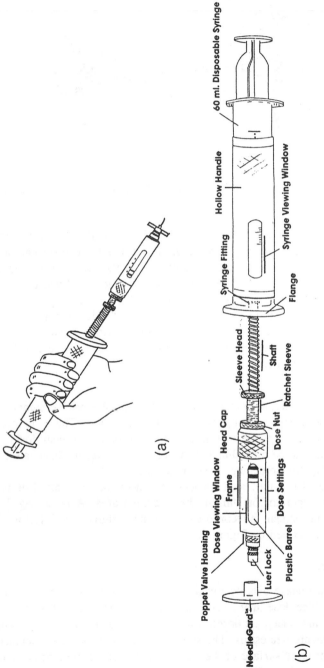

Figure 20 One-handed injection (a); Injecto-Stik (b). (Courtesy Ideal Instruments, Chicago, Illinois.)

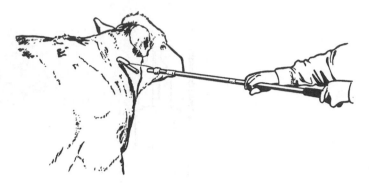

Figure 21   Stock Doctor medicating system. (Courtesy Syrvet Inc., Waukee, Iowa.)

The injection needle shaft features a beveled ring so that it is retained, while attached to the lightweight syringe, in the injection site. The syringe plunger is powered by an internal coil spring, activated by the pressure of needle insertion into the animal. A 20ft long retractable nylon tether, fitted to a pulley inside the pole handle, is attached to the collar of the syringe.

Upon impact of the pole mounted syringe needle the animal, the operator, or both, usually retreat from each other with the syringe in place. This uncoils the tether and, when the injection is complete, enables the operator to retrieve the syringe and needle.

8.   *Intraruminal*

The deposition of liquid anthelmintic directly into the bovine rumen is a hybrid technique because it "injects" an oral formulation. The technique eliminates the spillage and waste that may occur with oral or topical dosing. In addition, animals can be treated in a crowd alley or snake; the headgate or catch, often necessary for drenching, is not required.

The injector has a tube feed connection to a remote inverted drug reservoir carried by the operator or hung in the treatment area. A revolving dose volume change mechanism is incorporated into the design of the injector. The use technique is illustrated by Figure 22.

C.   Implants

The sustained release of a chemical dose can be accomplished by solid-form implantation. The dose may be released from one or a number of pelleted implants. Cattle and sheep are dosed with hormonal agents in this way to promote growth or synchronize estrus. The site of implantation is the subcutaneous tissue of the posterior surface of the ear. Implantation equipment is supplied in two parts:

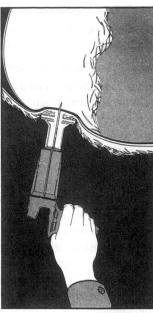

STEP 2: Press the injector into the animal's side. This causes the shroud to telescope into the injector while also compressing the layers of skin into a solid block against the ruminal wall.

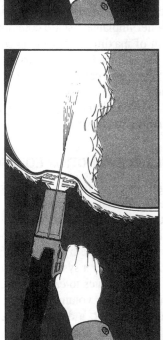

STEP 4: Pause for a second or two before withdrawing the injector to make sure the full dose is delivered. The shroud returns to its fully extended position as the injector is withdrawn.

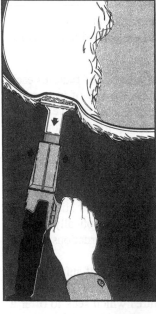

STEP 1: On the left side of the animal, place the shroud of the injector in the triangle formed by the last rib, the point of the hip and the shelf of the spinal column.

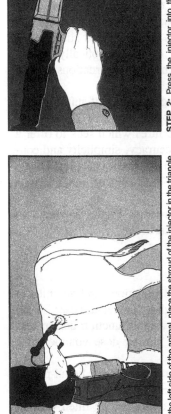

STEP 3: When the shroud is fully depressed, it activates the springloaded plunger. This pushes the needle into the ruminal cavity followed by the instantaneous delivery of the dewormer.

Figure 22 Four steps in the use of the intraruminal injector. (Courtesy Fort Dodge Animal Health, Overland Park, Kansas.)

1. Implants (Pellets or Silicone-Coated)

Growth-promoting implants are available in preloaded disposable belts or cylinder cartridges. Single-estrus synchronization implants are offered in foil packs.

2. Implanters (Also Known as "Guns")

Each manufacturer produces its own size and shape of pellets; this requires each to make available its own design of implanter. Because the site of implanting is harder to get at than others and the technique requires comparatively more administration skill, the design of the implanter is an important factor in the competition for growth promotion implant market share.

The basic implanting method is injection with a beveled needle, the internal diameter of which is only slightly larger than that of the implant pellet. After the needle has been inserted under the skin, the dose is deposited by pushing it out of the needle, into the deposit site, with a rod. The push rod slides along the needle bore. Each implant injection should be done as aseptically as possible, to reduce the risk of infection and its transmission. Instructions for the use of a single pellet implanter are given in Figure 23.

Multiple pellet cartridges, for individually dosing large numbers of animals, such as in a beef feedlot, require more sophisticated applicators.

Multiple pellet implanters share common design features which seek to offer a combination of maximum implanting speed with accuracy, simplicity and convenience. These include: forward trigger handle, to assist with accurate needle placement; easy load cartridge magazine; automatic dose counter; quick change needle mount.

Figure 24 shows an implanter with a cylindrical cartridge. Figure 25 shows an implanter depositing a six pellet dose from a belt cartridge.

VIII. MULTIPLE FUNCTION EQUIPMENT

The syringe principle delivers a wide variety of formulations and volumes to many different animal species and sites. Even though a "universal doser" is far from practical, there are circumstances that make a multifunction design feasible: for example, dosing the same species, with similar dose volumes of a variety of products made by one manufacturer.

Figure 26 illustrates a system which can be used to administer liquid preparations by various routes to cattle. The 25-ml capacity automatic syringe is fitted with variable dose volume adjuster; the barrel outlet can be fitted with an injection needle mount, a drench nozzle, or a topical applicator pipe. The syringe feed tube can be connected to a wide variety of containers; each change pack includes the accessories needed to complete a function transformation.

Figure 27 shows a dual-function automatic refill doser connected to a container. The drenching nozzle can be replaced by a needle mount, thus convert-

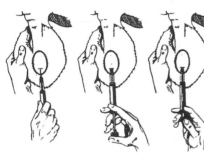
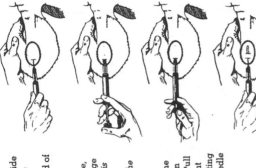

INSERT IMPLANT — With beveled side of needle away from the skin, pick up skin with the point of the needle and penetrate skin surface about one third of the way from the outer tip of the ear. *Do not puncture cartilage.*

With needle parallel to the ear surface, push needle between skin and cartilage (subcutaneously) *until all the needle is under the skin.*

Holding the needle *all the way in to the hub,* push the plunger all the way in; ejecting implant subcutaneously.

WITHDRAW NEEDLE — Withdraw the needle and make sure the implant is in place 1/4" to 1/2" from needle hole. Pull plunger back to remove empty implant sheath. Place used needle in disinfecting solution and replace with a sterile needle for next animal.

DISINFECT NEEDLES — Place needles in a shallow tray and cover with suitable disinfectant. Follow manufacturer's directions for proper disinfectant use.

RESTRAINT — Confine the animal in a squeeze chute or head gate. To reduce the risk of injury to the person doing the implanting, restrain the animal's head to limit movement.

SITE PREPARATION — Select an implant site on the outer surface of the ear. The site should be approximately midway between the tip and the fleshy base. Follow by clipping the hair, and scrubbing the site area with a brush and suitable disinfectant.

AFFIX STERILE NEEDLE — Unscrew nose piece half a turn, take a needle from the disinfectant solution, shake off excess solution, set it into the end of the implanter making sure it is seated all the way in. Retighten the nose piece to stabilize the needle.

LOAD IMPLANTER — First, remove the sheathed implant from its foil pack. *Be sure to keep the implant dry at all times.* Next, utilizing the implanter with sterile needle attached, pull the plunger back and place sheathed implant in slot. Then, release the plunger until it rests on the implant. To activate for implantation, push the implant out of the sheath a distance of 1/16th inch.

Figure 23 Instructions for use. Synchro-Mate B Implanter. (Courtesy Rhone-Merieux Animal Health Inc., Athens, Georgia.)

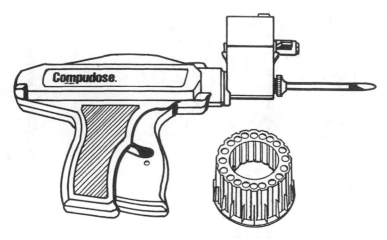

Figure 24 Implanter with 20-shot disposable cartridge. (Courtesy Elanco Animal Health, Indianapolis, Indiana.)

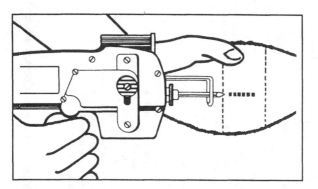

Figure 25 After inserting the needle to its full length, squeeze the trigger gradually withdrawing the needle. This should deposit the pellets in a straight line.

ing the instrument into an injection syringe. Doses of between 2.5 and 20 ml can be selected by adjusting the volume screw at the base of the handle. Multifunction equipment must be thoroughly cleaned and, if necessary, sterilized and lubricated between the administrations of different products.

IX. POWERED SYSTEMS

When a large-scale animal production operation requires routine medication of each animal, power-assisted delivery systems offer producers greater speed, lower fatigue, and reduced labor costs. Power can also force medication through

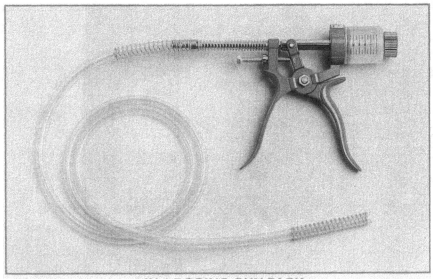

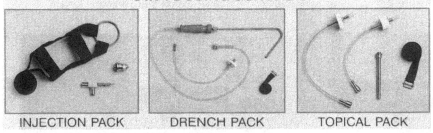

Figure 26 Three-in-one dosing gun. (Courtesy Schering-Plough Animal Health, Union, New Jersey.)

skin as well as propel it over a distance. The following examples illustrate ways in which power contributes to the efficient administration of animal medications.

A. Egg Injection

Biologically active compounds, such as vaccines, can be injected into fertile chicken eggs 3 days before they hatch. An injection rate of 20,000 eggs per hour can be achieved by two operators using an automated system powered by compressed air and electricity. Figure 28 illustrates the egg handling sequence (left to right).

The injection sequence is cleansed by a built-in sterilization wash to prevent the spread of infection. The entire system (Fig. 29) can be wheeled around the hatchery because it is narrow enough to pass through standard doorways.

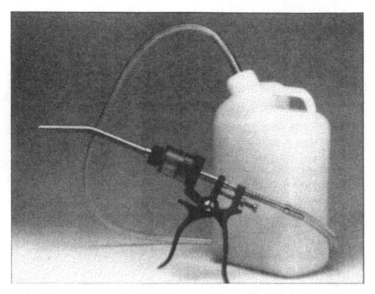

Figure 27   20-ml drencher/injector. (Courtesy Syrvet Inc., Waukee, Iowa.)

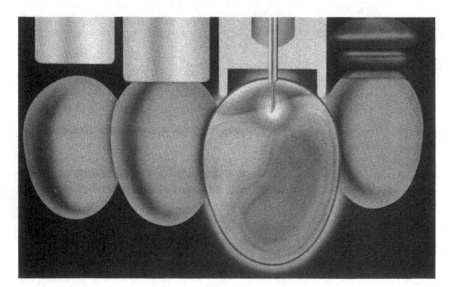

Figure 28   Inovojec egg injecting sequence. (Courtesy Embrex Inc., Research Triangle Park, North Carolina.) Left: Injection head lowered onto egg. Left Center: A small opening is made in the shell with a cutting tube. Right Center: An injection needle descends through the cutting tube, to a controlled depth, delivers a measured dose, and is withdrawn. Right: Eggs are moved to hatching baskets.

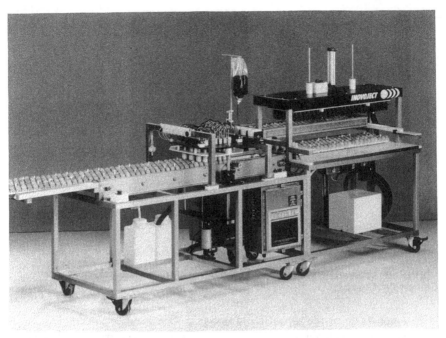

Figure 29 Inovojec egg injection system. (Courtesy Embrex Inc., Research Triangle Park, North Carolina.)

B. Chick Injection

Up to 3000 day-old chicks can be injected by one operator per hour, with between 0.07 and 0.4 ml Mareks disease vaccine, for example, using an automatic system powered by compressed air and electricity. Pressing the side of each chick's head against a shaped plate, fitted with touch sensitive buttons (Fig. 30), triggers a 20-g hypodermic needle injection into the subcutaneous neck tissue. The needle retracts through a hole in the baseplate after each injection.

Liquid vaccine is gravity-fed, by flexible tube, from a remote reservoir into the machine syringe. All syringe electronic and mechanical activation parts are contained in a portable box. Figure 31 illustrates a typical system measuring 19 × 8 × 5 in.

C. Spray Boxes

The immunization of day-old chicks against a number of poultry virus pathogens can be achieved with a coarse spray of vaccine. Between 40,000 and 60,000 chicks can be dosed by two operatives, in 1 hour, by passing standard trays of 100 chicks through a single box (Fig. 32).

Figure 30   Automatic chick injection plate. (Courtesy Select Laboratories Inc., Gainesville, Georgia.)

The underside of the top of the transparent box is fitted with two Venturi nozzles through which a total of 7 ml is dispensed to each tray of 100 chicks. Each dose-delivery sequence begins when the chick tray reaches the end of its slide travel into the box and trips a switch. Compressed air pumps the plunger

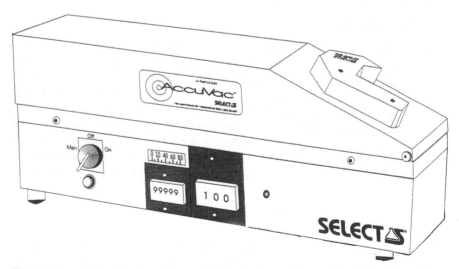

Figure 31   Chick injection system. (Courtesy Select Laboratories Inc., Gainesville, Georgia.)

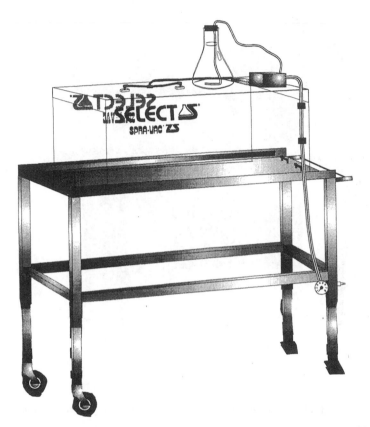

Figure 32 Chick vaccination spray box. (Courtesy Select Laboratories Inc., Gainesville, Georgia.)

of a sterile disposable syringe to fill the nozzle feed tubes; it also produces the spray. An average of 5 sec exposure to spray is required to immunize each 100 chicks.

Spray box makers often change basic design features, such as dose volume or number of spray nozzles, to meet users' particular handling needs and vaccine potency preferences. "In-line" systems of compressed air spray vaccination are more highly automated (Fig. 33). They detect the presence of chicks on a moving belt, by means of an electric eye, and begin spraying instantly. Such systems can be linked to automatic chick counting systems. No full-time operator is needed, only periodic inspection and addition of vaccine to the pressurized reservoir.

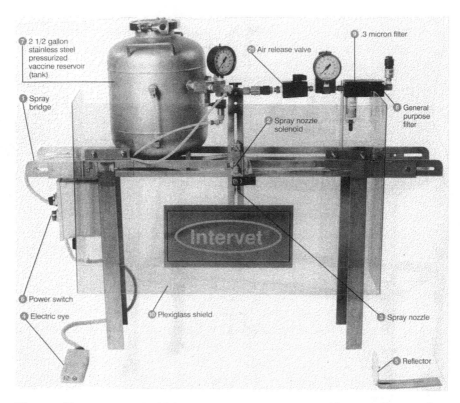

Figure 33 In-line and chick counter conveyer system. (Courtesy Intervet Inc., Millsboro, Delaware.)

D. Transdermal Jet

Since 1970 the "needleless" injection of humans with between 0.05 and 1 ml of liquid dose volumes has been possible by pressurizing medication with carbon dioxide gas. Sudden release, through a jet head pressed against the skin, forces the dose through the skin into the subcutaneous tissue. The depth of transdermal penetration is determined by the amount of gas pressure. Mass vaccination of humans, for example, requires about 300 psi. The advantages for human use are that it is painless; the dose does not pool, as with a needle, in the tissue, but disperses in a fan shape; repeated injections, as with insulin, produce less scar tissue. Figure 34 shows the jet head and pressure release mechanism adapted for veterinary use.

Jet injection of animals has not gained widespread favor largely because of the need to very carefully adjust the pressure for each combination of species and site of dose deposition. For example, intradermal injection of a chicken may require 200 psi while subcutaneous administration to a dairy cow may need 830 psi. A needle and syringe is more versatile and less expensive. Also the jet

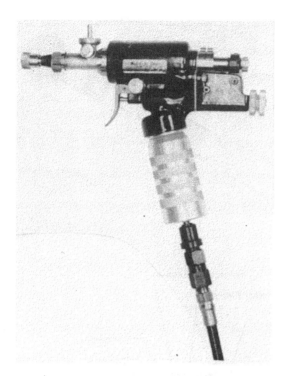

Figure 34 Pow'r-Ject needleless injector. (Courtesy Eidson Assocs Inc., Minneapolis, Minnesota.)

injector is slower, the pressure release noise frightens some animals, and the apparatus is less portable than a needle syringe. However, it is possible, as live-stock production intensifies and the development of low-dose performance en-hancers, for example, yields marketable formulations, that specific product/ species jet injectors could become the administration equipment of choice.

E. Missile Systems

From earliest times blowpipe darts and poisoned arrows have been used to hunt elusive animals, or those too dangerous to approach. Today, when such ani-mals need to be immobilized or treated, precise volumes of liquid drugs can be projected up to 75 yd and automatically injected by a special syringe. A missile system comprises two principal parts; the projectile and the projector. Figure 35 illustrates a reusable syringe and needle projectile.

Syringes, ranging in capacity from 1.0 to 15 ml, can be fitted with needles from ½ in. to 2-1/2 in. in length. Needles are collared or barbed; designs for intramuscular or subcutaneous injection are available. Selection of the size and design of needle is determined by the site of deposition, the thickness of the animal's hide or skin, and the intended method of syringe recovery. Colored

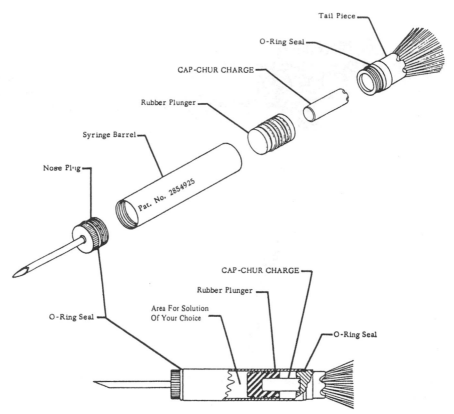

Figure 35   diagram of CapChur syringe. (Courtesy Palmer Chemical & Equipment Co., Douglasville, Georgia.)

tail pieces are useful indicators of different preloaded dosages and also aid in finding syringes in the field.

The barrel is filled, using a standard syringe, with the calculated volume of medication after the correct explosive charge and tail piece have been fitted. The appropriate needle nose plug is then fitted to complete the syringe loading process. Upon impact with the target site, the charge automatically explodes; this powers the plunger forward so that the entire content of the barrel is instantly expelled through the needle.

F.   Projectors

Cross or long bows can be used, with an arrow adapter, to propel syringes. Power projectors use either carbon dioxide gas or, for longer distances, powder charges.

Figure 36 shows a pistol-type projector, powered by carbon dioxide, which can propel each of about 20 syringes 40 ft using one gas bulb. A rifle-type projector can fire each of one dozen (approx.) 1-ml syringes about 30 yd using two gas bulbs. Gas projector performance is affected by the ambient temperature; low temperatures proportionally reduce projection distances. Rifle-type powder projectors (Fig. 37) can propel syringes up to 75 yards. Different 0.22 caliber blank loads are used to vary the range.

G. Dosers

Injection, liquid oral medication, and topical application can all be assisted by the application of power to a hand-held instrument. For example, in the U.S., cattle feedlot processing stations find that such assistance saves time and reduces fatigue.

The doser illustrated by Fig. 38 is powered by compressed air or carbon dioxide and can be fitted with a needle mount, drench nozzle, or pour-on wand. Dose volumes from 2 to 150 ml can be dispensed. Compressed gas enters the doser through the handle to the chamber behind the piston when the trigger button is pressed; this pushes the piston forward to dispense the dose. Releasing the trigger exhausts the gas, which allows the return spring to pull the piston back and refill the barrel. Compressed air operating pressure of between 50 and 90 psi (depending on the dose-volume delivery) requires at least a 3-HP electric or gasoline-powered compressor.

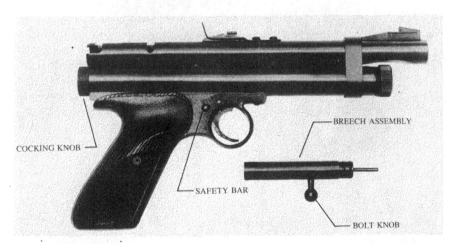

Figure 36 Short-range projector. (Courtesy Palmer Chemical & Equipment Co., Douglasville, Georgia.)

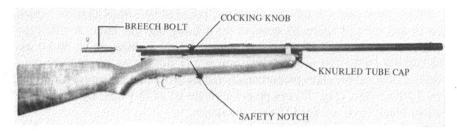

Figure 37 Extralong-range powder projector. (Courtesy Palmer Chemical & Equipment Co., Douglasville, Georgia.)

When air pressure is not available or when portability is required, compressed carbon dioxide from a tank, held in a back pack or placed close to the dosing area, can be used to power a doser. One pound of gas can propel about 500 doses, depending on dose volume.

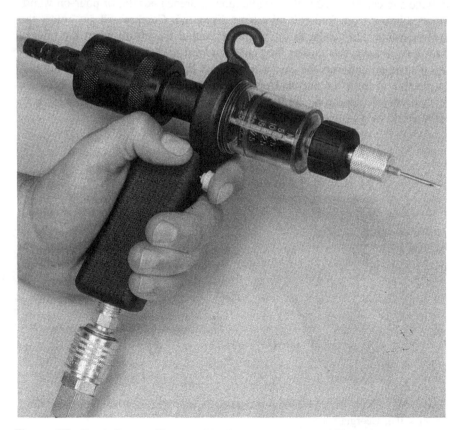

Figure 38 Pow'r Doser. (Courtesy MA Inc., Spring Valley, Wisconsin.)

A powered doser mechanism can also be permanently mounted close to the dosing area (for example, on a chute) and connected by pipes to the compressed air or gas. Flexible tubing from the doser allows the operator to apply topical preparations or give oral doses using a trigger on the "floating" wand or drench nozzle handle.

REFERENCES

1. Operating Instructions for the Auto-Medic II Liquid Proportioner, Product Brochure K4273A, ER. Squibb and Sons, Inc., Princeton, N.J. (1976).
2. Phillips, DSM., United Kingdom Patent 1,484,056 (1977).
3. Cortner, WC., Jr., U.S. Patent 3,902,461 (1975).
4. Sampson, DL., U.S. Patent 3,524,433 (1970).
5. Hanson, RL., U.S. Patent 4,060,083 (1977).
6. Oesterhaus, JH., U.S. Patent 1,325,699 (1910).
7. Brumfield, RE., U.S. Patent 1,868,308 (1932).
8. Stricklen, EA., U.S. Patent 2,170,599 (1939).
9. Wendt, DO., U.S. Patent 2,601,852 (1952).
10. Olson, AM., U.S. Patent 2,621,655 (1952).
11. Weil, JW. and Nutter, WL., U.S. Patent 2,650,593 (1953).
12. Klein, JH. and Tallentire, FL., U.S. Patent 3,238,941 (1966).
13. Mitchell, CN., U.S. Patent 3,238,941 (1966).
14. Corio, NN., U.S. Patent 3,934,584 (1976).
15. Adams, OR., Stomach tube passage in the equine, techniques and uses, Vet. Scope XV(1). (1970).
16. Flymort 24 Spray System, Crawley, West Sussex; Upjohn Ltd., Agricultural Veterinary Division.
17. Cooper's Improved Cattle Spray Race. Berkhamsted, Herts; Cooper, McDougall and Robertson, Ltd.
18. Zaffaroni, A., U.S. Patent 3,993,072 (1976).
19. Zaffaroni, A., U.S. Patent 3,993,073 (1976).
20. Robinson, TJ., U.S. Patent 3,916,898 (1975).
21. Groves, HH., U.K. Patent 1,471,465 (1977).
22. Fuchslocher, R., West Germany Patent 2,614,373 (1977).
23. Homm, R. and Katz, G., U.S. Patent 4,043,338 (1977).
24. Kimberly Clark Corp., West Germany Patent 1, 491,869 (1977).
25. Chvapil, M., Brazil Patent 7,608,701 (1977).
26. Sciarra, JJ., Pharmaceutical aerosols. In: Lachman, L., Lieberman, HA., Kanig, JL., eds. The Theory and Practice of Industrial Pharmacy. 2nd ed. Philadelphia: Lea & Febiger, 1976: 270.
27. Manuf Chem Aerosol News 1977, 48(12), 21.
28. Cohen, MJ., U.S. Patent 4,055,177 (1977).
29. Hurschman, AA., U.S. Patent 4,031,892 (1977).
30. Porter, RE., U.S. Patent 4,031,895 (1977).
31. Maury, JR., U.S. Patent 4,036,225 (1977).
32. Speer, SJ., U.S. Patent 4,040,420 (1977).

33. Guiney, AC., U.S. Patent 4,041,945 (1977).
34. Ampoules Inc., United Kingdom Patent 1,484,600 (1977).
35. Kobel, E., U.S. Patent 4,048,999 (1977).
36. Lindberg, RM. and Raghavachari, S.T., U.S. Patent 4,060,082 (1977).
37. Tischlinger, EA., U.S. Patent 4,059,109 (1977).
38. Pfizer Corp., Belgian Patent 0,738,082 (1970).
39. Lataix, G., U.S. Patent 4,067,440 (1978).
40. Zackheim, EA., U.S. Patent 3,494,359 (1970).
41. Silver, J. and Fuller, G., U.S. Patent 3,052,239 (1962).
42. Silver, J. and Fuller, G., U.S. Patent 3,052,240 (1962).
43. Silver, J., U.S. Patent 3,563,240 (1971).
44. Cheney, PE., U.S. Patent 3,685,514 (1972).

6

Specification Development and Stability Assessment

GARY R. DUKES and DAVID A. HAHN

Pharmacia & Upjohn, Inc., Kalamazoo, Michigan

I. OVERVIEW

For customers to be satisfied, each attribute of a product that is important to its intended use must be consistently acceptable every time the customer buys the product, and must remain within an acceptable range throughout the product's intended useful life. Before these attributes, or critical parameters, can be controlled (see Fig. 1):

1. The intended use of the product must be defined.
2. The critical parameters, the product attributes that are important to this use, must be identified and defined.
3. Approaches to measuring these critical parameters must be developed and validated to provide data that are relevant to the intended product use. The methods must be stability indicating, sensitive to important changes that might occur during storage and use.
4. Changes in the critical parameters with time as the product is stored and used must be studied.
5. A decision about the acceptable range of values for each critical parameter, as measured using the selected method, must be made.

Thus, a specification designed to control a critical parameter is made up of both a measurement method and the associated limits. Stated differently, a par-

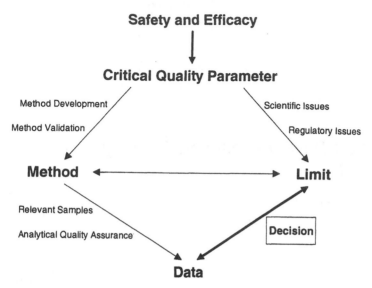

Figure 1   Specifications development process.

ticular set of limits is meaningful only when the critical parameter is measured using the associated method. For regulated products such as animal health products, the decisions about critical parameters, specifications, and the associated measurement methods are subject to review and approval by worldwide regulatory agencies.

Specifications for animal health pharmaceutical products help to assure safety, efficacy, and manufacturing consistency through the product shelf life. To accomplish these goals, the development process must be planned to provide the data and information needed to support good decisions. The set of critical parameters, the methods used to measure these parameters, and the associated limits normally evolve during the development process. This evolution largely occurs in response to stability information and changes in the manufacturing processes for the bulk drug and product as they are scaled up and finalized. Because not all safety and efficacy studies can normally be performed using final marketed product, the specifications also need to link the performance of marketed goods with the materials used during product development. Clearly, specifications are an integral part of the development process and will not serve their needed function if they are thought of as arbitrary controls established at the end of development. It is a truism that quality cannot be tested into a product.

The specifications that result from the development process constitute a critical element of the manufacturer's quality control and assurance systems. They are used to design and validate manufacturing processes to reliably produce acceptable product and to define the product packaging and storage conditions needed to maintain acceptable product through the shelf life. Specifications are an important legal commitment between the manufacturer and the licensing authorities. They can serve as elements of contractual agreements between customers and suppliers. Finally, the development of rational, scientifically justifiable specifications can help focus and integrate the formulation science, analytical science, and manufacturing aspects of product development.

This chapter focuses primarily on small molecule animal health pharmaceuticals. While the general considerations would apply equally to other types of products (i.e., biotechnology-derived products, biologicals, medicated feeds, feed supplements, etc.), specific considerations will vary among product types. Raw materials other than bulk drugs (i.e., excipients, packaging materials) are addressed elsewhere.

Product quality issues are generally the same for animal health and human health products, as reflected in the legal requirements in many markets. In each case, the need for specifications should be determined in light of the intended use. Although some differences can be identified, the differences for small-molecule pharmaceuticals are dwarfed by the similarities. For example, the central drug safety issue for animal health drugs is human food safety, whereas patient safety (i.e., target animal safety) is central for human health drugs. This difference is reflected in differences in specifications for parameters such as subvisible particulates in parenterals. The shelf life of some medicated feed supplements provides another example. Because of the way these products are stored and used, a shorter shelf life can be acceptable for medicated feed supplements than would be appropriate for most human health pharmaceuticals.

II. CRITICAL PARAMETERS

Specifications are meaningful only when they control critical parameters, which are attributes important to the intended use of the product. Efficacy and safety are the fundamental scientific and regulatory product quality issues for animal health pharmaceuticals, and most critical parameters are directly related to these parameters. For example, the critical parameters of a sterile parenteral product would normally include potency and sterility. In addition to critical parameters related to safety and efficacy, additional parameters are normally required to help assure that the product meets regulatory and market requirements for product quality and consistency. The set of critical parameters for a material or product normally evolves little during product development, in contrast with measurement methods and limits, which are addressed later in the chapter.

Importantly, consistency of critical parameters must primarily be assured by producing a material or product using validated, well-controlled processes. Quality must be "built in" to a product; it cannot be "tested in." Critical parameters are measured for materials and products, and the results are compared with associated limits to confirm that the validated process has performed as expected. This measurement and decision process is the subject of this chapter. Discussions of process validation and appropriate in-process testing for bulk drug and products can be found elsewhere (1).

A. General Considerations—Efficacy and Safety

Generally, the questions that need to be asked about a bulk drug, excipient, or product are straightforward:

Is the major component in the product the right thing?
How much (or what concentration) of the major component is present?
What *else* is there? How much?
Are the important physical properties acceptable?
Do the answers to these questions remain acceptable through shelf life?

These questions correspond with the FDA Center for Veterinary Medicine (CVM)'s "identify, stength, quality, and purity" (2), and the European Union (E.U.)'s "quality, activity, and safety" (3).

Efficacy is related to the amount of the therapeutic moiety in each dose of a product and may also be related to physical properties of the bulk drug or formulation which control drug release in vivo. Bulk drug particle size is an example of a physical parameter that can influence drug release. Additionally, physical properties of the bulk drug or formulation can affect the ability to consistently administer the correct dose, such as suspension resuspendability. Efficacy-related critical parameters must remain consistent with those of the batches used for pivotal clinical studies.

Safety requires assuring that the correct therapeutic moiety is present in the correct amount, that the chemical and microbiological purity is acceptable, and, for sustained-release formulations, that the physical properties of the bulk drug or formulation that control drug release are consistent with those of lots used in clinical or other relevant in vivo studies.

B. Bulk Drug

The critical parameters for a bulk drug must take into account the intended use of the material. The selected parameters should support anticipated critical quality parameters of the product. Parameters that might change during storage or use should be considered as stability studies are designed. The summary below includes the critical parameters usually considered for bulk drugs intended for use in common types of products. Normally, the critical parameters required by regulatory agencies are fully consistent with the scientific requirements. However, specific regulatory requirements and guidelines, appropriate pharmacopoeiae, and other sources should be consulted to assure that a complete set of critical parameters has been formed (3–6).

1. Identity

The primary purpose of an identification test is to discriminate against other drugs being used in the facility. While infrared spectroscopy is commonly used for bulk drugs, chromatographic retention matching may be sufficient for this purpose. If chromatographic retention matching is used, a more specific test (or another, alternative test), not necessarily intended for routine use, is often requested by regulatory agencies. Additionally, optical rotation is commonly specified for compounds with chiral centers to help assure that the correct isomer is present.

2. Appearance

This serves as a general check on identity and quality. Color of solution or color of powder is generally specified when color will be a critical parameter for a resulting product.

3. Potency

Major component content is measured and reported in terms of therapeutic moiety, not salt.

4. Optical/Isomeric Purity

For major components with one or more chiral centers, isomeric purity should be addressed. Isomeric purity is best addressed with an appropriate chromatographic assay. However, depending on the details of the synthetic process, specific rotation testing to assure that the correct isomer is present may be sufficient.

5. Chemical Minor Components

Process impurity and bulk drug degradation product levels, both individual and total, are almost always considered critical quality parameters. The specifications should include at least one degradation product that is characteristic of each relevant stressed degradation pathway. When specifications include these marker compounds, they may not need to include all degradation products. Limits are based on batch data, stability data, and drug safety evaluations.

Water and residual solvents are commonly measured. For salts, the counterion should be determined, unless process controls are considered sufficient to assure stoichiometry. Metals are normally controlled if they are used in the synthetic process.

6. Microbiological

Microbiological testing of bulk drug depends on the intended use. For drug to be used in a sterile product, microcount or sterility testing, as appropriate, is generally required as part of product sterility assurance. For bulk drugs to be used in oral products, microorganism count and pathogens should be characterized and the need for testing should be evaluated. Endotoxin (or pyrogen) testing is normally required for drug to be used in a parenteral product.

7. Physical Properties

Physical properties relevant to safety or efficacy in the eventual product are generally controlled. When the bulk drug is present as a solid in the product, particle size, specific surface area, and crystal form may be important physical parameters for assuring acceptable drug release, chemical stability, or physical stability.

C. Excipients

Critical parameters should be established for excipients to assure that they will consistently perform as desired in the product. This testing would normally include identification testing, pharmacopoeial testing if appropriate, and other specific testing based on the product. For example, the USPNF monograph for Peanut Oil includes: Identification, Specific Gravity, Refractive Index, Heavy Metals, Cottonseed Oil, Rancidity, Solidification Range of Fatty Acids, Free Fatty Acids, Iodine Value, Saponification Value, and Unsaponifiable Matter. Detailed discussions of excipients can be found elsewhere (7).

D. Products

Animal health products can take many different forms, as shown by the partial listing in the Table 1, and can be administered using many different devices and approaches. However, the critical parameters required to support product quality

Table 1 Dosage Forms

Pharmaceuticals	Medicated feeds
Solutions	Mashes
Gels	Meals
Suspensions	Pellets
Pastes	Crumbles
Powders	Blocks
Tablets	Liquids
Capsules	Drenches
Boluses	
Creams	
Ointments	
Timed-release mechanisms	

have much in common among these product types. The summary below includes the usual critical parameters, with some discussion about applicability to broad classes of products, rather than attempting an exhaustive catalog by product type. Normally, the critical parameters required by regulatory agencies are fully consistent with the scientific requirements. However, specific regulatory requirements and guidelines, appropriate pharmacopoeiae, and other sources should be consulted to assure that a complete set of critical parameters has been formed (3–6). The general requirements in the European Pharmacopoeia explicitly apply to veterinary products. Parameters that might change during storage or use should be considered as stability studies are designed.

1. Identity

The primary assurance that the correct active ingredient is used in the product comes from following standard operating procedures (SOPs) and Good Manufacturing Practices (GMPs). As for bulk drug, the purpose of an identification test is to discriminate against other drugs being used in the facility, and chromatographic retention matching is generally sufficient for routine use. A more specific test (or another, alternative test), not intended for routine use, is often requested by regulatory agencies. Additionally, identification tests for all excipients in the product are required in the E.U., unless the product is manufactured within the E.U.

2. Appearance

This serves as a general check on identity and quality. Color of solution is normally specified in the E.U.

3. Potency

Major component content should be reported in terms of therapeutic moiety, not salt—the same as for bulk drug. Usual limits are ± 5% at time of release. The lower limit through the end of shelf life can be lower, if justified by stability data. There is a general regulatory requirement to produce products at 100% of label potency at the time of release, with any overages included only to counter documented manufacturing losses.

4. Content Uniformity

This is clearly a critical quality parameter for all dosage forms other than solutions, regardless of whether they are single-dose or multiple-dose products. However, no explicit regulatory requirements are apparent for multiple-dose suspensions. Usual limits are based on pharmacopoeial requirements for single-dose products.

5. Degradation Products

Synthetic process-related impurities and excipient-related impurities do not need to be specified or qualified in the drug product, unless they are also degradation products or arise from drug-excipient interactions. Degradation product specifications should include at least one degradation product that is characteristic of each relevant stressed degradation pathway. When specifications include these marker compounds, they may not need to include all degradation products. Limits are based on batch data, stability data, and drug safety evaluations.

6. Microbiological Properties

Sterility testing is generally required for sterile products. However, because of the small number of units that can feasibly be tested, sterility testing provides little statistical sterility assurance and is primarily intended to detect gross contamination. Because sterility assurance arises primarily from process design, validation, and control (8), sterility evaluation can frequently be defended based solely on verification that the critical aspects of the sterilization process remained within the range of validation. This is referred to as parametric release. For a parenteral product, bacterial endotoxin testing is normally required. For oral products, microorganism count and pathogens should be characterized and the need for testing evaluated. The content of any preservative is a critical quality parameter, with the limits generally established to assure acceptable antimicrobial effectiveness testing results.

7. Physical Properties

Physical properties may be critical parameters and should be evaluated and controlled as appropriate. Drug release testing (or testing of parameters that control drug release) may be indicated for solid oral dosage forms, or for any dosage form that includes a solid-phase therapeutic moiety or which provides sustained release. Tablet weight variation, thickness, hardness, and tablet or capsule disintegration time are commonly specified parameters. For a suspension product, resuspendability, or ease of redispersion of the settled solids, is normally a critical parameter. For aqueous solution or suspension products, pH may be critical. For a solution or reconstitutable parenteral product, clarity and particulate matter testing may be indicated. For a reconstitutable product, reconstitution time testing should also be considered.

8. Other

Water content may be important for the stability or performance of some products, such as tablets and lyophilized powders. Volume of injection or deliverable volume is generally measured, where appropriate, to assure adequate fill volume.

III. MEASUREMENT ISSUES

A specification is the combination of the analytical procedure used to generate a result and the standard against which the result is judged. This link between method and limit is essential to the definition of specifications. The method must provide a meaningful measurement related to a critical quality parameter, and must be capable of supporting the associated limit. Conversely, when establishing a limit, the capabilities of the associated analytical method must be taken into account.

For example, microbiological and chromatographic methods to determine the potency of a product in fact measure slightly different things. The microbiological method measures the aggregate antimicrobial activity for a particular laboratory microorganism, which may include a significant contribution from related minor components. A chromatographic method for the same product will probably be specific for the major component. The meaning of a particular numerical limit will vary depending on which method is used, and limits must therefore be evaluated and established for a specific method. Additionally, the appropriate specification for the product potency will, of course, depend on the specific attributes of the product.

Method selection, development, and optimization must all be directed toward achieving a meaningful measurement each time the procedure is performed, regardless of day, laboratory, analyst, equipment, reagents, etc. Analytical method development is always constrained by the analyte, the amount or concentration of the analyte that is of interest, and the sample matrix (9–14). The method variability and the number of samples to be assayed will constrain the associated limits the method is capable of supporting. Appropriate sampling is also a critical part of this consideration (15). For a method to be part of a scientifically useful specification, it must measure the analyte in a way that is related to how the product will be used. For example, the potency of a multiple-dose suspension product can be measured in a meaningful way by assaying a dose prepared in the same way that the customer will prepare a dose. An assay of the total active ingredient contained in a vial would likely not be as meaningful in this case.

In method validation, the capability of the analytical method is characterized to support specification development. Validation sufficient to assure that the data are scientifically meaningful must be completed before data from the method are used for any purpose. Additionally, regulatory agencies require validation at a level appropriate to the stage of development before data from a method are used for clinical or marketing batch release or stability testing of a product or material. While appropriate validation is required for clinical supplies release, much more is normally known about the capabilities of analytical methods by the time an NADA or similar regulatory documents are being prepared. Additionally, the methods themselves may change or evolve during product development or after the product is on the market. This can occur because of changes in the chemical synthetic process or the formulation, evolution of the measurement objectives, changes in analytical technology, or because the method needed to be improved to meet the objectives. The need for revalidation should be considered when any change is made to the method or product.

This section discusses method validation and quality control for chemical assays. Method selection, development, and optimization are discussed extensively elsewhere. Useful starting points can be found in references 1–6, and guidance can be found in numerous analytical chemistry reference books, textbooks, and literature articles. While many general method validation considerations are similar for chemical, microbiological, and physical tests and assays, the specifics of method validation for microbiological tests and assays (16) for parameters such as microorganism count, pathogens, bacterial endotoxins, and sterility, and physical tests for particulates and related parameters (17) are beyond the scope of this chapter.

A. General Issues

1. Scientific Issues—Assays That SPARqle

Method validation can be viewed as a microcosm of the entire specifications development process, in that critical aspects of method performance are identified and controlled, appropriate methods are used to measure the performance of the assay, and the results are compared with limits to decide whether the analytical method is suitable for the purpose. Thus, method validation includes both measurement of method performance and associated acceptance criteria. Both statistical and practical significance must always be considered in comparing the validation data with the acceptance criteria. Neither statistical nor practical significance can meaningfully be examined in isolation. It is not uncommon for a statistically significant difference to prove to be too small to have any practical impact, considering the use of the data. Conversely, if the method variability is too great for the intended purpose, differences that are large enough to be important, considering the use of the data, can turn out to be statistically insignificant.

Critical assay performance characteristics and general validation criteria can be suggested for common types of assays, but in each case the fundamental question is whether the method provides a meaningful measurement. Importantly, the method must be capable of supporting development of a reasonable limit and subsequent quality-control measurements using that limit. Additionally, the set of validation experiments and the criteria used must be acceptable to regulatory agencies. The measurement objectives, the challenges that must be overcome to meet the objectives, and the regulatory environment, all can vary among products, resulting in legitimate differences in validation approaches and criteria. The information in this chapter, together with validation practices from the literature (18–22) and the general guidelines issued by regulatory and compendial agencies (23–29), should serve as starting points.

The assay performance characteristics of interest normally include specificity, precision, accuracy, and ruggedness. If the method is also quick and easy to perform, and has a linear response, there are clear operational advantages and ruggedness may be enhanced. This set of validation parameters can easily be remembered using the acronym "SPARqle," which standards for Specific, Precise, Accurate, Rugged, quick, linear, and easy.

The SPARqle acronym provides a convenient mnemonic both for the critical validation issues and for the appropriate chronological order in which the issues should normally be addressed. A method must be appropriately specific before it makes sense to evaluate any other dimension. If an assay method doesn't have the needed precision for the intended purpose, it isn't possible to assess accuracy. And if a method isn't suitably specific, precise, and accurate, the ruggedness is not of much interest. The other issues in the acronym are secondary. A method that is quick and easy may be more rugged. A method with a linear response will generally be quicker and easier because of the simpler calibration experiments. However, linearity is not required as long as the response function is known and understood scientifically, and appropriate calibration is performed.

Before the performance of the analytical method can be characterized, the method must be detailed in a written procedure. Ideally, this procedure should provide enough information that the method can be successfully run by a person who has never run it before, provided the person is already familiar with the analytical techniques that are used. Appropriately detailed written procedures are critical to method ruggedness.

Assay performance must be evaluated during validation in such a way that the results are representative of what will be obtained during normal use of the assay for decision making. This often means that the performance must be assessed in a lab other than the development lab (i.e., in the production quality-control laboratories). This can be achieved by involving other laboratories in the validation experiments, whenever possible, and by including appropriate analytical data quality-control checks in the assay procedure. Well-designed analytical system suitability tests, together with ongoing laboratory analytical data quality control programs such as use of laboratory standards, assure that all important aspects of the procedure are performing the same as during method validation.

Development of a rugged analytical method and method validation are *necessary* to assure that the analytical method is capable of providing valid results, which can support the needed decisions. However, method development and validation alone are *not sufficient* to assure that each result is actually valid. Representative samples are needed. System suitability tests are required to assure that all critical parts of the analytical method are performing acceptably each time the assay is run. Additionally, laboratory quality assurance and quality-control processes are needed to prevent errors whenever possible and to make sure that any errors that do occur are corrected before the results are reported.

a. Specificity. The analytical method should provide accurate results for the analyte of interest in the presence of potential interferences which might reasonably occur in the sample. It is also important to assure that the analyte signal has been correctly identified. Specificity is an important parameter for every analytical method, but is often addressed as part of the accuracy experiments for nonchromatographic assays. This specificity discussion will focus on chromatographic methods, but most of the points could easily be adapted to other analytical methods.

An analytical method must normally be capable of detecting expected changes in aged samples. Such a method is called stability-indicating. The major changes expected during stability studies include chemical degradation of the major component and placebo, possibly including interactions between the active ingredient and placebo components.

Potential interferences include potential process impurities and degradation products of the analyte of interest, other active or inactive ingredients in the formulation, and potential impurities and degradation products of the other components. For example, in chromatographic major component assays, potential process impurities and degradation products should be separated from the major component peak, and, in the case of impurity assays, also separated from the other impurities and degradation products. Potential interferences from the analytical reagents must be assessed by chromatographing a reagent blank.

Selectivity with respect to process impurities is normally evaluated by chromatographing samples of known process intermediates and other potential process impurities. Mother liquors and other process samples can also be used in this evaluation. Selectivity with respect to matrix components can be evaluated using samples of formulation components. In some cases, the formulation without the analyte of interest should be carried through the entire intended manufacturing process and used to assure that any products formed during manufacture do not interfere with the assay.

Potential degradation products are normally determined early in development using a variety of stressed degradation studies of bulk drug in solid and solution phases. Stresses such as heat, light, acidic and basic solutions, and oxidizing conditions are commonly used. Stressed degradation experiments should be broadened to include the formulated product at an appropriate stage of development, although not all stressed conditions are applicable to each possible formulation type. When the degradation pathways for the major component, formulation excipients, and formulation have been studied, selectivity with respect to known or expected degradation products can be evaluated using authentic samples if they are available. Aged samples of the formulation made without the analyte of interest should be used, as appropriate, to determine whether degradation products of any placebo components interfere with the analytes.

Evaluation of material balance in stability studies both at the intended storage temperature and at accelerated conditions can detect whether a significant degradation product (which does not coelute with the major component in the major component assay) is missed or inaccurately quantitated in the degradation products assay, although this approach is generally limited by the variability of the major component assay. The sorts of issues that material balance evaluation can discover include degradation products which are not eluted from the chromatographic column, or which have a lower response than that of the analyte of interest with the detection method being used.

Situations where a component does not elute in a useful region in the primary assay can be identified by comparing results with a second assay which uses a different separation mechanism. Thin-layer chromatography can be valuable in this regard because all components are available for detection, even those that fail to migrate or which migrate with the solvent front. Components that are not seen in the primary assay because of inadequate detection response can be found by substituting a different detection approach that is likely to provide a similar response for widely varying analyte structures (i.e., flame ionization detection, or thin-layer chromatography using spray reagents). Material balance can be assessed using each of the available minor component assay methods to seek discrepancies (30).

Experiments with authentic samples cannot definitively rule out coelution of an unknown or unexpected interference with an analyte of interest. To address this issue, analyte peak homogeneity is normally assessed in both fresh and degraded samples to provide additional confidence in method selectivity. One common approach to assessing peak homogeneity uses a detection method that provides structural information, such as mass spectrometry or UV diode array absorbance detection (31). The resulting data are assessed to determine whether they are consistent with a single component peak, often using chemometric methods. An alternative approach is to collect the peak of interest, either in its entirety or in several slices, and subject it to further analysis using a different analytical method. It is important to note that no approach can prove the homogeneity of a chromatographic peak, and that most experiments are not normally capable of finding small amounts (i.e., tenths of a percent) of structurally similar minor components under a major component peak. However, the overall set of selectivity experiments normally provides high confidence that there are no practically significant interferences with the analyte peak.

b. Precision. A number of different terms are used to refer to different aspects of method precision, or variability. The precision of the method is the closeness of agreement between a series of measurements obtained from multiple sampling of the same homogeneous sample. Variability is a broader term which includes both method precision and other sources of variability among the results for a product.

This section primarily addresses repeatability, which is the precision under the same operating conditions over a short interval of time. This aspect of precision must be addressed before meaningful accuracy experiments can be performed. Method repeatability is the lower limit for assay variability. Repeatability is studied by assaying a homogeneous sample a number of times on a single day without varying the method, reagents, analyst, or other method parameters. For procedures that are a combination of several operations, the repeatability of each operation should be assessed. If these are assessed independently, propagation of error can be used to estimate the repeatability of the final result. In any event, the repeatability of the final result should be compared with the anticipated limit to determine acceptability.

The overall method variability will not be smaller than the repeatability, but the overall variability will likely be much greater than the repeatability because of variation in sampling, sample storage and handling, sample preparation, reagents, equipment, analysts, and ancillary measurements (such as density). Additionally, variation in composition among samples can lead to high variability. The section on *ruggedness* addresses the aspects of variability that must be addressed to assure that meaningful results can be obtained for the life of the product in a quality-control laboratory setting. This longer-term variability is what is used to assess whether the analytical method is compatible with the desired limits.

The detection limit and the quantitation limit are largely determined by method precision. The limit of detection is often taken as the minimum level at which the analyte can be reliably detected with a particular degree of confidence, based on a statistical approach (32,33). The limit of quantitation is the lowest concentration or amount of analyte that can be quantitatively determined with suitable precision and accuracy.

A commonly used criterion for limit of detection is a signal:noise ratio of 3:1. This is often referred to as the decision limit (32):

$$L_D = x_{bl} + 3s_{bl}$$

where x_{bl} = the mean value of the blank response, and s_{bl} = the standard deviation of the blank (the RMS baseline noise). If the standard deviation is defined with many degrees of freedom, and if the noise is normally distributed, this definition leads to a probability of only 0.13% that a sample actually containing none of the analyte would be interpreted as having the analyte present. However, the probability of *not* detecting the analyte when it is actually present in the sample at L_D is 50%, which is not normally acceptable.

A better definition of L_D is given by the equation for detection limit in Reference 32:

$$L_D = x_{bl} + 6s_{bl}$$

Here, with the same assumptions as above, the probability of detecting the analyte when it is present at L_D is 99.87%, which is normally acceptable.

The limit of quantitation L_Q is given by the equation for determination limit in Reference 32:

$$L_Q = x_{bl} + 10s_{bl}$$

L_Q defined in this way, given the assumptions defined above, will result in a 10% relative standard deviation for a measurement made at the quantitation limit.

This approach based on statistical evaluation of baseline noise has its limitations. It was developed for two-level methods such as cuvette-based UV/visible absorbance measurements, where the average value for a single-level sample measurement is compared with the average value for a stable baseline. Additionally, variability is assumed not to be a function of signal level. For chromatographic assays, where peak area is normally the primary measurement, rigorous application of this approach to detection and quantitation limits is not straightforward because there will be some averaging of noise across the peak. Additionally, integration of peaks near the detection limit will likely be sensitive to baseline noise, and this is not considered in the statistical approach (34). However, use of peak height to estimate detection and quantitation limits is a reasonable approximation in most cases. Limits of detection should be confirmed using appropriate standards when this is important to the measurement objectives.

Limits of detection or quantitation are not absolute characteristics of the assay method, because baseline noise can vary with time, equipment, reagents, etc. This issue is normally addressed using appropriate system suitability tests, either checking detection of an appropriate standard at the detection limit, or assuring that the major component response is sufficiently high and baseline noise is sufficiently low.

c. Accuracy. Accuracy is the agreement between a measured value and the true value. Method bias should be negligible, considering the use of the assay. Specificity, or absence of interfering compounds, is a foundation of accuracy. Sampling to get a representative sample is also critical to accuracy.

Accuracy can be assessed, at least in principle, by comparing the results with results from an independent method which is well characterized and known to provide accurate results. While this is rarely a practical approach for initial method development, it can sometimes be useful as one element in accuracy evaluation when an existing method is being replaced with an updated or upgraded method. The conclusions that can be drawn from this approach are strongest if the methods rely on different analytical principles, and are as different as possible from the sample preparation step onward. The use of standard reference samples, when available, is a variation on this approach.

Accuracy includes complete (or consistent) recovery of the analyte from the sample matrix. Spiked recovery is normally performed to assess recovery. Complete absolute recovery should be sought during development (the usual limits are 98% to 102% recovery for pharmaceutical dosage form major component assays).

Spiked recovery is assessed over at least the range of concentrations or amounts required by the purpose of the assay. For example, a major component assay would normally be evaluated over at least 80% to 120% of the label amount, with appropriate replication. Spiked recovery data are normally presented as a graph of percent recovery versus amount spiked. Practical and statistical significance of any deviation from complete recovery is assessed. The deviations from complete recovery should not show any apparent patterns, and should appear to be randomly distributed.

Spiking is normally done on a weight percent basis. Impurities assays based on area percent calculations, with correction for relative detection response, do not provide results directly in weight percent because not all components are normally seen in the impurities assay (e.g., counterion, water). This should be taken into account in calculating recovery results.

Absence of influence of nonlinear effects from processes such as protein binding and adsorption to sample containers or analytical apparatus such as filters should be assessed. The variability of recovery should be addressed as a function of sample or matrix variability, reagent variability, analyst technique, etc., as part of ruggedness testing. Separate evaluation of extraction efficiency or other components of recovery can be useful in troubleshooting, but are not normally required for validation.

The matrix used for recovery studies should be carefully considered. For products such as compressed tablets, changes in the matrix during formulation may impact on recovery. Additionally, in some cases recovery may be reduced in aged stability samples. Alternatives to a spiked recovery study which may be more appropriate in some cases include dilution studies on real samples and the standard addition method. Use of radiolabeled analyte as an analytical development tool may help address recovery issues in some cases.

Acceptable material balance for stability and stressed degradation studies provides added assurance of selectivity for the potency and impurities assays used. Absolute recovery from column should be assessed, when that might be an issue. This can be done by comparing total detection response for the same sample injected with and without a column in the system (35).

Accuracy is often assessed at the same time as the method response function (i.e., linearity) is evaluated. It is important to note that accuracy does not necessarily require a linear response. However, if the method is expected to have a linear response, a nonlinear response is likely an indication of an accuracy problem. If the calibration approach used assumes linearity or any other method response function, the accuracy of the results depends on the reproducibility of the response function.

For methods that rely on a comparison to a reference material, the accuracy of the results is no better than the characterization of the reference standard. Thus, reference standard characterization is the foundation of accuracy. This is also relevant to impurities standards used to determine the impurity's response factor, the detection response relative to the major component. Response factors for impurities can be determined by measuring the response as a function of impurity concentration. Alternative detection methods likely to give a similar response for structurally dissimilar compounds can be applied to estimate impurity response factors when authentic samples are not available.

d. Ruggedness. A number of different terms are used to refer to different aspects of method variability. This section addresses robustness, intermediate precision, and reproducibility—the aspects of variability which must be addressed to assure that meaningful results can be obtained for the life of the product in a quality-control laboratory setting. *Robustness* is a measure of a procedure's capacity to remain unaffected by small but deliberate variations in method parameters. *Intermediate precision* measures within-laboratory variation: different days, different analysts, different equipment, etc. *Reproducibility* is between-laboratory precision. The section on precision addresses the aspects of variability that must be addressed before meaningful accuracy experiments can be performed. Good analytical data quality control is a foundation for method ruggedness, and is briefly addressed in a separate section below. A clear and complete assay procedure is needed before interlaboratory ruggedness experiments can be conducted. Sampling is a critical part of the measurement process, and sampling procedures should be defined during process validation.

Experiments should be designed to discover the critical variables in the assay which must be controlled to limit variability due to all issues except the product itself. Statistically designed experiments are recommended for ruggedness assessment (36). With an appropriate design, ANOVA analysis can be useful in sorting out the significance of different variables.

Ruggedness should be assessed for all procedures from sampling to final quantitation with respect to operational and environmental variables, including: variability among days; variability among analysts who may have slightly different technique or timing; age of the sample preparation; variation in reagents, including mobile phase composition; variations in chromatographic columns with time or among batches or manufacturers; differences among different instruments, such as differences in wavelength accuracy and bandwidth among HPLC absorbance detectors, or differences in accuracy of HPLC pump solvent gradient or flow rate; and differences between labs, which might be related to technology transfer issues, temperature, reagent source, or other factors. Additionally, ruggedness should be assessed with respect to sample attributes including sample age and condition, and sample or matrix variability.

Critical parameters for ruggedness identified in this testing should be controlled by careful drafting of the procedure and design of appropriate system suitability tests and other analytical quality assurance systems. For example, if the sample preparation is not stable for at least 24 h, a procedure limitation is normally needed.

The assay performance data reported in the regulatory documents should be as representative as possible of the results obtained in the product QC laboratories. Normally, the results obtained from a laboratory standard over a period of time provide the best available estimate of method variability for use in development of limits and regulatory documents. The laboratory standard is a material that is acceptably homogeneous, has a sample preparation procedure similar to that for actual samples, and is normally included in each potency assay run as a data quality-control tool.

e. Linearity The response characteristics should be studied over the range of responses that is relevant to the intended use of the method. For many systems, a linear response is expected over a useful range, meaning that the procedure is expected to provide test results that are directly proportional to analyte concentration. When this is the case, a single point calibration may be used. Nonlinear calibration should be used only when a nonlinear response function is expected for the technique used. For example, thin-layer chromatography using reflectance detection has an inherently nonlinear response, justifying use of a calibration curve.

The linearity is acceptable if the errors caused by using a linear calibration curve with zero intercept are not significant, considering both statistical and practical significance. Acceptable recovery over the intended operating range of the assay using a linear response function is an appropriate practical criterion for linearity.

Statistical approaches to evaluating linearity include correlation testing, examining the slope and intercept for agreement with expected results with a specific degree of confidence, lack-of-fit testing using a linear model, and assessment of the statistical significance of nonlinear terms from a fit to an alternative model. However, it is often difficult to interpret the practical importance of any deviations from linearity detected by such statistical approaches.

Linearity is normally assessed as part of the spiked recovery accuracy evaluation over the intended working range of the assay. For major component assays, linearity testing covering 50% to 150% of the target concentration is common. For area percent impurities assays, linearity is required from at least 0.05% to 110% of the target major component concentration. Otherwise, use of impurity external standards is recommended.

2. Regulatory Guidelines

The primary regulatory guideline addressing method validation is the FDA/ICH draft guideline (23,24). The main points of this guideline are given in Table 2. This guidance is in general agreement with the USP guidelines for validation of compendial methods (25), E.U. guidelines on analytical validation (26,27), and FDA/CVM analytical validation guidelines (28,29).

B. Specific Assay Issues

1. Reference Standard Characterization

Reference standard characterization is the foundation for accuracy of any method that relies on comparison with an external standard. The degree of characterization depends on the stage of development. At the earliest stages of development, a minimally characterized provisional standard is appropriate. Minimum testing for such a standard includes water and residual solvent content to estimate purity, hygroscopicity characterization to determine appropriate packaging and handling, and other testing as appropriate. By the time of clinical lot release, more extensive characterization should be complete. By the time of regulatory filings, the reference standard should be a well-characterized material of the highest available purity which has demonstrated stability relative to degradation and changes in water content for a minimum of 1 year after evaluation. Additionally, complete structural characterization is needed for regulatory filings.

Normally, a working reference standard is characterized for laboratory use. Ideally, this material should have a purity of at least 99%, because this makes accurate characterization easier; however, this is not always possible. Special synthesis or purification should be considered if needed. For chromatographic reference standards, the best salt can be selected for use as a reference standard based on purity or ease of handling, whether or not the same salt is used in the dosage form. An adequate supply of working standard is needed. When enough material to satisfy several years' projected needs can be obtained, the need for a replacement standard can be delayed. An ultrapure primary reference standard may be used as a tool in characterizing the working standard, but is not normally available for laboratory use.

Characterization of the material to support use as a working reference standard normally includes evaluation of chemical stability and hygroscopicity. These data define proper packaging, handling, and storage conditions. When possible, "as is" usage is preferable to requiring equilibration or drying before use. A hygroscopic material can be provided in single-use containers which are imper-

Table 2 FDA/ICH Draft Guideline on Methods Validation

Method characteristics (see text for definitions)	Type of analytical procedure			
	Identification	Impurities purity test		Assay: content/potency; dissolution: measurement only
		Quantitation	Limit	
Accuracy	–	+	–	+
Precision				
Repeatability	–	+	–	+
Intermediate precision	–	+[a]	–	+[a]
Reproducibility	–	–[b]	–	–[b]
Specificity	+	+	+	+[c]
Detection limit	–	+	+	–
Quantitation limit	–	+	–	–
Linearity	–	+	–	+
Range	–	+	–	+

Key: + Signifies that the parameter is normally evaluated. – Signifies that the parameter is normally not evaluated.
[a]In cases where reproducibility has been evaluated, intermediate precision is not needed.
[b]May be needed in some cases.
[c]May not be needed in some cases.

meable to moisture to allow as-is use under humidity conditions similar to those used for packaging the material. Homogeneity should also be assessed. Solubility in likely assay solvents should be determined, particularly if the salt selected for the reference standard is different from the bulk drug. A full structure proof is normally needed for the working standard.

The primary purity assignment is normally based on subtraction of minor components, because this approach results in smaller uncertainty in the final result than do most other approaches. The variability of each method used to determine impurities should be determined, and the resulting uncertainty in the purity assignment should be calculated. The following issues are normally addressed during reference standard characterization:

1. For materials isolated as salts, the quantity of the counterion is subtracted as a "minor component," based on experimental determination of the stoichiometry. This supports the usual reporting of major component content in terms of the therapeutic moiety, rather than the salt.

2. The methods used to quantitate impurities should be capable of detecting and quantitating the impurities expected from the synthesis of the reference material. Impurities similar in structure (synthesis or degradation-related) are usually determined by chromatographic methods. To ensure separation and detection of impurities, the use of closed-system methods, such as thin-layer chromatography, and detection schemes which provide similar response for a wide range of molecular structures should be considered. Multiple separation systems should be employed to help assure that all related impurities are found.

3. Impurities dissimilar in structure; for example, contaminating salts, water, residual solvents, residue on ignition, and organic/biological residues not otherwise determined are also subtracted. Methods used in these determinations should be capable of detecting and quantitating the impurities expected from the synthesis of the reference material.

4. Relative assay versus a primary standard, a highly purified and well-characterized standard that is available in only limited quantities, is commonly used to support the minor component test results for purity assignment.

5. Material balance should be assessed for several batches of bulk drug. Any discrepancies might indicate incomplete or inaccurate quantitation of minor components either in the reference standard or in the bulk drug batches.

6. Tests such as nonaqueous titration, thermal gravimetric analysis, differential scanning calorimetry, or phase solubility should be considered, when applicable, to support the proposed purity assignment with a direct estimate of purity. However, the limitations of these approaches should be considered carefully before they are applied.

The reference standard should be reevaluated periodically to assure that the purity assignment has not changed due to degradation, gain or loss of water or other volatile components, etc. Replacement standards are normally characterized mainly by comparison with the standard they are replacing. Additionally, it is prudent to repeat the key minor components assays to confirm material balance. Impurity reference standards for use in external standard impurities assays should be characterized to an appropriate extent. However, less extensive purity characterization is normally needed because the relative uncertainty in the assay result does not normally need to be as small as for a major component assay.

2. Major Component Assays

General points to consider for validation of major component assays for bulk drug or product are given in Table 3. Chromatographic methods are normally chosen for major component assays because they are capable of providing high selectivity with suitable precision. Liquid chromatography is the ususal choice because it can be performed at ambient temperature, limiting the potential for analyte degradation during the assay. Gas chromatography is also frequently used with compounds that are sufficiently volatile and stable. Thin-layer chromatography can also be used, although precision is frequently lower than for HPLC.

Major component assays normally use an external standard. Internal standards are not recommended in most cases because coelution with impurities or degradation products would bias the major component assay. Historically, internal standards were used to normalize injection volumes while HPLC autosamplers were in earlier stages of development. With the improved design and reliability of modern autosamplers, an internal standard is not normally required for this purpose. An internal standard can still be a useful part of some dosage form sample preparations. If an internal standard is used, the procedure should include a sample preparation made without internal standard to assure that nothing is hidden.

Microbiological assays are sometimes used for antibiotics and can be useful when the active ingredient is a mixture of several components. However, microbiological assays generally have higher variability, limiting the specification range and product shelf life that can be supported. Additionally, microbiological assays measure all components that have activity against the selected laboratory microorganism, and thus may not provide the degree of selectivity required for a stability-indicating assay, depending on the activity of process impurities and degradation products. It is important to note that this total microbiological activity is likely not clinically relevant, because the microorganism used in the assay is normally a convenient laboratory organism, not a relevant pathogen.

Titration assays are occasionally used and can provide higher precision than a chromatographic assay. However, suitable selectivity is normally hard to achieve because it depends on the absence of the titratable group in process impurities and degradation products.

Table 3 Major Component Assay—Validation Points to Consider

Validation attribute	Clinical stage	NADA stage
Specificity	Bulk drug and formulation: Synthesis precursor(s)	Bulk drug and formulation: Process intermediates Stress-degraded samples Major component peak homogeneity Contaminants from packaging materials Formulation: Formulation matrix components
Precision	Bulk drug and formulation: Chromatographic method; same day, analyst, reagents, etc.	Bulk drug and formulation: Chromatographic method and sample preparation method
Accuracy	Bulk drug and formulation: Provisional reference standard Formulation: Recovery from formulation matrix	Bulk drug and formulation: Fully characterized reference standard
Ruggedness	Bulk drug and formulation: Clear written procedure and appropriate system suitability tests	Bulk drug and formulation: Implementation data Laboratory standard data Clear written procedure and appropriate system suitability tests
Linearity	Bulk drug and formulation: response function characterized over limited range (80-120%)	Bulk drug and formulation: Response function characterized over sufficient range to support specifications; compare with theoretical response

3. Impurity and Degradation Product Assays

General points to consider for validation of minor component assays for bulk drug or product are given in Table 4. Assays for related impurities and degradation products are almost always chromatographic. Normal or reversed-phase liquid chromatography and size exclusion chromatography are all commonly used. Gas chromatography can also be used for sufficiently stable and volatile compounds. An area percent measurement based on major component peak area or total peak area is normally appropriate, provided that the method is linear over a sufficient range and relative detection response factors can be determined. Otherwise, the use of external standards for impurities is recommended.

Thin-layer chromatography is commonly used as a tool to investigate impurities and degradation products during method development because all compounds are present on the plate after development (unless they degrade), while highly retained compounds may not elute from an HPLC column within a useful time window. Additionally, all compounds can be detected using appropriate detection reagents, while differences in UV/visible absorbance can prevent detection of some components in HPLC with absorbance detection. Thin-layer chromatography can also be used for release assays. However, detection limits are frequently higher than in HPLC, chromatographic efficiency (and thus specificity) is normally poorer, and the precision may be poorer.

Optical purity is normally measured chromatographically whenever measurement of optical isomers is needed. Optical rotation is usually used to confirm that the correct enantiomer (or the racemate) has been obtained, and is sometimes used to estimate optical purity. For a compound with a single chiral center, this is a reasonable approach, although the specific rotation is relatively insensitive to low levels of the optical isomer. However, for compounds with multiple chiral centers, optical rotation provides little information on isomeric purity because the optical activity of each possible isomer is not normally known.

4. Selected Other Assays

a. Drug Release Testing. The need for in vitro dissolution testing for immediate-release animal health products should be determined in light of how soluble and well-absorbed the drug is, and whether dissolution has any relationship to bioavailability. For sustained-release products, drug release is normally a critical quality characteristic and is normally a requirement to gain regulatory approval. Although this section focuses on solid oral dosage forms, many of the concepts discussed in this section are also being applied to semisolid dosage forms such as topicals (creams, ointments, gels), transdermals, suppositories, and others. A properly designed dissolution test and specification minimizes the risk of bioinequivalence among batches both during formulation and process development and optimization and after approval. To meet these objectives, the in vitro dissolution test should be predictive of in vivo bioavailability (37).

The objective of dissolution test development and optimization is to obtain a test that can discriminate among formulations that differ in in vivo bioavailability (37). When this is not possible, discrimination is sought among formulations that differ in key variables such as excipient amounts and/or processing parameters. Optimum discrimination requires both maximum differences in dissolution rate among formulations and low variability for each particular formulation. Such a test will have the greatest probability of being related to in vivo bioavailability.

Initial test conditions for use during formulation development are selected based on the physicochemical properties of the drug, the formulation design, and the intended dose (37). Key drug properties include the pKa, the solubility as a function of the pH and surfactant concentration of the dissolution media, and the solution state stability of the drug as a function of pH. The dissolution medium and volume are selected such that USP "sink conditions" are met (i.e., the solubility of the drug is equal to or greater than triple the concentration of a completely dissolved tablet). Development issues are summarized in Table 5 (37).

The method is optimized using two or three lots that exhibit differences in in vivo absorption (if available) or lots that vary in formulation composition or manufacturing parameters. In these studies, the medium composition (pH, ionic strength, surfactant concentration and type) and/or the hydrodynamics (apparatus type or agitation, rotation speed) are modified in order to determine their effect on the dissolution rate of selected tablet lots (37). Because of the large number of possible combinations of variables, statistical experimental designs are often employed. The final drug release method may be different from the method used during formulation development, potentially requiring appropriate bridging experiments. The data from optimization studies are normally included in regulatory documents to justify the selection of the dissolution test conditions proposed for registration. In general, the approach to validation of a dissolution method is similar to that of any other method. Points to consider are summarized in Table 6 (37).

An in vivo/in vitro correlation relates a biological property produced by a dosage form, normally a pharmacokinetic parameter derived from the plasma concentration versus time curve, and a physicochemical characteristic of the dosage form, which is almost always derived from in vitro dissolution data (38). Because an in vivo/in vitro correlation is inherently an empirical relationship between two experimental observables, they should be considered formulation- (product-) specific. That is, there is no universal in vivo/in vitro correlation for

Table 4 Minor Components Assay—Validation Points to Consider

Validation attribute	Clinical stage	NADA stage
Specificity	Bulk drug and formulation: Available bulk drug process intermediates Formulation matrix components	Bulk drug and formulation: Bulk drug process impurities Stress degradation products Major component peak homogeneity Formulation: Formulation matrix components
Precision	Bulk drug and formulation: Chromatographic method Same day, analyst, reagents, etc.	Bulk drug and formulation: Chromatographic method and sample preparation method Detection limit and quantitation limit (baseline noise)
Accuracy	Bulk drug and formulation: Usually area percent assay with detection response assumed to be the same as for major component Formulation: Major component recovery from formulation matrix	Bulk drug and formulation: Detection response factors determined (area percent) or minor component reference standards Material balance Formulation Major component recovery from formulation matrix, column Minor component recovery from formulation matrix, column

Ruggedness	Bulk drug and formulation: Clear written procedure and appropriate system suitability tests	Bulk drug and formulation: Implementation data Laboratory standard data Clear written procedure and appropriate system suitability tests
Linearity	Bulk drug and formulation: Response function chracterized over sufficient range to support quantitation approach	Bulk drug and formulation: Response function characterized over sufficient range to support quantitation approach; compare with theoretical response

Table 5 Dissolution Test Parameters

Parameter	Typical range
Dissolution medium	Dependent on physiocochemical properties of the drug: aqueous buffers of physiologically relevant pH; may contain surfactant or bile salts to enhance drug solubility; deaeration if necessary
Apparatus	USP 1—Rotating basket USP 2—Rotating paddle USP 3—Reciprocating cell (BIO-DIS) apparatus USP 4—Flow-through apparatus
Agitation speed	50 rpm paddle (25–100 rpm) 100 rpm basket (50–150 rpm)
Flask size (medium volume)	1000 ml (500–900 ml)
Temperature	37°C (36.5–37.5°C)
Sampling method: volume	Manual: 15 ml (10–20 ml); automated: ≤ 3 ml
Filter	Disposable membrane filter ≤ 10 μm
Detection method	Rapid HPLC (t_R < 5 min) UV spectrophotometry

Source: Adapted from Ref. 37 with permission from Pharmaceutical Technology.

Table 6 Dissolution Assay Validation Points to Consider

Validation attribute	Comment
Specificity	Detection method must be free from excipient interference; specificity for process impurities and degradation products is not required
Accuracy (Linearity, Filter bias, and Recovery)	Detection method should be linear over the expected concentration range (i.e., 10–120% of label). Recovery of drug should not be significantly affected by filtration or the presence of placebo.
Precision/Ruggedness	Refers to the repeatability of the dissolution test across day, analyst, apparatus, and laboratory.
Automated sampling	Recovery should be assessed to ensure drug does not adsorb to the tubing. Also, dissolution profiles obtained by automated and manual sampling should be compared to ensure that there is no bias due to the automated sampling system.
Effect of dissolved gases	Dissolution profiles are compared using deaerated and nondearated media.
Stability	The room-temperature stability of the stock and working standard solutions should be determined.

Source: Adapted from Ref. 37 with permission from Pharmaceutical Technology.

a given drug. Strictly speaking, correlations are valid only for those formulations and manufacturing parameters for which there exist pharmacokinetic data.

An in vivo/in vitro correlation is sought because:

1. It enables a dissolution specification to be established that provides the maximum assurance of lot-to-lot consistency of in vivo product performance

2. It may allow the use of in vitro dissolution data in lieu of additional bioavailability data to justify a change in manufacturing site, raw material suppliers, or minor formulation/process changes

3. A correlation can be used for formulation development and optimization (37).

An in vivo/in vitro correlation is likely only if in vivo drug dissolution limits the rate of appearance of drug in the body. This can be appreciated from the following generalized kinetic scheme for in vivo release, absorption and disposition of a drug initially present in a solid-oral dosage form:

$$\text{Drug in dosage form} \xrightarrow{k_1} \text{Drug in solution} \xrightarrow{k_2} \text{Drug in body} \xrightarrow{k_3} \text{Elimination}$$

where k_1 is the dissolution rate; k_2 is the absorption rate; and k_3 is the elimination rate.

If in vivo dissolution is the rate limiting step to absorption ($k_1 << k_2$), then the dosage form controls the rate of input of drug into the body and a correlation with in vitro dissolution results is likely. An example of this would be a typical extended-release dosage form. On the other hand, if in vivo dissolution is much faster than absorption ($k_1 >> k_2$ or k_3), then absorption is the rate-limiting step to appearance of drug in the body, and it is highly unlikely that a correlation between in vivo results and in vitro dissolution will exist. This is normally the case with immediate-release dosage forms (37).

b. Water Analysis. The most commonly used water analysis method is Karl Fischer titration, which is applicable to most pharmaceutical samples that dissolve in a suitable solvent (39). Sampling, sample storage, and sample handling, while important for any assay, can be critical for water assays of hygroscopic materials. These issues must be addressed explicitly as part of method development and validation to help assure that the water result obtained represents the water content of the material of interest.

Precision is assessed by assaying several portions of a homogeneous sample using the proposed method, including the intended sample handling method. Accuracy and specificity validation is often performed by assaying samples with known amounts of water added either by spiking or by equilibration at elevated relative humidity. However, this approach does not rigorously address the two most significant potential sources of bias in the Karl Fischer titration—side reactions, and low or slow sample dissolution (39). Any of several alternative validation strategies can be used to address accuracy more rigorously (39). For example, the accuracy of the results can be investigated using an independent analytical method such as chromatography. The degree of validation of the independent method and the variability of both analytical methods limit this approach. Significant side reactions can be ruled out if a rigorously dried sample gives a water result not significantly different from zero. Unfortunately, this is a very difficult experiment to perform with most samples because of hygroscopicity and possible decomposition during drying. The linearity of the response of the Karl Fischer titration to varying amounts of moisture is generally accepted (39).

Methods that do not involve complete dissolution of the sample, such as loss on drying (LOD), are sometimes used to measure water content. These approaches provide an estimate of the water that is available to the analytical method, and results may not reliably correlate with the total water content of the sample. Use of such approaches requires careful characterization of the recovery of water from the sample matrix, and the ruggedness of the recovery with likely variation in matrix characteristics.

c. Residual Solvents. Residual solvents in bulk pharmaceutical chemicals are normally determined using a gas chromatographic method (40) which has been well characterized with respect to selectivity, precision, accuracy, and ruggedness. To assure accurate and reliable determination of residual solvents, the sample should completely dissolve in the assay solvent.

The first step in method validation is to assure that the selectivity of the underlying method is acceptable for all of the solvents used in the synthetic process. A representative batch of material is normally analyzed, residual solvents are identified by retention time matching with standards, and the chromatograms are examined for any artifacts. The solvents observed should make sense based on the synthetic process.

Another selectivity issue can arise if any sample degradation at the injector port and/or drug interaction with the dissolution solvent causes interferences that coelute with known solvents, and therefore cause high bias in the results. To evaluate such additive bias, the residual solvents originally present in a sample of bulk drug must be removed. This can be accomplished by first dissolving a sample in a solvent that is not seen in that lot and which will not interfere with the solvents that are seen. The solution is then blown to dryness under a stream of clean nitrogen to remove the solvent residues originally present in the sample. The dried drug residue is then dissolved in the assay solvent and analyzed. Since the solvents originally present in the sample should have been removed (or at least reduced), any response (other than for the dissolution solvent used in the experiment) is likely due to drug degradation and/or its interaction with the diluent.

Recovery of pertinent solvents from the drug/sample matrix is evaluated to ensure accuracy. To evaluate solvent recovery, samples are spiked with the solvents that may be found in the sample. Peak responses can be measured directly against injections of the standard preparation used to spike the bulk drug. Recoveries should normally be between about 90% and 110% near the intended specification level for each solvent. When acceptable recovery is found at one point, the linearity data for the underlying method can be used to infer linearity across the range of interest.

Methods that do not involve complete dissolution of the sample, such as head space gas chromatography, are sometimes used to measure residual solvents. These methods require careful characterization of the recovery of all potential solvents from the sample matrix, and the ruggedness of the recovery with likely variation in matrix characteristics.

C. Quality Assurance of Analytical Measurements

Method validation assures that the analytical method is *capable* of providing valid results which can support the needed decisions. Quality assurance of analytical measurements joins with method ruggedness and validation to assure that the assay system is in fact providing valid results each time the procedure is performed. Quality assurance includes system suitability tests to assure that all critical parts of the analytical method are performing acceptably each time the assay is run, and laboratory processes to prevent errors whenever possible and to make sure that any errors that do occur are corrected before the results are reported (41).

Useful results can be obtained only with control of the entire analytical process, which includes all activities between sampling the material to be tested and making a decision based on the test results, as shown in Figure 2. Some important parts of the process are not included in Figure 2, including method development, validation and implementation, clearly written analytical methods, standard operating procedures for each part of the process, procedure change control systems, meaningful system suitability tests for the analytical method, equipment qualification, and analyst training. Laboratory quality assurance and quality control build quality into each part of the process and assess the performance of the process on an ongoing basis to assure that it remains in statistical control. Process performance checks normally include system suitability tests, multiple standard determinations, replicate sample determinations, independent second-checking of records including calculations and transcriptions, and laboratory standard programs.

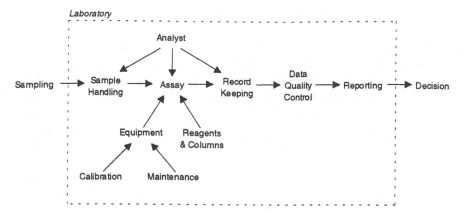

Figure 2   Analysis process.

Graphical control charts should be used to record the results of all tests performed on a routine basis to assure that the measurement process remains in statistical control. The use of control charts is recommended in addition to numerical databases because graphical charts can highlight developing problems before the method has actually failed.

1.   System Suitability Tests

System suitability tests (SSTs), in this context, are focused on the analytical method, which is part of the total assay "system" used to generate analytical results (see Fig. 3). Their goal is assuring that the analytical method is capable of generating data of sufficient accuracy and precision on a particular day, provided that the rest of the assay system is performing adequately. Appropriate

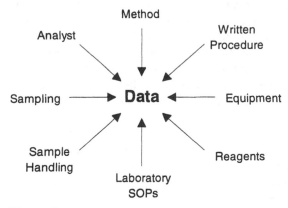

Figure 3   What is "the system"?

SSTs are normally needed to address each critical assay performance parameter, as determined during the validation and implementation of the assay. Additional SSTs may be needed to satisfy regulatory expectations, even if they do not directly address critical performance issues (42–44).

In general, the critical isocratic chromatographic potency assay performance parameters that are most sensitive to variations in assay components and environmental conditions are precision and selectivity. Precision is normally a critical assay quality parameter because the result must be compared with a limit. If the method variability is greater than was assumed during specifications development, the associated limit may not be sufficient to assure product quality. The assay precision must be assessed as part of each chromatographic run to assure that it remains acceptable. Standards or samples used for precision SSTs should be interspersed throughout the run, bracketing samples.

SSTs for selectivity in major component assays are required primarily to assure that critical impurities, which are known to elute close to the major component (or internal standard, if one is used), do not interfere with reliable quantitation. Additionally, the major component and any internal standard must be adequately resolved from each other to allow reliable quantitation. A resolution material containing critical impurities is used for this sort of system suitability test. The selectivity SSTs should be checked at both the beginning and end of a run. For minor component assays, it is also necessary to resolve critical impurities from each other. This is normally checked using a resolution material containing all critical impurities and degradation products. Information to guide peak identification can be provided by the same resolution material.

For most impurities assays, an SST is recommended to address detection limit either using a standard prepared at the detection limit or by setting limits on baseline noise and major component response. For some assays, additional chromatographic performance parameters may also be found critical to assay selectivity. For example, if integration ruggedness from chromatogram to chromatogram is found to be a limiting factor for precision, limits on appropriate chromatographic parameters, such as tailing factor or peak assymetry, may be prudent in addition to a direct precision determination. A capacity factor limit can help assure that the method is performing the same as during method validation, indicating that the selected resolution SSTs remain appropriate.

If necessary, directions for adjusting the chromatographic system so that the SSTs are met should be included in the procedure. For example, guidelines for adjusting the mobile phase composition to meet a selectivity SST could be given. Parameters that may be adjusted and acceptable ranges for adjustment should be determined during assay validation.

2. *Laboratory Analytical Data Quality Assurance*

The foundations of analytical data quality assurance include:

Personnel training
Written standard operating procedures
Instrument maintenance and calibration
Written analytical methods
Sample storage and handling facilities
Documentation of results
Systems to assure that the analytical system remains in statistical control
Documentation and data review before the data are used

These issues are important both to assure data quality (41) and for adherence to good manufacturing practice (GMP) requirements (45). For most product release and stability situations, good laboratory practices (GLPs) are not applicable.

Quality data require a number of practices that may be covered by laboratory standard operating procedures rather than specific assay methods. For example, use of replicate standard preparations and placing of reference standard preparations throughout an assay run may be required by operating procedures, rather than including these requirements in each analytical method. Additionally, the statistical control of the assay system over time can best be assessed using a laboratory standard program with control charting; this program is typically detailed in an operating procedure. The laboratory standard is a material that is acceptably homogeneous, has a sample preparation procedure similar to that for actual samples, and is normally included in each potency assay run as a data quality-control tool. Sampling procedures are normally detailed in operating procedures which may be specific to a type of product or material.

If the performance of the analytical method in the testing lab to be used was not characterized as part of method validation, a formal evaluation should be completed before data from the testing lab are used. This implementation or technology transfer process is critical to assuring that the method is continuing to perform as it did in the developer's laboratory during development. Assay implementation into a new laboratory should be the subject of a protocol prepared and agreed upon in advance. Critical implementation parameters normally include evaluation of the practical and statistical significance of any differences in mean and variability between the two labs for the same set of samples.

IV.   DRUG AND PRODUCT STABILITY

A.   Role of Scientific and Regulatory Studies

While it could be argued that any stability study is a scientific study, for the purposes of this discussion, a scientific study is defined as a one-time study that has been designed specifically to provide data to resolve a scientific issue. Regulatory stability studies, as the name implies, are conducted in response to a regulatory requirement. The primary difference is that scientific studies should be designed with the scientific objective as the only constraint—they may be conducted at any storage condition utilizing any assay schedule and set of tests and assays which will provide the necessary data. On the other hand, regulatory studies are constrained to specific storage conditions and assay schedules; a full complement of tests and assays and the number of batches are prescribed as given in regulatory publications. Due to the constraints within which regulatory studies must be conducted, they will not necessarily provide data to resolve the scientific issues. Therefore, prior to designing any stability study, it is very important to define the objectives, thereby determining which type of study should be conducted.

B. Chemical Degradation

1. Understand the Pathway

To develop a comprehensive stability database for a product, a knowledge of its critical performance characteristics is necessary. A crucial performance characteristic of any product is the manner in which it degrades. Without an understanding of the degradation pathways, it is impossible to develop analytical methods that can account for all significant degradation products. Without that assurance, the identity, strength, quality, and purity of the product cannot be assured. From another perspective, lack of knowledge of the degradation pathways leaves open the possibility that either late in the development process or even after approval, a previously undetected degradation product may appear. At best, a great deal of effort will be necessary to identify the new compound, verify (if possible) its occurrence in the clinical supplies, and demonstrate that it is not a toxicological concern. At worst, it could jeopardize the viability of the product.

An understanding of the pathways defines the key degradation products the analytical methods must be capable of measuring. (The point in the development time frame at which these activities should be accomplished is given in the section entitled "Stability Study Design.") For example, consider spectinomycin HCl, which is commonly formulated as an acidic sterile solution for injection. The degradation pathways are shown in Figure 4. In acid solution, the molecule is cleaved to yield actinamine and the hypothetical neutral sugar actinospectose. (Although actinospectose has not been isolated from this reaction, there is good evidence for its formation (46).) In basic solution, the incipient 1,2-diketone rearranges to the ring-opened actinospectinoic acid. Development of analytical methods to quantitate all three major degradation products would be very difficult, indeed, since actinospectose has no chromophore and is electrochemically inactive. However, since actinospectose is produced in a 1:1 molar ratio with actinamine, it is only necessary to develop a method for quantitation of actinamine.

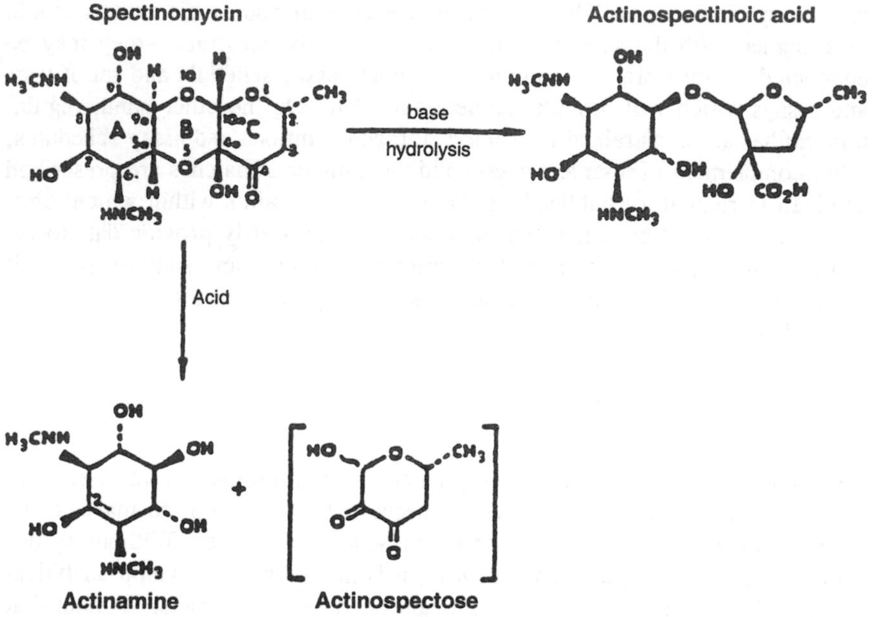

Spectinomycin   Actinospectinoic acid

base
hydrolysis

Acid

Actinamine   Actinospectose

Figure 4 Degradation pathways. (Reproduced with permission of the Japanese Antibiotics Research Association, Ref. 46.)

2. Implications for Specification Setting

Understanding the degradation pathways for a product can simplify the specifications-setting process. In the example given above of spectinomycin HCl sterile solution, it was indicated that it is only necessary to measure the actinamine content to be able to characterize the extent of degradation on that pathway. It logically follows that it is also only necessary to set a specification limit on the actinamine content, instead of both actinamine and actinospectose. The broader implication for the specifications-setting process is that an understanding of the degradation pathways increases the confidence that realistic specifications are being set which the product will be able to meet throughout its shelf life.

C. Relationship of Expiration Dating to Label Storage Temperatures

It is normally required that in addition to an expiration date, the product's label must state under what conditions it must be stored. But what is the relationship between the expiration dating period of a product and its label storage conditions? In practice, a given formulation and container/closure system may have different expiration dating periods assigned to them, depending on where, i.e., in which "climate zone" (47), they are being marketed. The relation between temperature and the rate constant for a chemical reaction can be approximated by the Arrhenius relationship.

1. Arrhenius Relationship

Many chemical reactions may be modeled by either zero- or first-order kinetics, especially for the first 10% to 20% of the reaction. For zero-order processes, the rate of disappearance of reactant A is constant and independent of its concentration:

$$-dA/dt = k$$

Solving the above equation yields:

$$A = A_o - kt \tag{1}$$

where A = the amount of A remaining at time t, A_o = the initial amount of A, and k = rate constant. For reactions following zero-order kinetics, a plot of A versus t yields a straight line whose slope is equal to $-k$.

For first-order processes, the rate of disappearance of the reactant A is proportional to the concentration of A at any time t.

$$-dA/dt = kA$$

Solution of the above equation yields:

$$\ln A/A_o = -kt$$

or

$$\ln A = -kt + \ln A_o \tag{2}$$

For stability data that follow first-order kinetics, a plot of $\ln A$ versus t will yield a straight line whose slope is $-k$.

The rate constant, k, is in almost all cases a function of the temperature, T. The rate constants for the degradation reactions of most pharmaceutical products increase as the temperature increases. This is the phenomenon described by the Arrhenius relationship, which states that for a given chemical reaction the empirical relationship between k and T may be written as:

$$k = b_o \, e^{-E/RT}$$

or

$$\ln k = \ln b_o - E/RT \tag{3}$$

where T = Kelvin temperature, E = activation energy, R = universal gas constant, and b_o = constant depending on the molecule of interest. If a particular reaction follows the Arrhenius relationship, then a plot of lnk versus 1/T will yield a straight line whose slope is E/R.

2. Applications to Stability Prediction

Equations 1, 2, and 3 have been found to be experimentally valid, over temperature ranges of interest, for many pharmaceutical formulations. If the degradation process for a particular dosage form follows zero- or first-order kinetic processes over the entire temperature range of interest, and reliable rate constants are shown to follow Arrhenius behavior, then it may be concluded that the mechanism of the reaction is the same over that temperature range and equations 1, 2, and 3 may be used to predict the dosage form's stability performance. Great care must be exercised, however, in extension of this treatment beyond temperatures and outside time periods used to establish the Arrhenius behavior.

As an example, consider a pharmaceutical product whose expiration dating period is limited by the degradation of the active ingredient. The loss of the active ingredient over time at 25°C has been shown to follow zero-order kinetics between 100% and 85% of the labeled amount. In addition, data from other studies at 15, 30, 45, and 60°C show the same zero-order dependence. If the rate constants from all these studies follow the Arrhenius relationship (Eq. 3), then from the slope of the plot of lnk versus 1/T for the active ingredient, the rate constant for any other temperature in the range of the data may be interpolated. Thus, knowing k for any temperature and the time to be spent at that temperature, the stability performance may be calculated by equation 1.

3. Climate Zone Concept

A more generalized application of the Arrhenius relationship is involved in the climate zone concept. Key to the climate zone concept is the calculation of a kinetic testing temperature (or mean kinetic temperature) (T_k). The kinetic testing temperature is defined as the isothermal temperature that corresponds to the kinetic effects of a time-temperature distribution. For a constant activation energy (E), k/b_0 can be determined from equation 3 for each temperature in a time-temperature distribution. These individual k/b_0 values are then weighted by the time at each temperature and summed as in equation 4 to yield an effective rate constant (k_{eff}). By substituting the effective rate constant back into the Arrhenius relationship (Eq. 3), the kinetic testing temperature (T_k) can be calculated (48).

$$k_{eff} = \frac{Sum\ k_1 t_1}{Sum\ t_1} \qquad (4)$$

where k_1 is the rate constant at temperature T_1, and t_1 is the amount of time at temperature T_1.

It has been shown in Reference 49 that the kinetic testing temperature changes only slowly with changes in activation energy for different reactions of interest (less than 2°C over the range 10 to 35 kcal/mole). Since most reactions of organic molecules exhibit activation energies close to 20 kcal/mole, that value is used for all subsequent calculations.

To simplify the development of storage conditions for each country in which it is desired to register a drug product, it is generally accepted that the earth can be divided into four climatic zones, as shown in Table 7 (50). Using the above approach, the countries of the world can be assigned to the four climate zones, and kinetic testing temperatures can be calculated for each zone (51,52). For climate zone I, the calculated kinetic testing temperature is 21°C; for climate zone II it is 25°C, and for climate zones III and IV it is 31°C. Climate zone I corresponds to northern Europe; climate zone II to southern Europe, the U.S., and Japan; climate zone III to the Sahara (Chad, Sudan, etc.); while climate zone IV corresponds to central Africa and the South Pacific (Kenya, Zaire, Indonesia, and the Philippines).

To test the assignment of the U.S. to climate zone II, data from a Pharmaceutical Manufacturers Association study on storage conditions in Dallas, Texas, were used to calculate kinetic testing temperatures for relatively extreme conditions (53). The data are from the hottest of a 3-year period in the early 1970s and are tabulated in 1-degree increments, and hours at each temperature for a year. For the case of an air-conditioned warehouse, any temperature below 20°C was assumed to be 20°C; any temperature above 30°C was considered to be 30°C. The calculated kinetic testing temperature is 25°C, If the warehouse were not air-conditioned (using the actual time-temperature distribution above 20°C), the kinetic testing temperature only increases 1 degree to 26°C.

4. Humidity

There are no generally accepted mathematical methods for calculating humidity-time-temperature relationships which are analogous to the Arrhenius approach for time-temperature distributions. So, an empirical approach has been used in at least two studies where relative humidity conditions from meteorological data

Table 7 Climate Zone Descriptions

Zone	Description
I	Temperate
II	Mediterranean and subtropical
III	Hot and dry
IV	Hot and humid

were paired with the kinetic testing temperature for each climate zone (50,51). The first regulatory body to embrace the climate zone concept was the Committee on Proprietary Medicinal Products (CPMP) of the EEC. On the basis of these studies, they assigned mean relative humidity conditions of 45%, 60%, 40%, and 70% for climate zones I–IV, respectively (54). These definitions have subsequently been incorporated into the ICH Stability Guidelines (47).

Obviously, sections of the same country could encompass different climate zones. For example, the northern part of the U.S. could be assigned to climate zone I and the southern part to zone II. Similarly, countries that are close trading partners may be assigned to different climate zones; e.g., the countries of the European Community are in both climate zones I and II. In a larger sense, the major pharmaceutical markets in the world (Europe, North America, and Japan) are all in zones I and II. Long-term stability studies conducted under storage conditions suitable for registration in zone II countries would also be suitable for registration in zone I countries. While it may be accurate, it is not very practical to distinguish between zones I and II for the purposes of conducting long-term stability studies. Therefore, the signatories of the ICH Guidelines (E.C., U.S., and Japan) recommended zone II conditions (25°C/60% RH) as the long-term storage conditions to support a room-temperature label storage statement for all countries in both zones I and II.

D. Release Limits

Of primary concern in the assignment of an expiration dating period to an individual product batch is the level of confidence that it will remain within its registered limits during that time period. For physical attributes such as appearance or dissolution, prior stability experience will indicate whether changes may be anticipated, but it is either difficult or impossible to develop a model to predict the change accurately. The level of confidence in these cases is based on experience at accelerated and label storage conditions as well as a sound process validation program. However, for changes that are governed by a chemical reaction, a quantitative degradation rate may be determined. From this rate a level of confidence can be calculated that the batch will remain within registered limits throughout its shelf life. For the purposes of this discussion, the following assumptions are used.

1. Decreasing potency is the stability-limiting factor.
2. The desired level of confidence, unless otherwise stated, is 95% (one-sided), which is generally accepted in the industry (55).
3. The manufacturing process has been validated and the underlying distribution of potency values is normal or can be approximated by a normal distribution.

4. The potency has a predictable rate of change.
5. The potency change is linear at least through the shelf life for all batches produced by the process. For a loss of potency of the order of 10% to 20%, this assumption is usually valid regardless of the order of the reaction.
6. The reaction mechanism is the same for all batches, and the true rate of change is a constant.
7. The assay is stability-indicating and sufficiently precise (it has been validated with regard to specificity, precision, accuracy, ruggedness, and other relevant parameters).

There are a number of factors that must be considered to ensure, with at least 95% confidence, that potency remains within its registered limits throughout the shelf life. For the simplest case, consider a product for which no changes are expected for any attribute. The only limitation in this case is the accuracy and precision of the release assay. Thus, to ensure that the potency will remain within registered limits over the shelf life, it is only necessary to determine, with 95% confidence, that the true potency upon release is within the registered limits. The lowest potency at which a batch could be released under these conditions may be defined as the lower release limit (56).

The calculation is very simple and straightforward, as shown in equation 1.

$$LRL = LR + t * S/\sqrt{n} \tag{1}$$

where LRL = lower release limit, LR = lower registration limit, S = assay standard deviation, DF = degrees of freedom for S, t = 95% confidence (one-sided) t-value with DF degrees of freedom, and n = number of replicate assays used for batch release.

A slightly more complex case is that given by a product, such as a tablet, which exhibits a significant degradation rate. For this case, a measure of the degradation rate and its associated variability is needed to calculate a lower release limit. To obtain these, the stability data are analyzed using standard regression techniques (57,58). The expiration dating period is determined by locating where the 95% confidence interval of the regression line crosses the 90% potency value. In addition, a test for poolability of the slopes from the individual batches is performed (59,60). If the slopes can be pooled, the average slope and its associated standard error are used for the release limit calculation. The average slope is a weighted average of slopes from separate regressions on each batch. The weighting factors are determined from the sum of the squared deviations of time values from the mean time value for each batch. The average slope is used since it is not influenced by the intercepts of each batch and, thus, is a better representation of the true rate of degradation than the pooled slope. (If the test for homogeneous slopes fails, a single slope may not be an appropriate representation of the true rate of degradation, and the resultant average slope and its associated standard error should only be used with caution.)

Since the release assay results and the average slope are independent measurements, the uncertainties due to the variation associated with the mean release assay result and the average slope may be added in quadrature as shown in equation 2 (61).

$$LRL = LR - EAC_T + t * \sqrt{S_T^2 + S^2/n} \tag{2}$$

where EAC_T = estimated attribute change = average slope of tablet change * shelf life; S_T = standard error of EAC_T = standard error of average slope * shelf life; and t = 95% confidence (one-sided) t-value with DF degrees of freedom. (*Note*: EAC_T is assumed to be negative. If it is positive, an upper limit could be calculated by equation 2, subtracting the error term.) The degrees of freedom may be calculated by the Satterthwaite Approximation (62) and all other terms are as defined for equation 1.

An example of this type of release limit calculation is given below.

Example: Consider tablet X with the following parameters.

LR = 90% of label
S = 1.1% of label
Average slope = −0.20% of label/month
Standard error of average slope = 0.03% of label
Shelf life = 24 months
EAC_T = −0.20 * 24 = −4.8% of label
S_T = 0.03% * 24 = 0.72% of label
n = 2
t = 1.67 (DF = 58)
LRL = 90 + 4.8 + 1.67*[(0.72)^2 + (1.1)^2/2]$^{1/2}$ = 96.6% of label

The result of this analysis is that for batches that assay at > 96.6%, it can be assured with 95% confidence that potency will meet the registered requirement after 24 months. From this example it should also be clear that the more precise the assay, the longer the expiration dating period that can be justified.

Calculation of the lower release limit without adding the variances in quadrature (i.e., simple addition of the uncertainty of the mean release assay result and the amount of degradation expected over the shelf life [at the 95% one-sided confidence level]) would result in an unnecessarily conservative value (97.3% of label for the example given above). The penalty for this incorrect calculation increases as the error terms become similar in magnitude.

A more complex case involves a powder that must be reconstituted prior to use, in which there are more independent factors (both fixed and variable) that must be considered.

Fixed factors	Variable factors
Formulation	Dry powder shelf life
Manufacturing process	Reconstituted solution shelf life
Assay method	Number of replicate release assay results
Stability data base	
Dry powder degradation rate	
Reconstituted solution degradation rate	
Variance of the mean release assay result	
Variance of the dry powder degradation rate	
Variance of the reconstituted solution degradation rate	

In similar fashion to the case described above, the release assay results and the average slopes of the dry powder and reconstituted solution degradation rates are independent measurements, allowing the addition in quadrature of the variances associated with these measurements. The lower release limit for this case may be calculated by equation 3:

$$LRL = LR - EAC_P - EAC_S + t * \sqrt{S_2^P + S_2^S + S^2/n} \qquad (3)$$

where EAC_P = estimated attribute change = average slope of dry powder change * shelf life (powder); S_P = standard error of EAC_P = standard error of average slope of powder * shelf life (powder); EAC_S = estimated attribute change of reconstituted solution = average slope of reconstituted solution * shelf life (reconstituted solution); S_S = standard error of EAC_S = standard error of average slope of reconstituted solution * shelf life (reconstituted solution); and all other terms are as previously defined.

It is readily apparent that the release limit represents a balance between all the variable factors. For a given shelf life, a change in one of the variables requires a counterbalancing change in one or more of the others. This is illustrated in the example given below.

Example: Consider reconstitutable product Y with the following parameters.

LR = 90% of label
Average slope of the dry powder = -0.15% of label/month
Average slope of reconstituted solution = -0.12% of label/day

Standard error of average slope of dry powder = 0.02% of label
Standard error of average slope of reconstituted solution = 0.02% of label
S = 1.0% of label
t = 1.67 (DF = 64)
n = 2

Various combinations of dry powder shelf lives and reconstituted solution shelf
lives will result in different release limits calculated by equation 3 as shown
below.

	Shelf life (reconstituted solution)		
Shelf life (dry)	7 days	14 days	21 days
24 months	95.9%	96.8%	97.7%
36 months	97.9%	98.8%	99.7%
48 months	100.0%	> 100%	> 100%

The results from this type of calculation may be combined with process capability data and marketing preferences to determine the optimal combination of
shelf life assignments. This calculation method can also indicate the impact of
changes in the fixed factors on the release limit and provide an objective means
for focusing attention on the relative benefits of increased assay precision, a more
robust formulation, or a more extensive stability database. Since release limits
are derived from a number of interdependent variable factors, they are inherently dynamic and should therefore be reevaluated on a regular basis.

E. Stability Study Design

1. General Considerations

A general definition of stability is the capacity of a drug product to remain within
its established specifications. In the development of a product with a well-characterized stability profile, there are a number of fundamental issues whose timely
and efficient resolution requires a well-designed, coherent stability program.
These issues include the following:

What tests and assays are necessary to monitor the quality of the product?
What are appropriate specifications for these tests and assays?
What packages are suitable (or unsuitable) for the product?
What storage conditions are appropriate?
What is an appropriate expiration dating period for the product?

There are both scientific and regulatory aspects to the resolution of these issues. The basis of the approach recommended here is to design the studies first to provide data to scientifically resolve these issues and then add whatever is necessary to satisfy regulatory requirements. The output from this process should result in a stability performance database which resolves all of the above issues.

As mentioned earlier, quality issues for small-molecule animal health and human health products are, in a general sense, identical. This is especially so for the determination of the stability performance of these products. Thus, the discussion in this section is equally applicable to both animal and human health products. It is important to understand that stability testing during pharmaceutical product development (both veterinary and human) should be viewed as an evolutionary process, which can be divided into a number of distinct, identifiable phases. Within each of these phases, differing pieces of information are sought that are important for building the stability performance database, and the progression from phase to phase gives rise to widely varying concerns and objectives. These widely varying concerns and objectives require a variety of stability study designs to be utilized. While it is possible to arbitrarily divide up the stability testing life cycle of a product into any number of phases, for consistency's sake, the model to be used here is the one given in a report published by the (then) PMA Joint QC-PDS Stability Committee in 1984 (63). The phases of stability testing in this model are given below.

Phases of Stability Testing for Product Development

Preformulation
Formulation development
Proposed product
New product
Established product
Revised product

For the purposes of this chapter, only the first three phases will be discussed.

2. Preformulation Phase

Each phase consists of one or more study topics. In the preformulation phase, which corresponds roughly to the pre-INADA period, there are two: drug substance reactivities, and toxicology supply stability. The objectives of the drug substance reactivity studies are, first, to profile the physical and chemical properties of the drug substance which is to be formulated into the product, and second, to establish the handling and packaging requirements for the drug substance. These objectives require a study design which involves short-term testing under accelerated conditions, looking for gross changes. The storage conditions are deliberately geared to promote degradation. The test schedule, number of intervals, and number of replicates are designed to allow estimates of rates of change of the attributes being monitored (at this stage, the basic attributes, such as potency, major degradation products, water content, and physical properties). Generally these studies are run in the analytical development laboratory as part of the analytical development process.

Studies of the drug substance are normally conducted in the solid state and in aqueous solution. In the solid state, the substance is subjected to extremes of temperature, humidity, light, and oxygen, while in solution it may be subjected to extremes of temperature, pH, oxidants, metal ions, and light. For example, a solution of a drug was studied at 70°C in the presence of a variety of transition metal ions (Fig. 5). The compound is obviously sensitive to the presence of transition metal ions and may indicate a necessity to minimize contact with metal surfaces during manufacturing and storage. Another common type of study involves measuring the reaction rate as a function of pH. Figure 6 illustrates the pH rate profile for an aqueous solution of spironolactone (64). It is clear that the degradation rate for this solution accelerates at either extreme

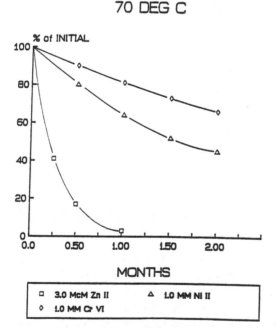

Figure 5 Effect of metal ions in solution.

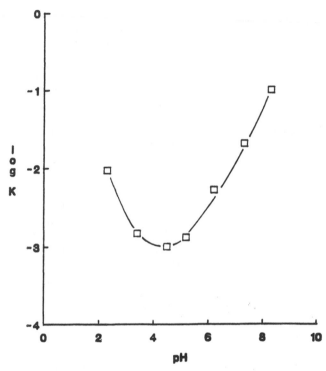

Figure 6   pH-rate profile of spironolactone at 40°C. (From Ref. 64.) (Reproduced with permission of the American Pharmaceutical Association.)

of pH and that the optimum pH for stability purposes is in the range from 3 to 6.

The results of all the tests are compiled into a profile of the physical and chemical properties of the drug substance. The profiles of two different drug substances are given below, in Table 8.

It is obvious that compound A is a fairly robust molecule, although it is sensitive to exposure to UV and flourescent light. Compound B, on the other hand, exhibits sensitivity to almost any type of stress. Based on these data, the handling and packaging requirements may be determined. Robust compound A, requiring only protection from light, can be safely stored at room temperature in an opaque container (e.g., a poly bag in a fiber drum). Compound B, on the other hand, should be stored at either refrigerator or freezer conditions in a container that will protect it from moisture pickup. A suitable container would be a poly bag in a foil-lined fiber drum with a rubber gasketed lever-lock metal closure. The profile of the chemical and physical properties of a drug substance is also useful during the formulation development phase, to be discussed later.

Table 8   Physical and Chemical Profile Data

Compound A		Compound B	
Conditions	Results (% of initial)	Conditions	Results (% of initial)
Solid state:		Solid state:	
70°C	No detectable	40°C/70% RH	56% in 3 weeks
25°C/75% RH	change after 28	60°C	23% in 5 days
UV light (RT)	days' storage	UV light (RT)	half-life about 1 year
Fluorescent light (RT)		Fluorescent light (RT)	No change in 28 days
Solution:		Solution:	(half-life at RT)
0.1 N HCl		0.1N HCl	230 days
0.1 N NaOH	No detectable	pH 3 buffer	70 days
pH 3, 5, 7, 9 buffers	changes after 42 days at 70°C	pH 5, 7, 9 buffers	40 days
		0.1N NaOH	<1 day
9 mM solutions of CU(II), Cr(II), Zn(II), Fe(III) in 0.1 N HCl	No detectable changes after 29 days storage at 25°C	Cu(II), Cr(III), Fe(III), Zn(II)	No effect
UV light (RT)	<10% in 29 days		(All solutions completely decomposed in <4 days at 70°C)
Fluorescent light (RT)	74% in 42 days		

    The objectives of the toxicology supplies stability studies are very different. The toxicology supplies represent the first crude formulations and the objectives are to provide data to document potency and levels of degradation products throughout the course of the toxicology trials, and to develop supporting data for subsequent formulation development work. These objectives require a different study design which involves short- to medium-term testing under label storage conditions monitoring potency, significant degradation products, and significant physical properties. The length of the studies is directly related to the length of the toxicology trials. The samples to be tested should be representative of each formulation and its packaging. The testing schedules should be established as necessary in order to monitor the supplies—more frequent testing should be done for less stable supplies. If the trial is of short duration, the stability testing may consist only of entrance and exit assays. The point is that the data should allow the correlation of toxicology results to the potency and level of degradation products during the trial.

3. Formulation Development Phase

The next phase is formulation development, in which there are two major study topics—formulation comparisons and clinical supplies stability. These are fundamentally different types of studies which, in practice, may be conducted almost simultaneously. The most fundamental difference is that the formulation comparisons studies are scientific studies as opposed to the clinical supplies stability studies which are mandated by regulatory agencies. Thus, the formulation comparison study design is driven entirely by the knowledge base which has been accumulated during the preformulation phase. For example, if during preformulation testing the drug substance exhibited high reactivities (compound B in Table 8), the formulator may decide to conduct excipient compatibility testing prior to designing experimental formulations. This testing is typically performed on binary mixtures of the drug substance and possible excipients and is of short duration at accelerated conditions. The data may be useful in focusing formulation development efforts. If, on the other hand, the drug substance reactivities are low (compound A in Table 8), excipient compatibility testing is not likely to provide cost-effective information.

The objectives of the formulation comparison studies are:

1. To determine any experimental formulation reactivities and any significant drug substance degradation products
2. To compare the stability characteristics of various experimental formulations
3. To collect preliminary information regarding possible stability limiting factors.

It is important to note that to accomplish these objectives, it is imperative that analytical methods development must be tightly linked to formulation development efforts. A useful means for efficiently screening a series of formulations is to compare the effects of short-term accelerated conditions studies on their physical properties. After elimination of those that are physically less stable, the remaining candidates may then be screened for relative chemical stability. Chemical stability is usually assessed by testing at accelerated conditions for periods ranging from a few weeks to a few months. The storage conditions should be deliberately chosen to stress the formulations. Since these are scientific stability studies, specific test schedules and storage conditions are not given here or in any regulatory guidelines; they must be derived from the development data in hand at the time the studies are designed.

On the other hand, the objectives of the clinical supplies stability studies are driven primarily by regulatory requirements and are first, to determine the values of all critical quality parameters used to establish safety and efficacy as a function of age; second, to ensure that only satisfactory material is used in the clinical setting, and; third, to estimate an approximate shelf life for the eventual product.

To ensure that only satisfactory material is used in the clinical setting and to be able to correlate clinical response to critical quality parameters, clinical supplies stability studies must be conducted at label storage conditions for at least the length of time the supplies are in use. At least one batch of each formulation and associated container/closure systems used in the clinic should be placed on stability. The tests and assays should include potency, significant degradation products, and other attributes relevant to the specific dosage form. The test schedule and duration of the studies are determined by the nature of the clinical trial.

4. Proposed Product Phase

When the final formulation has been chosen, product development enters the proposed product phase. The first types of stability studies at this point in the development process are generally scientific studies conducted to specifically resolve any remaining issues.

For example, assume the issue is to determine the impact of moisture content on the stability performance of a tablet. The most straightforward means to resolve the issue would be to put unprotected tablets (e.g., open bottle) on stability at various constant humidity levels (e.g., 30%, 45%, 60%, and 75% RH) at label storage temperature for 6 to 12 months, monitoring stability indicating parameters. A common pitfall in this approach is that very careful sample handling is necessary to obtain an accurate measure of the moisture content of the tablets. One means of removing this pitfall is to determine the equilibrium moisture sorption isotherm of the tablets at the label storage temperature (e.g., 25°C for controlled room temperature). From this measurement, the equilibrium moisture content of the tablets at the various constant humidity storage stations is known and can be correlated to any changes in measured parameters. These data may then be used to provide a rationale for a water content specification. If it is determined that there is no dependence of chemical or physical degradation on water content, then the data can be used to justify the absence of a water specification, the elimination of testing for water content on stability, and the elimination of broached container testing.

Regulatory stability studies are conducted during this phase specifically to accomplish the following objectives: first, to provide data to aid in setting specifications; determine any stability limiting factors; second, to establish an initial expiration dating period for the NADA filing; and third, to serve as the basis for the post-approval stability protocol.

As mentioned earlier, there does not seem to be a compelling reason for major differences between the development of a product destined for the veterinary pharmaceutical market as opposed to the human pharmaceutical market. For the development, specifically, of the protocol for the primary stability studies to support the filing of a regulatory application, a particularly valuable resource is the ICH (International Conference on Harmonization) Harmonized Tripartite Guideline for Stability Testing of New Drug Substances and Products, which was accepted for implementation by the regulatory agencies (human health) in Japan, the E.U., and the U.S. in 1993 (47). They will be hereafter referred to as the ICH guidelines. Until the VICH (Veterinary International Conference on Harmonization) process has resulted in VICH stability guidelines, it will be important to consult country-specific guidelines for details on how to conduct studies for particular markets.

The primary stability studies should be conducted using samples that are representative of the following:

Drug substance to be used in the marketed product
Formulation to be marketed
Manufacturing process to be used for the marketed product
Packages to be marketed

In general, at least three lots are required to be placed on stability. The ICH guidelines state that two of the three lots should be at least pilot scale and the third lot may be smaller. Pilot scale is defined as follows:

> The manufacture of either drug substance or drug product by a procedure fully representative of and simulating that to be applied on a full manufacturing scale. For solid oral dosage forms this is generally taken to be at a minimum scale of one tenth that of full production or 100,000 tablets or capsules, whichever is larger.

The tests and assays should be justified using the data generated during the preformulation and formulation development phases. The ICH guidelines provide that:

> . . . the testing should cover those features susceptible to change during storage and likely to influence quality, safety and/or efficacy. . . . The range of testing should cover not only chemical and biological stability but also loss of preservative, physical properties and characteristics, organoleptic properties and where required, microbiological attributes. Preservative efficacy testing and assays on stored samples should be carried out to determine the content and efficacy of antimicrobial preservatives.

The selection of the packages to be studied (i.e., marketed) involves a blend of scientific and business considerations. The scientific considerations include what kind of protection the formulation requires and how much protection is needed as well as the need to avoid product/package interactions. Data generated during the earlier development phases should be available to determine whether the product needs protection from moisture, light, or oxygen. If not, additional specific, short-term scientific studies should be designed and conducted. Data from the formulation comparison studies as well as stability studies of clinical formulations should provide the basis for selection of packaging materials that will not interact with the drug product. The storage conditions to be used for the long-term primary stability studies will depend on the label storage condition chosen for the product. The choice of the label storage condition involves, again, a blend of scientific and business considerations. For example, the data to date may indicate that the product will most likely have a short shelf life at room temperature (12 to 18 months), as opposed to 30 to 36 months at refrigerated conditions. If it is considered more important to have a longer shelf life than a room temperature label storage condition, then the label storage condition for the product will be 2 to 8°C.

For a product that is to be stored at room temperature, the stability storage conditions in the ICH guidelines are 25°C/60% RH. The storage conditions for accelerated studies should be at least 15°C higher than those for the long-term label conditions studies. Thus, for a room temperature product which would be studied long term at 25°C/60% RH, the accelerated conditions are 40°C/75% RH unless it is a product that could be adversely affected by low relative humidity, in which case the relative humidity at the accelerated conditions should be on the order of 10% to 20%. For a refrigerated product, 25°C/60% RH is generally considered to be the accelerated condition primarily from the practical standpoint of the availability of that condition in the stability storage facilities.

The testing schedule most widely recognized for the long-term studies is quarterly the first year, semiannually for the second year, and annually thereafter. No testing schedule for studies at accelerated conditions is specified in the ICH guidelines, but, to be able to determine a degradation rate by statistical analysis of the resulting data, a minimum of four data points is advisable.

F. Specialized Stability Studies

Some stability studies are hybrids. They are one-time studies designed to resolve a specific issue, but they are required by most regulatory agencies. The most common of these are photostability, reconstituted stability, and broached container stability studies.

1. Photostability

The photostability performance characteristics of bulk drugs and their formulations should be determined to be able to design packaging components that afford suitable protection from light. (*Note*: It is important to note that the testing being discussed here is separate from the stress testing done as part of the preformulation studies given in Sec. E of this chapter.) At the time of this writing, there are no specific guidelines from any of the major regulatory agencies (U.S., Europe, or Japan), but there is a draft ICH guideline (65).

While there is no agreement on the details of the light sources or how the data are to be used, there does appear to be consensus on the broader aspects of photostability testing. The approach recommended in the draft ICH guideline was developed in part on the basis of an industry consensus position compiled by EFPIA (European Federation of Pharmaceutical Industry Associations). It represents a systematic approach in which the extent of testing is established by assessing whether or not acceptable change has occurred at the end of light exposure testing (Fig. 7). Since a definition for acceptable change has not been agreed, a rationale for the definition used in the individual instance should be developed for inclusion in the regulatory file. Testing should be carried out on

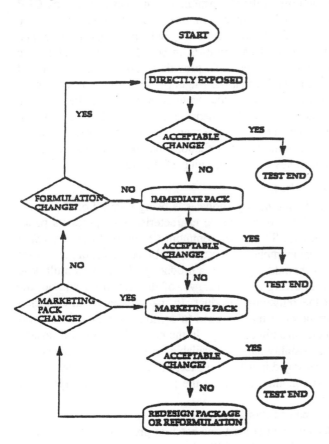

Figure 7 Decision flow chart for photostability testing of drug products.

one batch of the drug substance and the product. There is general agreement that exposure levels should be not less than 1.2 million lux-hours visible illumination (400 to 800 nm) and an integrated near UV (320 to 400 nm) energy of not less than 200 watt hours/m^2. Light intensities used in industry are generally in the range that will require continuous exposure for about 50 to 60 days to reach these exposure levels. To be able to separate thermal effects from photo effects, it is strongly recommended that a second sample suitably protected from light (e.g., covered with aluminum foil) be used as a dark control and placed alongside the authentic sample.

2. *Reconstituted Stability*

Some products require reconstitution prior to use. For any label statement other than one which instructs immediate use after reconstitution, stability data are necessary to justify the shelf life of the reconstituted solution as required by CFR 211.137 (c). Usually the study is conducted on one batch of the product. Since the resulting data are to be used to determine the shelf life of the product and may also be used to calculate a release limit (as in Sec. D of this chapter), the study should be very carefully designed. To provide an appropriate amount of statistical power, it is recommended that the length of the study be at least twice the desired or expected shelf life of the reconstituted product and that the data be collected at a minimum of six to eight scheduled intervals. Where the amount of sample for the entire study requires more than one vial to be reconstituted, the variability introduced by multiple reconstitutions may be minimized by pooling of the reconstituted samples at the beginning of the study.

3. *Broached Container Stability*

Increased attention from the regulatory agencies is being focused on the determination of the in-use stability performance characteristics of products packaged in multidose containers. These are commonly referred to as *broached container studies*. As the name implies, the purpose of broached container studies is to establish a period of time during which a product packaged in a multidose container may be used following the withdrawal of the first dose without affecting the integrity of the remaining product. In many cases, data collected during earlier development work may be used to construct a rationale to demonstrate that further studies are unnecessary. In the case of a solid oral dosage form, for example, where stability data are available for the product when stored in an open container or packaged in a PVC blister, no further knowledge would be gained by a broached container study on the product when packaged in a 500-count HDPE bottle.

For cases where this type of rationale cannot be constructed, a study should be designed to be conducted using at least one batch of product. It should cover the period of time in which the product will normally be expected to be in use after the first dose is withdrawn, monitoring chemical (e.g., sensitivity to oxidation), physical (e.g., increase in particulates, stopper failure), and microbiological (e.g., failure of the preservative system) attributes.

V. DEVELOPMENT OF LIMITS

The goal of the set of methods and limits making up the product specifications is primarily to assure product safety and efficacy throughout the shelf life by addressing each of the product's critical parameters. Additionally, specifications are often established for product parameters which contribute to the customer's perception that the product is of high quality, or which satisfy a regulatory agency expectation. For product to be practically produced and released for marketing, the limits established for each parameter must also be consistent with the capability of the associated analytical method and the capability of the manufacturing process.

Limits are interactive with methods. Method precision constrains the limits that can be supported by any method. The results frequently depend on the measurement method selected. For example, different particle-size methods do not necessarily yield comparable results, and particle size limits must be linked with a specific method. For some assays, the measurement method evolves based on the results to seek a correlation with an outside variable. For example, in drug release testing, the in vitro dissolution test is developed to seek a correlation with in vivo biopharmaceutical differences. The in vivo results then drive development of limits for the in vitro test.

The tight link between method and limit requires that limits be evaluated for appropriateness whenever a method is changed. It might be appropriate, for example, to establish a limit for an impurity in a simple solution dosage form using an HPLC method which quantitates that impurity as a percentage of the total integrated area in the chromatogram. If a complex solid dosage form is developed at a later time, it is likely that excipients will be detected in the chromatogram. Even if the impurity is completely resolved from the excipients and the major component, it would clearly be inappropriate to use a limit based on the total area in the chromatogram. Instead, a limit based on a determination of the impurity with respect to the area of the major drug peak, or perhaps with respect to an independent standard, would be more appropriate.

Differences in methods used throughout development need to be considered as the total database of development information is assembled for determining limits, and for regulatory review and approval. For example, if the solution conditions used in a tablet dissolution method are changed part way through development, dissolution data from early tablet batches might not be comparable to data from later batches. Of course, when earlier batches can be retested using the new method, a consistent database can be maintained. However, this is not always possible when development time frames span several years.

The link between method and limit must also be considered when comparing drugs or products from different sources. A specification from one manufacturer might not provide an equivalent level of quality control to that from a different manufacturer if the methods differ. Similarly, if a customer analyzes a product purchased from a vendor, it is important that the supplier's method be used, if the result is to be judged against the supplier's specification.

This section focuses primarily on product testing for release and stability evaluation. Process validation and in process testing are not specifically addressed.

A. Approaches to Limits

Typical approaches to setting limits are listed in Table 9 for several bulk drug parameters which are commonly the subject of specifications. A similar listing for common product parameters is given in Table 10.

1. Limits Based on Process Capability

All limits must be consistent with the capability of the manufacturing process and the analytical method. Otherwise, an unacceptable fraction of lots will have assay results outside the limits. When there are no safety, product performance, or regulatory issues, a statistical approach based on data for batches that are representative of the final process is commonly used to determine appropriate limits. The statistical approach is discussed in detail below.

The applicability of process capability-based limits will vary, depending on unique characteristics of each drug and formulation. Additionally, this approach will not be appropriate for all parameters, since issues of safety, product performance, and regulatory acceptability must also be addressed. Whatever approach is taken to establishing limits, the data should be presented so that the relationship is apparent between the distribution of results and the proposed limits. This can be achieved by presenting the lot data as a histogram, together with the corresponding normal distribution and the proposed limits. An example is provided in Figure 8.

2. Limits Based on Safety, Performance, or Regulatory Acceptability

If safety, performance, or regulatory acceptability require tighter limits than those based on process capability, one must either improve the process or be willing to accept higher than normal lot rejection rates. Thus, process capability must be considered even when the limit will be based on another consideration.

Limits based on product performance can be decided based on the range over which the critical parameter has varied in successful product batches. The range of successful experience can be defined either in scientific studies where the

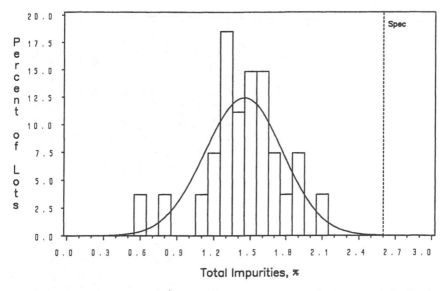

Figure 8 Bulk Drug Impurities Data Histogram. The curve is the normal distribution that most closely represents the data.

variables are intentionally varied, or by retrospective analysis of the development batch database.

Regulatory expectations are expressed in regulations, pharmacopoeiae, and guidelines. Additionally, these expectations can frequently be clarified by meeting with regulatory agencies.

3. Limits Based on Product Quality and Consistency

Occasionally, specifications will be adopted that are designed to assure pharmaceutical elegance but which are not directly related to product safety or efficacy. These vary for different types of products. For example, a color of solution specification is usually an example of an aesthetic specification.

B. Use of Batch Data—Process and Analytical Method Performance

Inevitably, there is variability in a product made using any process. This variability is related to differences in raw materials from batch to batch and differences in the details of how the process is performed, in addition to random variation. The process capability is defined by the mean result for critical product parameters and the variability in the results for the parameters.

When several batches of the product are available, the variability in measurable critical product parameters can be characterized statistically, and limits can be calculated which will provide a particular level of confidence that fu-

Table 9 Issues for Bulk Drug Limits Development

Test	Safety an issue?	Effect on performance an issue?	Common approach to limits development
Identity	Yes	Yes	Agreement with reference chromatogram or IR spectrum
Color	No	Yes (product aesthetics)	For color, approach based statistically on process and method capability is reasonable
Potency	Maybe upper limit, but unlikely to be basis for limits	Yes, but unlikely to be basis for limits	Approach based statistically on process and method capability is reasonable; may be limited by analytical method unless total impurities exceeds 1%
Impurities (incl. stereoisomers)	Yes	Total impurities may impact degradation rate	Approach based statistically on process and method capability is reasonable
Water	No	Stability could be an issue	Limits should be based on either scientific studies or range of successful experience

Residual solvents	Yes	Possibly	Generally based upon compendial limits
Heavy metals, residue on ignition	Yes	No	Compendial limit
Specific optical rotation	Unlikely	Unlikely	Approach based statistically on process and method capability is reasonable
Endotoxins, microcount, pathogens	Yes	No	Compendial limit
Particle size/surface area and other physical tests	No	Yes	Scientific study or range of successful experience; approach based statistically on process and method capability may be useful

Table 10 Issues for Product Limits Development

Test	Safety an issue?	Effect on performance an issue?	Common approach to limits development
Identity	Yes	Yes	Agreement with reference chromatogram or IR spectrum
Appearance, color	No	Yes (aesthetics)	For color, approach based statistically on process and method capability and stability data is reasonable
Potency	Maybe upper limit, but unlikely to be basis for limits	Yes, but unlikely to be basis for limits	Usually ± 5% of label content at time of release; a lower limit at the end of shelf life may be justified by stability data
Content uniformity	Yes	Yes	Compendial limits
Degradation products	Yes	Total impurities may impact degradation rate	Approach based statistically on process and method capability and stability data is reasonable
Sterility, endotoxins	Yes	No	Compendial limit
Physical properties	Yes	Yes	Scientific study or range of successful experience; approach based statistically on process and method capability may be useful

ture lots produced by the same process will meet the limits. Analytical method variability should also be considered in this calculation.

Limits based on process capability should be calculated for all numerical parameters. These limits should be compared with the limits required to assure safety, range of successful experience, and anticipated regulatory acceptability. If the limits based on process capability are the same as or tighter than the limits based on other issues, the limits based on process capability should be considered for filing.

The limits based on process capability may be broader than the limits based on other issues. This can occur either because of a mismatch between the capability of the process and the needs of the product, or because only a small number of batches are available for statistical analysis. In these cases, the calculations can be rearranged to predict the risk of lot rejection associated with the needed limits. In this situation, the costs of the expected lot rejection rate should be compared with the anticipated costs of the alternatives in reaching a decision.

The anticipated rejection rate could be decreased if it is possible to modify the process to reduce the range for the parameter of concern. It might be possible to make the process less sensitive to process variables, or to reduce the range within which some process variables are expected to vary during normal production. If the broad limits are due to limited degrees of freedom, preparation of additional batches may help better define the needed limits.

Alternatively, studies could be performed to justify higher levels of impurities or extend the range of successful experience, or attempts could be made to defend a deviation from anticipated regulatory acceptability. Of course, each of these alternatives would likely increase development costs or risk delaying regulatory approvals.

It is valuable to include a statistician in the process of calculating limits based on process capability. Data sets that fail a normality test and data sets with only a few numbers above a reporting limit or quantitation limit present difficulties that do not appear to have a general solution. A discussion of these issues is beyond the scope of this chapter; consultation with a statistician is recommended.

1. Representative Batches, Number of Batches, Etc.

Specifications should be based on lots made using the intended production process, which are representative both of what the process can make and of the variability in the process that is expected during normal production. Specifications intended to reflect process capability should ideally be based on a substantial number (> 20) of regular production lots. In this case, process variability is characterized by random sampling. However, this is not normally possible at the time specifications are initially being developed.

When enough pilot or production lots have been made to provide confidence that the normal variability of the process has been observed, lot selection should consider the following points:

1.   The process used should be the same as planned for routine production.

2.   Selection should, in general, be based on input characteristics and not based on results. Lots should not be excluded without a failure that is traceable to an assignable cause.

3.   Lot data should be examined for distinguishing effects that would make the pooling of lots unjustified (i.e., scale, site, etc.). If any such effects are detected, additional process characterization is indicated.

When only a few lots have been made using the intended production process, it is unlikely that the entire range of regular production process variability has been sampled. If a sufficient number of production lots is not available, process variability can be characterized using laboratory experiments. For example, results could be obtained for laboratory lots intentionally produced throughout the anticipated range of normal process variability. Although it may not be possible to use data from laboratory batches as part of the regulatory database, knowledge of the variability or ruggedness of the process allows assessment of both the suitability of the limits calculated from the regulatory batches and the degree of control required in the manufacturing process to routinely meet the specifications.

2.   Evolution of Analytical Methods—Replicates

Either of these approaches assesses the capability of a composite process that includes both the manufacturing process and the analytical measurement process. If the lots used for limits development have not been assayed using the same assay and the same number of replicates as are planned for lot release, the effect of assay variability on the distribution width should be examined. The method for calculating tolerance intervals for derived measures, which is discussed in the section below on Use of Stability Data, can be used to calculate appropriate tolerance intervals which include both process and analytical variability.

3.   Statistical Tools—Tolerance Intervals

The general recommended approach for limits based on estimated process capability is use of statistical tolerance limits (66). Tolerance limits include a specified portion of a normally distributed population with a specified degree of confidence. For example, one can calculate the limits within which 99.5% of the population is expected to fall with 95% confidence. This is referred to as 99.5% population coverage. Like confidence intervals, tolerance intervals are calculated by multiplying the estimated standard deviation by a statistical factor derived from the degrees of freedom in the standard deviation measurement.

However, unlike confidence intervals, tolerance intervals take into account the uncertainty in both the standard deviation and the mean, while confidence intervals take into account only the uncertainty in the standard deviation. Tolerance intervals are either one- or two-sided. One-side limits are normally appropriate for impurities, water, and residual solvents. Two-sided limits are normally appropriate for potency and optical rotation. At least eight lots are recommended for the use of statistical tolerance intervals, although four to seven lots can be used with caution. One-sided and two-sided tolerance interval multipliers for 95% confidence and various population coverages are given in Tables 11 and 12, respectively.

Before calculating tolerance intervals, appropriate lots must be selected, the range of coverage and level of confidence (usually 95%) must be decided, truncated data should be addressed, and the normality of the distribution should be tested. A statistician should normally be consulted when truncated data or a nonnormal distribution are involved.

a. What Range of Population Coverage Is Appropriate? Limits based on process capability should generally be set to predict NMT 2% *overall* lot failure rate. Because well-controlled processes are generally expected to result in less than 1% to 2% product rejection rate, at least 98% of the population should

Table 11 One-Sided 95% Tolerance Limit Factors (K) for a Normal Distribution

No. of obs.	P (% coverage)				
	95.0	98.0	99.0	99.5	99.9
5	4.203	5.121	5.741	6.313	7.502
6	3.708	4.516	5.062	5.566	6.612
7	3.399	4.141	4.642	5.103	6.063
8	3.187	3.884	4.354	4.787	5.688
9	3.031	3.695	4.143	4.556	5.413
10	2.911	3.550	3.981	4.378	5.203
15	2.566	3.136	3.520	3.874	4.607
20	2.396	2.933	3.295	3.628	4.318
25	2.292	2.810	3.158	3.478	4.142
30	2.220	2.725	3.064	3.376	4.022
40	2.125	2.613	2.941	3.242	3.865
50	2.065	2.542	2.862	3.157	3.766
100	1.927	2.380	2.684	2.963	3.539
200	1.837	2.276	2.570	2.839	3.395
300	1.800	2.232	2.522	2.787	3.335
400	1.778	2.207	2.494	2.757	3.300
∞	1.645	2.058	2.326	2.576	3.090

Table 12   Two-Sided 95% Tolerance Limit Factors (K) for a Normal Distribution

No. of obs.	P (% coverage)				
	95.0	98.0	99.0	99.5	99.9
5	5.115	6.071	6.722	7.325	8.587
6	4.436	5.265	5.830	6.353	7.448
7	4.022	4.774	5.286	5.761	6.753
8	3.742	4.442	4.918	5.360	6.283
9	3.540	4.202	4.652	5.070	5.943
10	3.386	4.019	4.450	4.849	5.684
15	2.956	3.509	3.885	4.234	4.963
20	2.753	3.268	3.618	3.943	4.622
25	2.632	3.124	3.459	3.769	4.418
30	2.550	3.027	3.351	3.652	4.281
40	2.445	2.902	3.213	3.501	4.105
50	2.379	2.824	3.126	3.407	3.994
100	2.233	2.650	2.934	3.198	3.749
200	2.143	2.543	2.816	3.069	3.597
300	2.106	2.499	2.767	3.016	3.535
400	2.084	2.474	2.739	2.985	3.499
∞	1.960	2.346	2.576	2.812	3.291

be included within the proposed limits (98% coverage). It is straightforward to apply this criterion to each specification separately. However, applying this criterion to each specification separately would likely lead to rejection of more than 2% of production lots, because there are several specifications and not all are correlated.

If there are n independent (uncorrelated) specifications, and the probability of passing each separate limit is p, the probability of simultaneously passing all n limits is p^n. For example, if there are three independent specifications for a particular product, and a 2% rejection rate is predicted for each, the overall rejection rate is expected to be:

$$(1 - 0.98^3) \times 100 = 6\%$$

For p^n to be 0.98 (i.e., 2% predicted *overall* rejection rate), each independent limit must cover the following fraction of the population (p values), depending on n:

n:	1	2	3	4	5
p:	0.98	0.99	0.993	0.995	0.996

For bulk drug, use of 99.5% coverage for each individual limit is suggested. Not all parameters are independent. For example, anhydrous and solvent-excluded potency will be correlated with total impurities content, but will likely be independent of sample water content. In general, five groups of related bulk drug parameters can be identified:

Purity-related: anhydrous and solvent excluded purity, impurities, counterion, and specific rotation;
Drying-related: water and solvents;
Extraneous material: gravimetric insoluble material and particulates;
Microbiology: microcount, pathogens, and endotoxins;
Physical properties: crystal form, particle size, and specific surface area.

Limits based on process capability would normally be set for at least some parameters in at least two or three of these groups (see Table 9), suggesting use of $n \geq 2$. Because the items within each group are not completely correlated, use of the coverage associated with four independent limits is suggested.

For product, use of 99.0% coverage for each individual limit is suggested. Four groups of related product parameters can be identified:

Potency-related: potency, content uniformity;
Chemical purity-related: degradation products;
Microbiology: sterility, endotoxins;
Physical properties: drug release, tablet parameters, suspension resuspendability, solution pH.

However, limits based on process capability would normally be set only for parameters in one or two of these groups (see Table 10). This suggests use of $n = 2$, because the items within each group are not completely correlated.

4. Example: Bulk Drug Impurities at Time of Release

As an example of the use of batch data and tolerance intervals to calculate limits, a limit for an impurity in a hypothetical bulk drug has been calculated. The ICH Document on Impurities in New Drug Substances (67) states that:

> Limits should be set no higher than the level which can be justified by safety data, and, unless safety data indicate otherwise, no lower than the level achievable by the manufacturing process and the analytical capability. In other words, where there is no safety concern, impurity specifications are based on data generated on actual batches of the new drug substance allowing sufficient latitude to deal with normal manufacturing and analytical variation, and, the stability characteristics of the new drug substance.

While the ICH guidelines do not apply directly to animal health pharmaceuticals, they represent current regulatory expectations about the general approach and format for impurities limits. For this example, we assume that there are no safety concerns within the limits which would be proposed based on process capability.

The impurities data for 27 batches are shown in Figure 8 as a histogram with an overlaid normal distribution. These data have a mean of 1.45% and a standard deviation of 0.32%, and pass the W-test for normality. The tolerance interval approach yields a limit of:

Limit = average + (k * standard deviation)
   k = 3.457   (interpolated from Table 11: 27 lots (26 degrees of
                    freedom), 95% confidence, 99.5% coverage, single-sided)
   Average = 1.45%
   Standard deviation = 0.32%
Limit = 1.45% + (3.457 * 0.32%)
Limit = 2.6% at the time of release

C. Use of Stability Data

1. Statistical Tools—Tolerance Intervals for Derived Measures

Measures with more than one source of variability, such as the sum of two separate results, are called derived measures. Calculation of a joint standard deviation for derived measures is straightforward (61). However, calculation of a joint confidence or tolerance interval for derived measures is a difficult statistical problem which is not discussed in most applied statistics books, except for the limiting case where each measurement is defined with the same degrees of freedom. The usual approach to this problem is first to calculate the joint standard deviation and to estimate the joint degrees of freedom. Once the degrees of freedom have been estimated for the derived measure, multipliers such as t (confidence intervals) or k (tolerance intervals) are applied to the joint standard deviation.

The Satterthwaite approximation is probably the most commonly used approach to estimating the degrees of freedom for a derived measure, and is used in a published release limits calculation (56,68). In the Satterthwaite approximation, the appropriate number of degrees of freedom for calculating the confidence or tolerance interval from the standard deviation on the sum is calculated using:

$$DF = \frac{(V_a + V_b)^2}{\left[\dfrac{V_a^2}{DF_a} + \dfrac{V_b^2}{DF_b}\right]}$$

where $V_a = s_a^2$; $V_b = s_b^2$; $DF_a = n_a - 1$; and $DF_b = n_b - 1$.

2. *Example: Bulk Drug Impurities at End of Shelf Life*

Extending the example from above, assume that a stability study with a total of 14 data points showed an average slope of $+0.04\%$/month, with a standard error of 0.02%/month. If the proposed shelf life is 24 months, this corresponds to an expected increase of:

Increase = 0.04%/mo × 24 mo = 0.96%

Standard error = 0.02%/mo × 24 mo = 0.48%

The joint standard deviation, accounting for the variability in the batch data at the time of release (from above) and the variability in the stability data is given by:

Joint standard deviation = $(0.32\%^2 + 0.48\%^2)^{\frac{1}{2}} = 0.57\%$

The degrees of freedom on the joint standard deviation is estimated using the Satterthwaite approximation (see above):

Degrees of freedom = 23

A limit for the end of the product shelf life can be calculated as above, except that 95% coverage is normally used to calculate limits for the end of shelf life:

Limit = average at time of release + average increase on stability +
   (k * standard deviation)
k = 2.33 (interpolated from Table 11: 23 degrees of freedom,
   95% confidence, 95% coverage, single-sided)
Sum of averages = 2.41%;
   Joint standard deviation = 0.57%
Limit = $2.41\% + (2.33 * 0.57\%)$
Limit = 3.8% at the end of shelf life

D. Use of Process Performance Data—Process Development and Validation

Making consistent product, batch after batch, is the main goal of process development and validation. A product will typically be made starting with small research batches, and proceeding through one or more scale-up steps to full production scale. During this process development, the sensitivity of the identified product critical parameters to changes in raw materials and processing details is studied. The critical aspects of the process are determined, and appropriate process controls and in-process testing and limits are specified. The production process is then validated to assure that the process is capable of consistently producing acceptable product. Revalidation should be considered when any changes are made to a production process. Controls on raw materials used in the process could be considered a special case of in-process testing and limits.

Detailed consideration of process validation, process controls, and in-process testing and limits is beyond the scope of this chapter (1).

E. Use of Product Performance Data

1. Efficacy, Pharmacokinetic, Bioavailability, Drug Safety Data

Limits based on product performance, such as in vivo bioavailability (69), can be decided by measuring the critical parameters in the lots that have been successful and basing the limit on the range over which the critical parameter has varied in successful product batches. The range of successful experience (RSE) can be defined either in scientific studies where the variables are intentionally varied, or by retrospective analysis of the development batch database.

The lots at the low and high ends of the RSE should be reassayed appropriately to accurately and precisely define the RSE. The variability of the assay normally used for lot release should then be considered in recommending a limit.

2. Example: Bulk Drug Particle Size

Bulk drug particle size/specific surface area is generally considered an important physical parameter for drug substances, except for solution formulations, and is an example where RSE logic would normally be used. Particle size can critically impact on: dissolution, solubility, and biopharmaceutics; content uniformity in the final product; critical process steps; suspension physical properties; and stability. The range of particle size observed in formulated bulk drug lots that have been successful in each of these areas would be considered the range of successful experience.

For example, several bulk drug lots with a range of particle sizes or specific surface areas might be formulated and used in a bioavailability trial. If all the lots were bioequivalent (no unsuccessful experience), the limits should be set to provide 95% confidence that all successful experience will be accepted, considering the variability of the normal release assay.

On the other hand, if the lot formulated with the lowest surface area bulk drug is not bioequivalent to the other lots, this would represent an unsuccessful experience. In this case, the specification should be set to provide 95% confidence that a similar unsuccessful experience will be avoided, considering the variability of the normal release assay.

ACKNOWLEDGMENT

The authors would like to thank James E. Freeman for helpful discussions, Jackson P. Scholl and Brian C. Hoff for developing the SPARqle acronym, and Richard Gaines for generating the tolerance interval tables.

REFERENCES

1. Berry IR, Nash RA, eds. Pharmaceutical Process Validation. 2nd ed. New York: Marcel Dekker, 1993.
2. FDA Center for Veterinary Medicine. Drug Stability Guideline. 4th Rev. 1990.
3. European Pharmacopoeia. 2nd ed. 19th fascicule, 1995. Preparations for parenteral use, p. 520; capsules: p. 16; tablets, p. 478; liquids for oral administration, p. 672; topical semisolid preparations, p. 132.
4. Commission Directive 92/18/EEC of 20 March 1992 modifying the Annex to Council Directive 81/852/EEC on the approximation of the laws of the member states relating to analytical, pharmacotoxicological, and clinical standards and protocols in respect to the testing of veterinary medicinal products.
5. Committee for Veterinary Medicinal Products. Note for Guidance: Quality of Prolonged and Controlled Release Dosage Forms for Veterinary Use. Approval before release for consultation Dec. 13–15, 1995.
6. United States Pharmacopoeia XXIII. 1088. In Vitro and In Vivo Evaluation of Dosage Forms, 1924 (1995).
7. Handbook of Pharmaceutical Excipients. Washington D.C.: American Pharmaceutical Association, 1994.
8. Mortimore S, Wallace C. HACCP: A Practical Approach. London: Chapman and Hall, 1994.
9. Gilpin RK, Pachla LA. Pharmaceuticals and related drugs. Anal Chem. 1995; 67:295R.
10. Munson JW, ed. Pharmaceutical Analysis: Modern Methods. New York: Marcel Dekker, 1981.
11. Schirmer RE. Modern Methods of Pharmaceutical Analysis. 2nd ed. Boca Raton, FL: CRC Press, 1991.
12. Szepesi G. HPLC in Pharmaceutical Analysis. Boca Raton, FL: CRC Press, 1990.
13. Wainer IW. Liquid Chromatography in Pharmaceutical Development: An Introduction. Springfield, OR: Aster, 1985.
14. Snyder LR, Glajch JL, Kirkland JJ. Practical HPLC Method Development. New York: John Wiley, 1988.
15. Chowhan ZT. Sampling of particulate systems. Pharm Tech. 1994; April: 48.
16.. Hewitt W, Vincent S. Theory and Application of Microbiological Assay. San Diego: Academic Press, 1989.
17. Borchert SJ, Abe A, Aldrich DS, Fox LE, Freeman JE, White RD. Particulate matter in parenteral products: a review. J Parenter Sci Technol. 1986; 40:212.
18. Hokanson GC. A life cycle approach to the validation of analytical methods during pharmaceutical product development. Part I. The initial method validation process. Pharm Tech. 1994; 118 (Sept).
19. Hokanson GC. A life cycle approach to the validation of analytical methods during pharmaceutical product development. Part II. Changes and the need for additional validation. Pharm Tech. 1994; 92 (Oct).
20. Taylor JK. Validation of analytical methods. Anal Chem. 1983; 55:600A.
21. Clarke GS. The validation of analytical methods for drug substances and drug products in UK pharmaceutical laboratories. J Pharm Biomed. Anal. 1994; 5:643.
22. Carr GP, Wahlich JC. A practical approach to method validation in pharmaceutical analysis. J Pharm Biomed Anal. 1990; 8:613.

23. International Conference on Harmonization. Validation of Analytical Procedures—Draft Consensus Text, Step 2 of the ICH Process. Oct. 26, 1993.

24. International Conference on Harmonization; Draft Guideline on Validation of Analytical Procedures for Pharmaceuticals; Availability. Fed Reg. 1994; 59:9750.

25. United States Pharmacopoeia XIII. 1225. Validation of Compendial Methods, 1982 (1995).

26. EEC Draft Notes for Guidance on Analytical Validation, III/844/87-EN. Rev. Dec. 9, 1988.

27. European Union CVMP Quality Guideline Number 1.5a. Analytical validation. Adopted January 1992.

28. FDA Kansas City District Office. Analytical method validation. Presented at GMP Compliance Workshop for Animal Drug Manufacturers, Nov. 21, 1991.

29. FDA Center for Drug Evaluation and Research. Reviewer Guidance: Validation of Chromatographic Methods. Washington DC: November 1994.

30. Kirschbaum JJ. Synergistic use of multiple assays and achievement of mass balance to validate analytical methods. Trends Anal Chem. 1988; 7:16. (1988).

31. Seaton GGR, Marr JGD, Clark BJ, Fell AF. Chemometric methods for the validation of peak homogeneity in HPLC. Anal Proc. 1986; 23:424.

32. Massart DL, Vandeginste BGM, Deming SN, Michotte Y, Kaufman L. Chemometrics: A Textbook. Amsterdam: Elsevier, 1988, Chap. 7.

33. Analytical Methods Committee. Recommendations for the definition, estimation, and use of the detection limit. Analyst 1987; 112:199.

34. Grushka E, Zamir I. Precision in HPLC. In: Brown PR, Hartwick RA, eds. High Performance Liquid Chromatography. New York: John Wiley and Sons, 1989:529.

35. Kischbaum BJ, Perlman S., Joseph J, and Adomovics J. J Chromatogr Sci. 1984; 22:27–39.

36. Mulholland M. Ruggedness testing in analytical chemistry. Trends Anal Chem. 1988; 7:383.

37. Skoug JW, Halstead GW, Rohrs BR, Theis DL, Freeman JE, Fagan DT. Strategy for the development and validation of dissolution tests for solid oral dosage forms. Pharm Tech. 1996; 58 (May).

38. USP Subcommitee on Biopharmaceutics. In vitro/in vivo correlation for extended-release oral dosage forms. Pharm Forum. July/Aug. 1988:4160.

39. MacLeod SK. Moisture determination using Karl Fischer Titrations. Anal Chem. 1991; 63:557A.

40. Foust DW, Bergren MS. J Chromatogr. 1989; 469:161.

41. Taylor JK. Quality Assurance of Chemical Measurements. Chelsea, MI: Lewis Publishers, 1987.

42. Wiggins DE. How to set realistic system suitability criteria. LC-GC Magazine. 1989; 7:851.

43. Darnowski RJ. Quantitative chromatographic system suitability tests revisited. Pharm. Forum. 1985; Nov-Dec:941.

44. Wahlich JC, Carr GP. Chromatographic system suitability tests—what should we be using? J Pharm Biomed Anal. 1990; 8:619.

45. US Code of Federal Regulations, 21. Parts 210 and 211.

46. Rosenbrook W. Jpn J Antibiotics. 1979: XXXII(suppl):s211–s227.
47. Stability testing of new drug substances and products. ICH Tripartite Guideline Endorsed by the ICH Steering Committee at Step 4 of the ICH Process. 1993; Oct. 27:11.
48. Haynes JD. J Pharm Sci. 1970; 60:927.
49. Dukes GR, Bibart CH. Drug Information J. 1989; 23:441–447.
50. Futscher N, Schumacher P. Pharm Ind. 1972; 34:479–483.
51. Grimm W, Schepky GN. Pharmaz Bet. 1980; 28:335–372.
52. Grimm W. Drugs Made Ger. 1986; 29:39–47.
53. Fusari SA, Hostetler GL. Reference thermal exposures and performance of room temperature stability studies. Pharm Tech. 1984; 8:48–54.
54. Stability tests on active substances and finished products. Notes for Guidance Concerning the Application of Section F of Part I of the Annex to Directive 75/318/EEC III/66/87-EN Rev. 4. Brussels, 1988:7.
55. Carstensen, JT, Nelson E. J Pharm Sci. 1976; 65: 311.
56. Allen PV, Dukes GR, Gerger ME. Pharm Res. 1991; 8:1210–1213.
57. Dunn OJ, Clark, VA. Applied Statistics: Analysis of Variance and Regression. New York: John Wiley and Sons, 1974; 307–322.
58. Carstensen JT. Drug Stability: Principles and Practices. New York: Marcel Dekker, 1990:226-247. (1990)
59. Neter J, Wasserman W. Applied Linear Statistical Models. Homewood, IL: Richard D. Irwin, 1974; 703–713.
60. Draper NR, Smith H. Applied Linear Regression Analysis. 2nd ed. New York: John Wiley and Sons, 1981; 58–59.
61. Taylor JR. An Introduction to Error Analysis. Mill Valley, CA: University Science Books, 1982:40–80.
62. Snedecor GW, Cochran WG. Statistical Methods. 7th ed. Ames, IA: Iowa State Press, 1980: 97–98.
63. PMA Joint QC-PDS Stability Committee. Stability concepts. Pharm Technol. 1984; 8(6):42–48.
64. Pramar Y, Gupta VD. Preformulation studies of spironolactone: Effect of pH two buffer species, ionic strength and temperature on stability. J Pharm Sci. 1991; 80:551–553.
65. Tripartite guideline for the photostability testing of drug substances and products. Draft Step 2 document of the ICH Process. July 31, 1995.
66. Wadsworth HM, Stephens KS, Godfrey AB. Modern Methods for Quality Control and Improvement. New York: John Wiley, 1986; 408–419.
67. Boehlert JP. ICH document—impurities in new drug substances. Sixth draft. March 15, 1994.
68. Winer BJ. Statistical Principles in Experimental Design. New York: McGraw-Hill, 1962:36–39.
69. FDA Center for Veterinary Medicine. Bioequivalence Guidance, 1996. Jan. 1996.

7

Bioavailability Bioequivalence Assessments

MARILYN N. MARTINEZ and MELANIE R. BERSON
Center for Veterinary Medicine, United States Food and Drug Administration, Rockville, Maryland

I. INTRODUCTION

Characterization of a drug's absorption, distribution, metabolism, and excretion (ADME) is an important goal in the development of new chemical entities. These pharmacokinetic variables should be considered when comparing drug bioavailability among target animal species, routes of administration, or enantiomeric forms.

Relative bioavailability comparisons are an assessment of comparability among drug products based on a recognized relationship between drug pharmacequivalence studies is to determine whether differences in product manufacturing and formulation will affect the rate and extent of drug absorption. The fundamental assumption of all bioequivalence testing is if the rate and extent of drug absorption are comparable, then the products will be medically indistinguishable and therefore interchangeable.

This chapter provides a review of scientific principles underlying the development of bioavailability/bioequivalence studies. These principles were utilized in the development of the Center for Veterinary Medicine (CVM)/Food and Drug Administration (FDA) Bioequivalence Guidance.* Due to the broad range

---

*To obtain the current CVM Bioequivalence Guidance, write to the Communication and Education Branch, HFV-12, Center for Veterinary Medicine, Food and Drug Administration, 7500 Standish Place, Rockville, Maryland 20855. The reader is also referred to the review of the 1993 Veterinary Drug Bioequivalence Workshop (1).

of topics covered in this chapter, discussions are limited to the fundamental points governing each issue. An extensive set of references are provided if greater detail is needed on specific topics.

II. PRODUCT BIOEQUIVALENCE: REGULATORY BACKGROUND

A. History of the GADPTRA

The basis for bioavailability/bioequivalence (BA/BE) study interpretation is that the rate and extent of drug appearance in the central compartment (sampling compartment) correlate with the corresponding clinical response profile. This concept is the scientific foundation for the generic animal drug legislation.

The Generic Animal Drug and Patent Term Restoration Act (GADPTRA) signed into law on Nov. 16, 1988 [Public Law 100-670], amends section 512 of the Federal Food, Drug, and Cosmetic Act (FFDCA) [21 U.S.C. 360b]. GADPTRA provides statutory authority to approve abbreviated new animal drug applications (ANADAs) for generic copies of off-patent animal drugs approved for safety and efficacy.

Prior to GADPTRA, the FDA could approve abbreviated applications only for copies of animal drugs approved prior to 1962 and found to be effective under the DESI (drug efficacy study implementation) program. The intent of GADPTRA is to encourage competition and lower animal drug prices by allowing abbreviated applications for copies of previously approved drugs, without the generic drug sponsor duplicating the safety and efficacy studies that were required for the original new animal drug application (NADA) approval.

The FFDCA, as originally enacted in 1938, required only that new drugs be adequately tested to show safety for the labeled conditions of use. The 1962 amendments to the Act required the sponsor to submit substantial evidence to show that a new drug product was effective as well as safe for its intended uses. The 1962 amendments required the FDA to evaluate effectiveness of every product on the market between 1938 and 1962.

FDA enlisted the National Academy of Sciences/National Research Council (NAS/NRC) to accomplish the efficacy review. FDA requested the holders of New Drug Applications (NDAs) to submit data to the Agency to allow evaluation of effectiveness claims for the drug products. FDA provided the data to NAS/NRC. NAS/NRC reviewed the data and made recommendations to FDA regarding the effectiveness of therapeutic classes of products.

FDA reviewed the NAS/NRC recommendations and published DESI notices in the Federal Register, rating products as to their effectiveness. Sponsors were required to supplement their NDAs to conform to conditions imposed by the DESI notices for any products that were rated less than effective. From 1962 to 1988, ANADAs could be submitted only for copies of products approved prior to Oct. 10,1962 (the effective date of the Drug Amendments of 1962 [Public Law 87-781]) and rated as "effective" by the DESI review. The 1968 Animal Drug Amendments to the FFDCA first defined new animal drug and consolidated into one place the various parts of the Act that related to regulation of animal drugs.

GADPTRA was enacted on Nov. 16, 1988, but no approvals could be granted under the new law until Jan. 1, 1991. GADPTRA provides the legislative authority for approval of duplicates and certain related versions of animal drugs previously approved by FDA for safety and effectiveness and listed in the agency publication of approved animal drug products. GADPTRA was preceded by the Drug Price Competition and Patent Term Restoration Act of 1984 (Public Law 98-417), which was very similar legislation to provide for abbreviated new drug applications (ANDAs) for prescription human drugs. GADPTRA directs the Agency to follow the provisions of the human generic drug regulations [21 CFR 314.55 and 21 CFR 320] until the Agency issues regulations to implement GADPTRA.

Certain drug categories are ineligible to be copied under the provisions of GADPTRA. Drugs manufactured primarily using biotechnology, drugs subject to a notice of hearing (NOH) on a proposal to withdraw approval, and drugs withdrawn from sale for safety or effectiveness reasons cannot be copied through generic applications. In addition, generic approvals cannot be granted for copies of products until the relevant marketing exclusivity periods granted under the provisions of GADPTRA or patents issued by the Patent and Trademark Office (PTO) have expired.

GADPTRA defines the requirements for approval of an ANADA. In general, the data requirements necessary for approval of an ANADA are limited to bioequivalence studies and, in the case of drugs for food-producing animals, tissue residue withdrawal studies. The law requires that the generic product be shown to be bioequivalent to a previously approved pioneer drug product.

GADPTRA permits an ANADA for an animal drug product that is the same as an animal drug product listed in the approved animal drug product list published by the Agency (listed new animal drug) with respect to conditions of use recommended in the product labeling, active ingredient(s), dosage form, strength, and route of administration. An ANADA applicant may petition the Agency under section 512(n)(3) of the Act for permission to file an ANADA for a new animal drug product with certain defined changes from an approved pioneer product. The changes are limited to the following:

1. Change of one active ingredient in an approved Ian combination product
2. Change in dosage form
3. Change in dosage strength

4. Change in route of administration
5. Change in one active ingredient in a feed mixed combination

A generic sponsor may file a suitability petition to request that the specific change in the pioneer product be allowed. The petition must follow the format and content described for a citizen petition [21 CFR 10.20] which is a public document filed with the FDA. The FDA response to the petition is also a public document. The Agency determines whether to deny or approve the petition.

Petitions are approved if the FDA determines that the requested change in the pioneer product would not require investigations to establish the efficacy, target animal safety, or human food safety of the new product. An approved suitability petition gives the petitioner permission to file an ANADA for a generic product with the requested change in the pioneer product. However, the ANADA cannot be approved unless the application includes a successful demonstration of bioequivalence between the pioneer product and the generic product with the allowable change from the pioneer product. For products for food-producing animals, a tissue residue withdrawal study is generally required.

CVM has processed numerous suitability petitions requesting certain changes in approved new animal drugs (pioneer products). Examples of requests that have been approved by CVM include the following: change in dosage form from an oral paste to a powder delivered in the feed; change from two tablet strengths to four tablet strengths; change in dosage form from a soluble powder to a liquid concentrate for drinking water solutions; substitution of one approved salt for another in one of the active ingredients in a combination product; change in strength of an oral solution.

Examples of requests that have been denied include: substitution of one active ingredient for another in a single active ingredient medicated feed; change in dosage form to provide microencapsulation of the active ingredient in an implant product; change in strength, dosage form, and inactive ingredients in an injectable product; and change in dosage form from a lyophilized powder to be reconstituted to an aqueous injectable suspension (pioneer product) to an oil-based, ready-to-use injectable suspension (generic product). In many of the examples of denied petitions, the Agency determined that the proposed innovation in the generic product presented target animal safety issues that could not be addressed through bioequivalence studies, and therefore, the requested change could not be approved as a generic product.

B. Definitions of Bioavailability and Bioequivalence

Systemic drug absorption depends on the physicochemical properties of the drug, the nature of the drug formulation, and the physiology of the absorption site. There are four types of equivalence to be considered (2):

1. Chemical equivalence: two or more dosage forms which contain the same quantity of a particular drug
2. Clinical equivalence: the same drug in two or more dosage forms which provide identical in vivo effects as measured by some pharmacological response or by the control of the symptoms of a disease
3. Therapeutic equivalence: structurally different chemicals that yield the same clinical effect
4. Bioequivalence: a drug contained in two or more similar dosage forms reaches the general circulation at the same relative rate and to the same relative extent

Based on human drug regulations (21 CFR 320.1), bioavailability can be defined as the rate and extent to which the active ingredient or active moiety is absorbed from a drug product and becomes available at the site of action. For drug products that are not intended to be absorbed into the bloodstream, bioavailability may be assessed by measurements intended to reflect the rate and extent to which the active ingredient or active moiety becomes available at the site of action.

GADPTRA states that a new animal drug shall be considered to be bioequivalent to the approved new animal drug if:

1. The rate and extent of absorption of the drug do not show a significant difference from the rate and extent of absorption of the approved new animal drug referred to in the application when administered at the same dose of the active ingredient under similar experimental conditions in either a single dose or multiple doses, or

2. The extent of absorption of the drug does not show a significant difference from the extent of absorption of the approved new animal drug referred to in the application when administered at the same dose of the active ingredient under similar experimental conditions in either a single dose or multiple dose, and the difference from the approved new animal drug in the rate of absorption of the drug is intentional, is reflected in its proposed labeling, is not essential to the attainment of effective drug concentrations in use, and is considered scientifically insignificant for the drug in attaining the intended purposes of its use and preserving human food safety, or

3. In any case in which the Secretary [of HHS] determines that the measurement of the rate and extent of absorption or excretion of the new animal drug in biological fluids is inappropriate or impractical, an appropriate acute pharmacological effects test or other test of the new animal drug and, when deemed scientifically necessary, of the approved new animal drug referred to in the application in the species to be tested or in an appropriate animal model does not show a significant difference between the new animal drug and such

approved new animal drug when administered at the same dose under similar experimental conditions.

The second definition refers to certain sustained-release (SR) dosage forms. The third definition refers to the hierarchy of preference for determination of bioequivalence. Ultimately the selection of the method for demonstrating product bioequivalence will depend on the purpose of the study, the availability of analytical methodologies, and the nature of the drug product.

The CVM Bioequivalence Guidance provides the preferred hierarchy of bioequivalence studies as (in descending order of sensitivity): a blood concentration-time profile study; pharmacologic end-point study; and clinical end-point study. When absorption of the drug is sufficient to measure drug concentration *directly* in the blood (or other appropriate biological fluids or tissues) and systemic absorption is relevant to the drug action, then a blood (or other biological fluid or tissue) level bioequivalence study should be conducted. The blood level study is generally preferred above all others as the most sensitive measure of bioequivalence. The generic animal drug sponsor should provide justification to CVM for choosing either a pharmacologic or clinical end-point study over a blood profile (or other biological fluids or tissues) study.

When the measurement of the rate and extent of absorption of the drug in biological fluids cannot be achieved or is unrelated to drug action, a pharmacologic end point (i.e., a drug-induced physiologic change that is related to the approved indications for use) study may be conducted. Lastly, in order of preference, if drug concentrations in blood (or fluids or tissues) are not measurable or are inappropriate, and there are no appropriate pharmacologic effects that can be monitored, then a clinical end-point study may be conducted, comparing the test (generic) product to the reference (pioneer) product and a placebo (or negative) control.

Whenever possible, the comparison of blood concentration-time profiles for the active drug moiety is recommended. However, clinical or pharmacologic end-point studies are used to test product bioequivalence when the drug is not systemically absorbed. The assumption is that the rate and extent of drug delivery to the site of action will dictate its pharmacologic effects. This assumption may not be correct if the effect lies on the plateau portion of a dose-response curve. Moreover, the variability associated with a pharmacodynamic effect is generally greater than that associated with drug pharmacokinetics (3). Therefore, these alternative test procedures tend to be less sensitive measures of product relative bioavailability.

C. Human Food Safety

ANADA requirements, as defined by GADPTRA, include information to show that the withdrawal period for the generic animal drug is consistent with the tolerances established for the approved new animal drug. Alternatively, if the withdrawal period is proposed to be different, the sponsor must provide information showing that the residues of the generic animal drug at the proposed different withdrawal period is consistent with the tolerances established for the approved new animal drug.

While blood level comparisons are generally adequate to confirm product comparability with respect to target animal safety and efficacy, they may not accurately reflect the relative drug tissue concentrations at the innovator's approved withdrawal time. In particular, potential differences in drug terminal disposition may not be quantifiable in the blood due to limitations associated with the analytical method (1). In these situations, products could successfully meet standard blood level bioequivalence criteria despite markedly different drug concentrations within the target animal's edible tissues. Consequently, the Center has adopted the recommendation of the Panel on Human Food Safety at the 1993 Veterinary Drug Bioequivalence Workshop (1) that a tissue residue depletion study would generally be required for the approval of a generic animal drug product in food-producing animals.

Certain drug products may be exempt from the requirement to conduct a tissue residue depletion study. Exemptions include products for which a waiver of in vivo bioequivalence testing is granting, and products for which the assay method used in the blood bioequivalence study is sensitive enough to measure blood levels of the drug for the entire withdrawal period assigned to the reference product (e.g., drug products in which the innovator is assigned a zero withdrawal time). Other requests for waiver of the tissue residue study will be considered on a case-by-case basis.

III. CHEMISTRY AND MANUFACTURING ISSUES

A. Product Dissolution

Several factors can affect the rate and extent to which a drug is absorbed: in vivo disintegration; in vivo dissolution; local blood flow (perfusion rate-limited); and drug movement across biological membranes (diffusion rate-limited). The impact of local blood flow and drug permeability can be seen by the effect of injection site on parenteral drug bioavailability (4–6).

More commonly, however, the rate and extent of drug absorption is controlled by the process of in vivo dissolution (7). This is true both for oral dosage forms (8,9) and for parenteral products (10,11). Factors that can modify the dissolution characteristics include formulation, surface area, particle size, drug crystalline state, compression force (solid oral dosage forms), and manufacturing procedure (e.g., wet vs. dry granulation, blending time) (12–16).

Some drug-excipient interactions may be overlooked if not tested under appropriate conditions. For example, due to the pKa of the excipients and/or the drug molecule itself, dissolution properties may be highly dependent on the pH of the dissolving fluid (17). In these situations, products may appear to exhibit similar in vitro dissolution profiles when tested under one set of conditions, but very different profiles when tested under other in vitro conditions. Alternatively, interactions may exhibit a time dependency (18). These interactions may not be distinguishable shortly after product manufacture but would alter product performance over time. Such effects often result in a decrease in product shelf life. The effect of formulation on the bioavailability of pharmaceuticals intended for veterinary use has been reviewed by Pope (11).

B. Scale-Up

Production scale-up can affect product bioavailability (19). Therefore, pivotal target animal studies (clinical/bioequivalence trials, residue depletion studies, and target animal safety studies) as well as stability studies should be conducted on batches that represent at least 10% of the largest proposed production batch (Manufacturing Guideline, Docket No. 92D-0039). The test batch must be produced on equipment comparable to that ultimately used for production purposes.

C. Bioequivalence Waiver Policy

For certain generic drug products, the Agency may waive the requirement for demonstration of bioequivalence to the pioneer product. In general, the bioequivalence between solutions is considered to be self-evident for solutions with the same active and inactive ingredients in the same concentrations and with the same pH and physicochemical characteristics as the pioneer product.

Categories of products that may be eligible for waivers include parenteral solutions intended for injection by the intravenous, subcutaneous, or intramuscular routes of administration; oral solutions or other solubilized forms; topically applied solutions intended to produce local therapeutic effects; and inhalant volatile anesthetic solutions.

When a generic drug sponsor seeks the approval of a multistrength solid oral dosage form product, waiver of in vivo bioequivalence study requirements may be granted if: in vivo bioequivalence has been demonstrated between the highest strengths of the generic and pioneer products; the lower dosage strengths exhibit in vitro dissolution characteristics comparable to that of the corresponding strength of the pioneer product; and the generic product formulations are dose-proportional with respect to the ratio of active and inactive ingredients. In cases where the active drug has poor water solubility, manufacturing and tableting conditions may alter product bioavailability, and in vivo bioequivalence studies of both the highest and lowest dosage strengths may be needed. A bioequivalence waiver may be granted on the basis of in-vitro dissolution if minor changes in the composition or manufacturing method of an approved product are proposed (20–22).

D. Correlating In Vivo Bioavailability with In Vitro Dissolution Characteristics

Correlations between in vivo bioavailability and in vitro dissolution have been identified for many human drug products (21–23). However, for some products, in vitro dissolution does not correspond with in vivo bioavailability. The inability to correlate in vivo bioavailability with in vitro product performance may reflect the selection of inappropriate in vitro test conditions. Problems in methodology may be associated with the rotational speeds of the dissolution apparatus (24) or the choice of media within which product dissolution rates are being tested (17,25–27). Alternatively, a lack of correlation can occur when in vivo dissolution is not the rate-limiting step in product absorption or when the in vitro test fails to account for physiologic factors affecting drug absorption (28,29).

Some excipients can directly exert a positive or negative effect on in vivo drug absorption. For example, mannitol, an "inert" excipient, can cause a significant reduction in drug bioavailability by decreasing product in vivo transit time (30). Conversely, excipients may enhance product bioavailability by affecting physiological factors which can impede drug absorption.

Low oral bioavailability may be associated with poor membrane permeability, presystemic biotransformation, or drug degradation within the lumen of the gastrointestinal (GI) tract. Examples of formulation approaches for overcoming these bioavailability problems include (31):

1. Coadminstration of metabolic inhibitors to reduce the extent of presystemic drug metabolism
2. Formulation of polar compounds with a counterion to increase drug lipophilicity
3. Inclusion of nonsurfactant membrane permeation enchancers such as EDTA, bile salts, fatty acids, and salicylates
4. Use of surfactants to enhance drug solubility, prolong GI residence time, protect the drug from luminal degradation, enhance membrane permeability, and protect the drug from gut wall metabolism.

The impact of these formulation approaches on in vivo drug bioavailability may not be reflected in their effect on in vitro product performance. The use of absorption enhancers (such as bile salts, anionic and nonionic detergents, medium-chain glycerides, salicylates, acyl amino acids, acylcarnitines, lysolecithin, ethylenediaminetetraaceticacid, and particulate carriers) can enhance the bioavailability of polar compounds without significantly affecting product in vitro dissolution characteristics (32–34). Conversely, surfactants used to increase the rate of in vitro dissolution of lipophilic compounds may have the opposite effect on in vivo bioavailability (35,36).

Drug bioavailability can also be markedly affected by the rate and extent to which a drug is delivered to specific portions of the GI tract. Physiological factors that can affect the site of drug absorption include surface area, membrane permeability, presence of specialized transport mechanisms, local blood and lymph flow, and presystemic intestinal metabolism (28). Presystemic intestinal drug loss can result from biotransformation processes associated with bacterial gut flora, digestive enzymes secreted into the gut lumen, and metabolizing enzymes contained within the mucosal cells lining the gut wall (37,38).

Gut wall metabolism can be a significant source of drug loss for a variety of compounds, including antibiotics (39). Enzymatic activity resides primarily in the mucosal epithelial cells, is greatest in the villous tips, and decreases progressively toward the crypts (37,38). It also tends to be greater in the duodenum and jejunum than in either the ileum or colon (37,38), although exceptions do occur (38). A multitude of enzyme systems have been implicated in these biotransformation processes including esterases, deacylases, alcohol dehydrogenase, acetyltransferase, cytochrome P-450, sulfotransferase, glucuronosyltransferase, monoamine oxidase, epoxide hydrolase, β-glucuronidase, peptidases, and hydrolases (37,38,40). An extensive list of cytochrome P-450 superfamily members identified with eukaryotic and prokaryotic species has recently been published (41).

The process of presystemic (intestinal) drug loss may be facilitated by p-glycoprotein-mediated countertransport mechanisms (42–44). It has been suggested that by slowing the mucosa-to-serosa transport of drugs, these proteins can increase the intracellular residence time of the drug and increase drug exposure to intracellular enzymes, thereby increasing the extent of intracellular drug biotransformation (45). Although these proteins have been implicated in the development of multiple drug resistance (46,47), they have only recently been considered important in determining the bioavailability of certain compounds (42).

Intestinal degradation of xenobiotics can decrease the bioavailability of a wide variety of compounds (38). Therefore, efforts are under way to identify methods for minimizing this bioavailability problem. Accordingly, there exists a rapidly growing list of factors that can significantly affect presystemic (gut and hepatic) drug metabolism. These include vitamin A (decrease P-450 activity) (48), excess dietary cholesterol (increase P-450 activity) (48), iron or selenium deficiency (decrease P-450 activity) (48), starvation (decrease P-450 activity) (49), grapefruit juice (decrease P-450 activity) (50,51), presence of systemic infections (decrease P-450 activity) (52), polyethylene glycol derivatives (decrease p-glycoprotein activity) (53), and possible vitamin E (decrease p-glycoprotein activity) (54). A list of substrates, inhibitors and inducers of the cytochrome P-450 and p-glycoprotein system are provided elsewhere (55–57). It should be noted that differences may exist in the effect of regulators on the P-450 systems of the liver and the intestine (48).

Drug absorption is a complex process reflecting the interaction of animal physiology and drug formulation. Due to the complex nature of these interactions, comparative product in vitro dissolution data may not adequately reflect differences in product in vivo performance. Consequently, for approval of the majority of oral and parenteral (not intravenous) dosage forms, product bioequivalence must be confirmed on the basis of in vivo test methods.

IV. BIOEQUIVALENCE STUDY PROTOCOL CONSIDERATIONS

The use of comparative blood profiles as the basis of generic drug approval requires an assumption that comparable blood concentrations will yield comparable clinical responses. This assumption will hold true only if:

1. The active molecule passes goes through the systemic circulation to reach its site of action.
2. Blood is the body compartment that provides the most sensitive discriminator of product inequivalence.
3. The study design accurately reflects product relative bioavailability under conditions of clinical use.

This last point is critical in determining the appropriateness of a bioequivalence study design.

A. Study Design

The fundamental assumption in bioequivalence testing is that clearance remains constant across study periods (crossover design) or between subjects (parallel design). This follows from the equation:

$$AUC = FD/Cl$$

where F is the fraction of the administered dose (D) which reaches the systemic circulation, and Cl is the systemic clearance of the drug (58). By substituting FD/Cl for AUC, the systemic availability comparison can be described as:

$$[FD/Cl]_{test}/[FD/Cl]_{ref}$$

To provide an unbiased assessment of product relative systemic availability (F), clearance and dose must be constant.

B. Parallel Versus Crossover Study Design

Bioequivalence studies are most frequently conducted as crossover trials since this design offers the advantage of allowing treatments to be compared within subjects. Since intrasubject variability tends to be much smaller than intersubject variability, crossover designs tend to be more powerful than parallel designs for identifying statistically significant treatment effects.

The simplest crossover design is a 2×2 factorial in which all subjects receive both treatments (Fig. 1). In this design, half the subjects are randomly assigned to sequence 1 (receiving treatment 1 in period 1 and treatment 2 in period 2) while the other half is assigned to sequence 2 (receiving treatment 2 in period 1 and treatment 1 in period 2).

Individual cell means are termed y_{12}, y_{21}, y_{11}, and y_{22}, respectively. The difference between the cell means for period 2 – period 1 equals:

$$y_{12} - y_{11} = (\text{TRT B} - \text{TRT A}) + (\text{PER 2} - \text{PER 1}) + CO_{A:B} \tag{1}$$

and

$$y_{22} - y_{21} = (\text{TRT A} - \text{TRT B}) + (\text{PER 2} - \text{PER 1}) + CO_{B:A} \tag{2}$$

where TRT = treatment, PER = period, and CO = carryover effects (59,60). Upon subtracting these values and multiplying the differences by 0.5, the resulting comparison of the individual cell means equals:

$$0.5 \times \{(y_{12} - y_{11}) - (y_{22} - y_{21})\} = (\text{TB} - \text{TA}) + 0.50\,(CO_{A:B} - CO_{B:A}) \tag{3}$$

Based on equation 3, the difference between treatment effects within any given sequence is a function both of the actual difference between treatment means and the difference in the carryover effects from periods 1 and 2. Note that period effects have been subtracted out of the model. Thus, the presence of a statistically significant period effect does not invalidate the treatment comparison.

	PER 1	PER 2
GRP 1	TRT A (y_{11})	TRT B (y_{12})
GRP 2	TRT B (y_{21})	TRT A (y_{22})

Figure 1   General layout of a 2×2 crossover study design.

However, to achieve an unbiased treatment comparison, one of two conditions must be met:

1. $CO_{A:B} = 0$ and $CO_{B:A} = 0$ or
2. $CO_{A:B} = CO_{B:A}$

To avoid the possibility of unequal carryover effects, CVM recommends that 99.9% of the assayed moiety be eliminated from the body prior to administering any subsequent treatments in a single-dose bioequivalence trial. This translates into a drug washout interval equal to 10 times the apparent terminal elimination half-life. Additional time may be needed if physiologic carryover effects are anticipated (e.g., enzyme induction, electrolyte imbalance, etc.). When carryover effects are anticipated or if the duration of the washout interval risks changes in drug clearance, the investigator is advised to consider the use of a parallel study design.

C. Multiple-Dose Studies Versus Single-Dose Studies

Whereas single-dose studies are generally adequate for demonstrating product bioequivalence for most immediate-release dosage forms, it may occasionally be necessary to obtain blood level comparisons after multiple dosing. Unless otherwise contraindicated by the product label, multiple-dose studies should be carried out to steady state (dosing for 4 to 7 terminal elimination half-lives) (61). Three consecutive C_{min} (trough) values should be obtained (generally, predose blood concentrations) to confirm that steady-state concentrations have been achieved. Multiple-dose studies may be appropriate in the following situations:

1. Nonlinear Kinetics

Multiple-dose studies should be conducted when a drug exhibits nonlinear (saturation) kinetics since differences observed under single-dose conditions may be inflated during actual conditions of use (i.e., when drug accumulation results in enzyme saturation). Nonlinear kinetics refers to time (e.g., autoinduction) or dose dependencies in pharmacokinetic parameters (61–63). Nonlinear kinetics can be associated with absorption factors (such as saturable carrier systems [e.g., oral absorption of amoxicillin, cyclacillin, cephalexin, cefroxadin, cephradine, and cefadroxil], dose-related changes in GI transit time, presystemic drug metabolism), saturable binding to plasma proteins, blood cells or extravascular tissue, saturable excretion mechanisms and end-product inhibition (i.e., the formation of metabolites which compete for parent drug elimination).

Based on a literature review, Ludden (63) concludes that of the hundreds of marketed human drug entities, fewer than 40 have nonlinear kinetic properties which require that dose adjustments occur during clinical use. Similarly, for the majority of veterinary drug products, single-dose bioequivalence trials (using the highest approved dose) will be adequate to confirm product interchangeability.

2. Long Elimination Half-Life, Large Intersubject Variability

A sponsor may choose to conduct a multiple-dose crossover study in lieu of a single-dose parallel study when a drug product has a very long elimination half-life and intrasubject variability is markedly less than intersubject variability. In this situation, the treatments are immediately crossed over, thereby eliminating the use of a traditional washout interval (64–66). Thus, rather than waiting for 10 elimination half-lives between study treatments (time for 99.9% drug elimination), the time interval between blood profile determinations will equal the time to reach 95% steady-state concentrations (five times the apparent terminal elimination half-life). Although this design does not reduce the total time associated with the study, it will significantly reduce the time interval between blood samples and thereby reduce the likelihood of physiological changes (size, maturity, health) which could bias the treatment comparison.

3. Sustained-Release (SR) Preparations

Multiple-dose studies are generally needed to confirm the equivalence of SR products labeled for repeated administration. Assuming linear kinetics, dosing to steady state will enable a more accurate assessment of product relative bioavailability since the extent of product bioavailability (area under the curve [AUC]) equals AUC measured from time zero to time infinity (AUC_{0-inf}) (61,67). In addition, since AUC over a dosing interval at steady state ($AUC_{0-\tau}$) equals AUC_{0-inf}, fewer samples are needed to quantify the extent of drug absorption. Steady-state studies will also provide accurate assessments of peak (C_{max})-to-trough (C_{min}) ratios (65), a measure that may be critical in the clinical response to a drug.

When evaluating the bioequivalence of SR formulations (or when evaluating new SR formulations), blood profiles should be obtained over both the initial and the final dosing intervals (68–70). The initial dose provides an estimate of product absorption rate since absorption and elimination processes are highly confounded upon multiple dosing (71). Relative C_{max} values and steady-state peak-to-trough ratios may be incorrectly predicted when based solely on single-dose data (65,72).

4. Reducing Variability, Tightening C_{max} Confidence Intervals

Multiple dosing strategies may reduce the intersubject variability associated with blood level profiles (66,72), reduce the apparent differences in C_{max} values, and reduce the width of the corresponding confidence interval associated with the C_{max} comparison (65,73). The degree to which multiple dosing studies can affect bioequivalence decisions is a function of the relationship between the absorption and elimination rates, the possible presence of saturation kinetics, and the random noise associated with the concentration-time estimates (inherent variability and experimental error).

5. Accommodating Low Assay Sensitivity

If assay sensitivity is inadequate to measure single-dose blood levels to three terminal elimination half-lives beyond t_{max}, a multiple-dose study may be chosen if substantial accumulation is predicted when dosed at the maximum approved dose. To determine whether repeated administration will provide the necessary drug concentrations, the sponsor could estimate the accumulation factor, R:

$$R = 1/[1 - e^{-(\lambda n)\tau}]$$

where λn is the terminal disposition rate constant [where $t_{1/2} = 0.693/\lambda n$] and τ is the dosing interval (61). This equation can be used to describe the relationship between dosing rate, half-life, and drug accumulation (Table 1). Alternatively, the sponsor may choose to perform a single-dose study at a dose exceeding that approved for the pioneer product.

D. Fed Versus Fasted Conditions

Diet can significantly alter the bioavailability of certain orally administered drugs (74–76). Food effects can be related either to the drug entity itself (formulation-independent effects) or to a particular formulation (e.g., that seen with SR products). Food has been shown to both increase (77–79) and decrease (80–82) systemic bioavailability. Food effects have been associated with both hepatic metabolism (e.g., propranolol) (83) and renal excretion (84).

Table 1 Predicting Drug Accumulation at Steady State

τ (expressed as multiples of $t_{1/2}$)	Accumulation index
0.25	6.29
0.5	3.41
0.75	2.47
1.00	2.00
1.25	1.73
1.50	1.55
1.75	1.42
2.00	1.33
2.50	1.21
3.00	1.14
5.00	1.03

For bioequivalence investigations, concerns focus on potential formulation-specific food effects (79–85). Therefore, to assure product bioequivalence under conditions of clinical use, studies should be conducted in accordance with the approved product label. If the pioneer product label indicates that the product is to be administered in either the fed or fasted state, then the bioequivalence study should be conducted accordingly. If the bioequivalence study parameters pass preagreed confidence intervals, then the single study is acceptable as the basis for approval of the generic product.

Food consumption can result in an unintentional surge in the amount of drug released from an SR dosage form (dose dumping). Dose dumping may be associated with the interaction of product formulation and the feed contents (68,72,86,87). Therefore, unless otherwise indicated on the product label, for SR products intended for use in monogastric species/classes, bioavailability should be examined under postprandial conditions. Since the presence of food in monogastric species can increase the variability associated with drug absorption (88), the sponsor may choose to conduct two trials. One trial is conducted under fasted conditions, and confidence interval criteria are applied. A second, smaller study, in the presence of food, can be conducted and is evaluated solely on the basis of product means (C_{max} and AUC).

Investigations have also demonstrated that diet can affect the plasma concentrations of several oral anthelmintics (fenbendazole, albendazole, and ivermectin) in ruminants (89–91). In animals fed restricted rations, the bioavailability of the drug was enhanced, as compared with animals on full rations. Bioavailability of the drug was also higher in housed animals fed hay and concentrates than in grazing animals on pasture. Therefore, both plane of nutrition and feed type should be considered in the design of bioequivalence studies in ruminants. Although rumen fill and feed type may not be expected to exert formulation-dependent effects on bioavailability, it may affect the magnitude of variability associated with bioequivalence comparisons.

V. DETERMINING WHICH MOIETIES TO MEASURE

A. Total Versus Free Drug Concentrations

If the percentage of free versus bound drug is constant over the therapeutic dosing range, product relative bioavailability will be accurately determined by measuring total (free plus bound) drug concentrations. When less than 90% of the total drug fraction is bound, protein binding generally has little impact on drug activities (92). Drugs that bind to serum albumin generally do not exhibit nonlinear protein binding due to the high concentration of albumin in the blood.

To identify the conditions in which nonlinear protein binding can bias the bioequivalence determination, a drug's elimination and distribution characteristics must be considered (7,61,93). The impact of nonlinear protein binding will depend on its effect on volume of distribution (Vd) and clearance (Cl). Since the elimination half-life equals $(0.693 \times V_d)/Cl$, changes in the free fraction can potentially alter a drug's terminal elimination half-life. These changes are summarized in Table 2 (7,93).

For drugs with low extraction efficiency, clearance is a function of the free fraction of drug. Conversely, drugs with high extraction efficiency are relatively insensitive to the extent of protein binding since blood flow serves as the rate-limiting step in its elimination processes. In contrast, an increase in free fraction will have little effect on the V_d of a small V_d (0.1 L/kg) drug (e.g., polar compounds), but will increase the V_d of large V_d (1.0 L/kg) compounds (lipophilic agents). These relationships can be used to predict the changes in total drug versus free fraction which might occur in the presence of saturation kinetics.

If an increase in free fraction causes an increase in either V_d or Cl, total drug concentrations will decrease. We can therefore conclude that in the presence of nonlinear protein binding, the total concentrations of drugs B, C, and D will provide an adequate assessment of product relative bioavailability. One may argue that the total drug concentrations of low V_d high Cl compounds will not be affected by an increase in the free drug fraction and therefore both free and total drug concentrations will need to be assessed. However, compounds resembling drug A would appear only transiently in the blood, being very rapidly extracted from the circulation. Therefore, type A-like compounds would most probably exist only as pro-drugs, in which case product bioequivalence could be based on concentrations of the active metabolite rather than of the parent compound.

Table 2 Effect of Altered Protein Binding on Drug Cl and V_d

	Character of drug			Changes in		
Drug	Change in protein binding	V_d (L/kg)	Cl	V_d	Cl	$t_{1/2}$
A	Decrease	0.1	High	No change	No change	No change
B	Decrease	1.0	High	Increase	No change	Increase
C	Decrease	0.1	Low	No change	Increase	Decrease
D	Decrease	1.0	Low	Increase	Increase	No change

It should be noted that when comparing the systemic availability of low-Cl drugs that are associated with nonlinear protein binding, the assumption of a constant Cl may no longer be valid. The comparison of AUC values may reflect both differences in systemic availability (F) and Cl.

B. Parent Drug Versus Active Metabolite(s)

The rate of metabolite appearance may be primarily determined by the rate of drug absorption (k_a) or the rate of metabolite formation (k_m), depending on which of these two rate constants is the smaller. The smaller the k_m relative to the k_a, the less of an impact the parent compound absorption rate has on the metabolite C_{max} values. However, when $k_m > k_a$, the rate of appearance of the parent compound and the metabolite should be similar (94).

For linear systems, the extent of metabolite systemic availability (AUCm) is insensitive to product absorption rate (61) as described by the equation:

$$AUC(m) = \frac{fm * Fh(m) * fa * d}{Cl(m)}$$

where fm = the fraction of drug converted to metabolite, Fh(m) = the systemic availability of the metabolite, fa = the fraction of the administered dose that is absorbed, d = dose of the parent compound, and Cl(m) = the clearance of the metabolite. Note that the absorption rate constant, k_a, does not appear in this equation (94).

Based on these considerations, product bioequivalence can be based solely on concentrations of the parent compound in the majority of cases. Exceptions to this would be if the parent compound is not measurable (e.g., Ref. 95), or if nonlinear elimination is associated with an active metabolite but not with the parent compound (e.g., Refs. 96,97).

C. Chiral Compounds

Enantiomers, or chiral compounds, are molecules that have identical chemical and structural formulas but which differ with respect to the arrangement of the four functional groups about an asymmetric carbon atom. The resulting asymmetry frequently gives rise to different pharmacokinetic and pharmacodynamic properties. Although these differences are well recognized (98–101), most drug compounds are marketed as racemates (102) because of the high costs associated with preparation of pure enantiomers (99).

One published report demonstrated that the results of a bioequivalence study varied when the assessment was based upon blood levels of the racemate (nonstereospecific assay) versus the individual enantiomers (stereospecific assay) (103). When considering the confidence intervals for the parameters AUC and C_{max}, one would conclude that the two products were bioequivalent based on nonstereospecific blood level data. However, the data for the enantiomers failed to demonstrate product bioequivalence. This discrepancy appeared to be attributable to greater variability associated with the blood levels of the enantiomers than that associated with the racemate (i.e., resulting in wider confidence intervals for the four enantiomers). The mean estimate of the differences in product rate and extent of absorption were comparable.

Stereospecificity in the assessment of human drug bioequivalence for racemic mixtures has been a subject of debate (104,105). Although enantiomer-specific formulation effects are rare, they may occur if there exists stereospecific nonlinear kinetics (105–107). Therefore, the need for a stereospecific determination of human drug bioequivalence is assessed on a case-by-case basis (105).

Currently, CVM does not require stereospecific assays for the determination of product bioequivalence. However, animal drug sponsors should recognize that stereospecific differences in enantiomer bioavailability may occur across animal species or when changing from an oral to parenteral dosage form (98).

VI. DEFINING THE STUDY POPULATION

Bioequivalence studies are generally conducted in healthy animals representative of the species, class, gender, and physiological maturity for which the drug is approved. The bioequivalence study may also be conducted with a single gender for which the pioneer product is approved. The question of whether age, disease, or gender could affect product relative bioavailability under conditions of clinical use may be considered.

A. Age and Disease Considerations

Aging (108–110), disease (7,111), and maturation (112–116) can significantly affect drug pharmacokinetics. However, in the case of bioequivalence trials, the fundamental question is whether these variables can affect product interchangeability.

There currently is little information comparing product bioequivalence across old versus young or healthy versus diseased populations (117). For human generic drug approvals, the assumption made is that if two products are shown to be bioequivalent in young, healthy adults, the products will perform comparably across all potential patient populations (118). However, observations to the contrary have been reported (119).

Age-related changes occurring in the gastrointestinal system of monogastric species can significantly alter drug bioavailability (109). These changes include a decrease in gastric hydrochloric acid secretion with a corresponding increase in gastric pH (achlorhydria) and increased gastrointestinal transit time. Therefore, if one of the two formulations in a bioequivalence trial is pH-dependent, product inequivalence may occur under conditions of achlorhydria, but not in normal healthy subjects. This disparity was observed with two formulations of diazepam in humans (117).

Regarding parenteral dosage forms, age-related and maturation-related changes in product relative bioavailability may occur since reductions in muscle mass, interstitial fluid, and tissue perfusion rates could have a significant effect on the dissolution rate of suspensions (108,120). Ultimately, the effect of age, maturation, or disease on parenteral product bioavailability will depend on whether the rate and extent of drug absorption are limited by the rate of particle dissolution, blood perfusion, or capillary permeability.

B. Gender Effects

Although gender-related differences in drug pharmacokinetics are known to occur (121,122), statistically significant gender-by-formulation interactions have yet to be described (123).

VII. PHARMACOKINETIC DATA ANALYSIS

CVM recommends the use of noncompartmental methods for the analysis of bioavailability data. Estimates of rate and extent of product absorption should be based on observed rather than fitted data.

A. Measuring the Extent of Drug Absorption

Noncompartmental methods are based on statistical moment theory and linear systems theory (8,61,124). AUC represents the zero statistical moment. Assuming that the last quantifiable drug concentration (C_{last}) occurs during the terminal elimination portion of the curve, adding the extrapolated area from C_{last} to time infinity ($AUC_{last-inf}$) to AUC calculated from time zero to C_{last}(AUC_{0-last}) will provide an estimate of the total extent of product bioavailability following a single administration (AUC_{0-inf}). Under steady-state conditions, the area estimated during a single dosing interval ($AUC_{0-\tau}$) equals AUC_{0-inf}.

CVM does not recommend using AUC_{0-inf} estimates as the pivotal parameter in a bioequivalence trial due to errors introduced when determining the extrapolated area. Instead, the extent of product absorption should be compared on the basis of AUC_{0-last} since this parameter provides bioavailability comparisons (based on the 90% confidence interval approach) which are comparable to those obtained if the true AUC_{0-inf} values were known (125).

The CVM Bioequivalence Guidance suggests that AUC values be estimated by the linear trapezoidal method, but states that other methods of estimation may also be acceptable. In cases where sampling times are separated by relatively long time intervals, the log-linear method of AUC determination may provide the more accurate estimate of product bioavailability (126,127).

B. Measuring the Rate of Drug Absorption

In addition to C_{max}, a number of metrics have been proposed for assessing rate of drug absorption. These include mean residence time (124,128–130), maximum entropy (131), C_{max}/AUC (132,133), partial AUCs (134), Wagner-Nelson plots (135), or center of gravity (which may be particularly important in its application to drugs with multiple absorption maxima) (136). The applicability of these various metrics to bioequivalence assessments has been investigated for both immediate-release and sustained-release dosage forms (71,137,138). It has been determined that the quality of a particular measure is highly dependent on the particular bioavailability scenario (139).

Given the problems associated with the determination of product absorption rate, C_{max} may not be the metric of choice for all bioequivalence determinations (140,141). In part, this may be because C_{max} is a function of both rate and extent of drug absorption (61,137–139). Despite these apparent difficulties, the CVM Bioequivalence Guidance recommends C_{max} as the metric of choice for comparing rates of absorption for most bioequivalence comparisons. The choice of C_{max} reflects the clinical significance associated with peak drug concentrations, its ease of use, and the vast regulatory experience acquired with C_{max} in the human drug arena (66). CVM recognizes that the release characteristics of certain products (e.g., SR formulations) may be more appropriately defined by alternative metrics. Animal drug sponsors are encouraged to explore these alternatives with CVM.

VIII. STATISTICAL ANALYSIS OF BIOAVAILABILITY DATA

Sponsors should identify the method of data analysis (including the use of data transformation) prior to initiating the in vivo bioequivalence study. The importance of an a priori determination of the specific question being addressed and corresponding method of data analysis is illustrated in the following example (142):

A study was conducted in two hospitals to determine whether one hospital tended to retain its patients for a longer period of time. There were two patient categories examined: obstetrics patients and geriatric patients. The results of the investigation were:

Patient category		Hospital 1					Hospital 2				
Obstetrics		2	2	2	3	3	2	2	2	3	4
		3	3	4	4	4					
	mean			3.0					2.6		
	n			10					5		
Geriatric		20	21	21	20		19	22	22	21	
							20	20	23	22	
							21	21	20	21	
	mean			20.5					21.0		
	n			4					12		
Unweighted (LS) mean				11.75					11.80		
Weighted (arith) mean				8.0					15.59		

Clearly, the conclusions derived from this investigation depend on whether the weighted or unweighted (least-square)/means are considered. A determination of which mean is "correct" depends on the question being addressed. If the question is whether hospital B is more conservative in in patient retention time than hospital A, then the type of patient should be factored into the analysis and the unweighted (LS) mean is the choice for the comparison. Conversely, if one wishes to know whether patients tend to stay longer at hospital B than A, regardless of the reason, comparisons should be generated on the basis of the weighted (arithmetic) mean.

A. Sequence Effects in the 2 × 2 Crossover Design

At an α value of 0.10, 10 of 100 studies using the 2 × 2 crossover study design are expected to show sequence effects when no unequal residual effects present. Unfortunately, this study design does not allow the investigator to ascertain the origin of the sequence effect (143–145).

A limitation associated with the use of a standard two-period, two-treatment, two-sequence crossover design is that true sequence effects (error associated with the process of subject randomization) are confounded with carryover effects (physiologic or pharmacologic) and treatment-by-period interactions (when the assessment of the treatment effects are dependent on the period in which the comparisons are generated). Although a true sequence effect does not bias the treatment comparison, a statistically significant sequence effect can indicate a problem with the use of a standard two-period, two-treatment crossover study design.

If the effects attributable to sequence are statistically significant (at $P < .10$ since it is an intersubject comparison) but no statistically significant period effects are observed ($P < .05$), the data from both periods of the crossover study can be included in the statistical analysis. In this case, it is assumed that the sequence effect is benign and will not affect the outcome of the comparison. However, since treatment effects are confounded with period-by-sequence interactions (59), the presence of statistically significant period and sequence effects could indicate the presence of unequal residuals or a treatment-by-period interaction. Therefore, when statistically significant period and sequence effects are observed, the treatment comparisons must generated solely on the basis of period 1 data (i.e., the data must be handled as if generated in a parallel design study).

B. Use of the Analysis of Variance

Hypothesis testing using an analysis of variance (ANOVA) requires certain assumptions: homogeneity of variances, normality, independence of the main effects (additivity), and the absence of a subject-by-treatment-interaction (146–149). If these assumptions are not met, data transformation or nonparametric procedures should be considered.

An appropriate statistical model to describe the observations generated in a standard two-treatment, two-period, two-sequence crossover design (146,149,150) can be written as follows:

$$Y_{ijk} - \mu + seq_k + subj_{i(k)} + period_j + trt_{(j,k)} + error_{ijk}$$

where Y_{ijk} = the observation associated with the ith subject (nested within the kth sequence) during the jth period, μ = the population mean for the measure of interest, seq_k = the kth sequence, $subj_{i(k)}$ = the ith subject nested within the kth sequence, $period_j$ = the jth period, $trt_{(j,k)}$ = treatment associated with the jth period and the kth sequence, and $error_{ijk}$ = the unexplained variability associated with the ith subject (nested within the kth sequence) during the jth period. This error estimate determines the width of the 90th confidence interval.

If a parallel study design is used, the model reduces to:

$$Y_{im} = \mu + trt_{(m)} + error_{im}$$

where Y_{im} = the observation associated with the ith subject and the mth treatment, μ = the population mean for the measure of interest, trt_m = the mth treatment, and $error_{im}$ = the unexplained variability associated with the ith subject and the mth treatment. This error estimate determines the width of the 90% confidence interval.

C. Data Transformation

The use of data transformation reflects the investigator's belief that the assumptions of the ANOVA are better met when the data are presented on a transformed scale (151). Numerous kinds of data transformations are possible (152,153). However, the logarithmic transformation is the data modification of choice when describing biological systems (148,154). Reasons for this include:

Pharmacokinetic models are multiplicative and therefore considered by some not to be in compliance with the assumption of additivity
Logarithmic transformation stabilizes the variances
Many biological systems are associated with log-normal distributions
Bioequivalence comparisons are generally expressed as ratios rather than differences
Other types of data transformation will be very difficult to interpret

D. Discordant Observations

The classification and handling of extreme statistical outliers remain a controversial issue both from administrative and scientific perspectives (1). Although data should not be discarded without prior knowledge of the distribution characteristics of the population, most bioequivalence studies do not contain a sufficient number of subjects to define that population. Therefore, what appears to be a discordant observation may, in fact, be a subject representative of a subpopulation within which the formulations are not bioequivalent. Therefore, sponsors are strongly advised to seek FDA advice prior to deleting subjects from their statistical analyses.

Another type of "outlier" results from error encountered during sample analysis. For example, the assay of a particular sample may produce a value that appears inconsistent with the values of flanking samples or seems inconsistent with a predetermined pharmacokinetic model. To deal with the type of discordant observation, sponsors should have a standard operating procedure which clearly states the criteria for sample reanalysis and the method by which the final sample concentration will be determined.

E. Use of an Interval Hypothesis for Testing Product Bioequivalence

The use of an interval hypothesis as the basis of product approval is founded on the assumption that there exists some boundary of acceptable differences within which the products can be expected to have comparable safety and efficacy. Currently, the Agency utilizes the two one-sided tests procedure for evaluating product bioequivalence (155). Unlike the Student's t-test, which is based on the null hypothesis of no treatment difference, the 90% confidence interval approach is based on the hypothesis that two products are in fact bioinequivalent. With this test procedure, product bioequivalence is confirmed only when the null hypothesis is rejected (155).

1. Algorithms for Calculating Confidence Intervals (140,155,156)

a. Untransformed Data

$$\text{Lower limit} = \frac{(T-R) - SE * t_{0.95(v)}}{R}$$

$$\text{Upper limit} = \frac{(T-R) + SE * t_{0.95(v)}}{R}$$

where T = mean value for the test product, R = mean value for the reference product, SE = standard error for the estimate of the differences between the means, and v = the error degrees of freedom.

b. Log (LN) Transformed Data

$$\text{Lower limit} = EXP^{[(T-R) - SE*t(0.95(v))]}$$

$$\text{Upper limit} = EXP^{[(T-R) + SE*t(0.95(v))]}$$

When sequence groups of a crossover study contain unequal numbers of subjects, the differences between the observed mean values do not provide an unbiased estimator of the differences between the population means. Therefore, the confidence intervals should be calculated using the estimate of the difference between the least-square (LS) means and the standard error (SE) associated with that estimate (150,155).

F. Sample Size Determinations

Drs. Liu and Chow (157) have estimated the sample size needed for Schuirmann's two one-sided tests procedure based on an untransformed data set. These estimates are a function of the variability in the estimate of the difference and the magnitude of the difference in treatment means (Table 3). As seen in Table 2, sample size estimates have also been developed for log-transformed data sets (158,159).

For both Tables 3 and 4, n represents the number of observations per treatment. Therefore, if a study is conducted as a standard crossover trial, n also equals the number of study subjects. If, however, a parallel design is chosen,

Table 3 Sample Sizes (n) for Schuirmann's Two One-Sided Tests Procedure at 80% Power and a Bioequivalence Limit = 0.2* μ_R at the 5% Nominal Level

		Difference, T–R			
		0	5	10	15
%CV	10	8	8	16	52
	12	8	10	20	74
	14	10	14	26	100
	16	14	16	34	126
	18	16	20	42	162
	20	20	24	52	200
	22	24	28	62	242
	24	28	34	74	288
	26	32	40	86	336
	28	36	46	100	390
	30	40	52	114	448
	32	46	58	128	508
	34	52	66	146	574
	36	58	74	162	644
	38	64	82	180	716
	40	70	90	200	794

the necessary number of study subjects equals 2n (# observations for treatment 1 + # observations for treatment 2).

IX. PROBLEMATIC ISSUES IN BIOEQUIVALENCE DETERMINATIONS

A. Multiple Absorption Maxima

The comparison of product bioavailability characteristics may be confounded by the presence of multiple peaks. Potential origins of multiple peaks include variability in the rate of gastric emptying and intestinal flow (160), enterohepatic recycling (161,162), bladder reabsorption (163), and product formulation (164).

Generally, the highest peak is designed as the C_{max}. For immediate-release products and for SR formulations which are associated with an initial loading dose, C_{max} is generally designated as the initial peak in the concentration/time profile. However, some formulations may provide a slow and steady release throughout the dosing interval. When this results in multiple maxima, the definition of product absorption rate becomes unclear (165,166). Methods of handling such data sets are under investigation.

Table 4 Approximate Sample Sizes to Attain a Power of 80% in the Case of the Multiplicative Model

		Ratio test/Reference							
		0.85	0.90	0.95	1.00	1.05	1.10	1.15	1.20
%CV	5.0	12	6	4	4	4	6	8	22
	7.5	22	8	6	6	6	8	12	44
	10.0	36	12	8	6	8	10	20	76
	12.5	56	16	10	8	10	14	30	118
	15.0	78	22	12	10	12	20	42	170
	17.5	106	30	16	14	16	26	58	230
	20.0	138	38	20	16	18	32	74	300
	22.5	172	48	24	20	24	40	92	378
	25.0	212	58	28	24	28	50	114	466
	27.5	256	70	34	28	34	60	138	564
	30.0	306	82	40	34	40	70	162	670

B. Use of Composite Curves to Establish Bioequivalence

When only a single blood sample can be obtained from each study subject, blood level profiles represent a composite of both intrasubject and intersubject variability. The need to generate composite curves is encountered when assessing the pharmacokinetics of drugs intended for use in fish, chicks, and turkey poults. Since the two sources of variability cannot be differentiated, estimates of variance in population pharmacokinetic parameters tend to be inaccurate (167–170).

Mathematical methods have been proposed for conducting t-tests on AUC values based on composite data (171–173). Using this estimated t value, confidence intervals can be generated about the difference in treatment means (Martinez, unpublished observation). However, the regulatory applicability and implications of this method of analysis remain unresolved.

C. Products with Nonzero Baselines

When evaluating the relative bioavailability of products in which the active substance is an endogenous compound, sponsors must decide whether it is more appropriate to compare total drug concentrations, correct values by subtracting a pretreatment baseline measurement, or use a statistical model which includes a covariate (174,175). The choice of procedure should be based on whether or not the baseline values are correlated with an inherent physiologic effect which determines how the drug is handled. If this is the case, the use of a covariate in the statistical model may be appropriate.

The relationship between levels of endogenous and exogenous drug should also be considered. If the endogenous levels constitute a significant percent of the total drug concentrations, subtraction of baseline values may be appropriate. However, subtraction of background levels may not be appropriate if endogenous drug concentrations are highly variable within an individual or if the presence of exogenous drug alters the endogenous levels of the compound under investigation.

Regardless of how the background levels are handled, it is inappropriate to simultaneously use more than one correction procedure. Methods of handling these kinds of data sets should be discussed with CVM prior to initiating the study.

X. BIOAVAILABILITY STUDIES IN NEW ANIMAL DRUG APPROVALS

In addition to the bioequivalence studies already discussed, bioavailability studies can be utilized in the NADA process. Pharmacokinetic/bioavailability data can be used to predict appropriate dosage regimens for a drug or to provide a rationale for interspecies extrapolation of doses.

Relative bioavailability studies may be used to evaluate comparability (e.g., new salt forms, change in route of administration, or manufacturing changes for previously approved products), as distinguished from bioequivalence studies, which focus strictly on interchangeability. In relative bioavailability studies, the criteria used for data evaluation may be adapted to the product category a priori, especially if the pharmacokinetic/pharmacodynamic relationship is well known for the particular drug. The relative bioavailability study may identify differences in the rate and extent of drug absorption, but a judgment on its impact on target animal safety and efficacy may determine whether additional confirmatory studies are needed.

Pharmacokinetic/pharmacodynamic relationships have been well described for many antimicrobial compounds and may provide the basis upon which dose ranged veterinary drug labels are developed. The pharmacokinetic/pharmacodynamic relationship of antimicrobial agents may fall within one of three general categories (176,177):

1.   Agents with little concentration-dependent bactericidal activity. These compounds tend to have little or on postantibiotic activity and drug concentrations must be maintained above the MIC for the majority of the dosage interval (time-dependent killing). This group includes most β lactams and glycopeptides.

2.   Compounds exhibiting marked concentration-dependent killing activity. Examples include the aminoglycosides and the fluoroquinolones.

3. Agents that are predominantly bacteriostatic. Examples include erythromycin, tetracycline, and chloramphenicol.

Whereas pharmacokinetic/MIC relationships are generally useful in estimating an efficacious dose (178), the large number of variables influencing drug activity in the animal necessitates clinical validation (179).

CVM supports a professional flexible labeling (PFL) concept which encourages animal drug sponsors to provide a dose range rather than a point dose for prescription new animal drugs. PFL has been the topic of two workshops (180,181) cosponsored by the AAVPT (American Academy of Veterinary Pharmacology and Therapeutics), FDA/CVM, the AHI (Animal Health Institute), and the AVMA (American Veterinary Medical Association). Although the emphasis for PFL has been on prescription antimicrobial drugs, the concept also has potential relevance to other classes of veterinary drugs.

Bioavailability information has been identified as an important element of flexible labeling to facilitate dose selection by the practitioner. Blood concentration profiles generated at the low and high ends of the approved dose range, coupled with MIC data, would help the veterinary practitioner select the appropriate dose to use for the particular disease, organ system, and bacteria involved. The dose range may be defined by a clinically confirmed dose at the low end, and target animal safety (and human food safety for food-producing species) at the upper end of the dose range.

An important goal of approving dose-ranged products is to increase the therapeutic options available to practitioners by providing labels with more clinically relevant information than is currently available on animal drugs. More effective use of bioavailability data in the NADA process may prove to be a valuable tool in increasing animal drug availability, facilitating approvals for minor species through interspecies extrapolation, and expanding the utility of drug labeling for the practitioner.

REFERENCES

1. Martinez MN, Riviere JE. Review of the 1993 veterinary drug bioequivalence workshop. J Vet Pharmacol Ther 1994; 17:85-119.
2. DiSanto AR. Bioavailability and bioequivalence testing. In: Remington's Pharmaceutical Sciences. 18th ed. Easton, PA: Philadelphia College of Pharmacy and Science, 1990:1452-1458.
3. Levy G, Ebling WF, Forrest A. Concentration- or effect-controlled clinical trials with sparse data. Clin Pharmacol Ther 1994; 56:1-8.
4. Rutgers LJE, Van Miert ASJPAM, Nouws JFM, Van Ginneken CAM. Effect of the injection site on the bioavailability of amoxycillin trihydrate in dairy cows. J Vet Pharmacol Ther 1980; 3:125-132.
5. Firth EC, Nouws JFM, Driessens F, Schmaetz P, Peperkamp K, Klein WR. Effect

of the injection site on the pharmacokinetics of procaine penicillin G in horses. Am J Vet Res 1986; 47:2380–2384.

6. Marshall AB, Palmer GH. Injection sites and drug bioavailability. Trends Vet Pharmacol Ther 1980; 6:54–60.
7. Rowland M, Tozer TN. Clinical Pharmacokinetics: Concepts and Applications. Philadelphia: Lea and Febiger, 1995.
8. Riegelman S, Collier P. The application of statistical moment theory to the evaluation of in-vivo dissolution time and absorption time. J Pharm Biopharm 1980; 8:509–534.
9. Selen A. Factors influencing bioavailability and bioequivalence. In: Welling PG, Tse FLS, Dighe SV, eds. Pharmaceutical Bioequivalence. New York: Marcel Dekker, 1991:117–148.
10. Toutain PL, Raynaud JP. Pharmacokinetics of oxytetracycline in young cattle: comparison of conventional vs long-acting formulations. Am J Vet Res 1983; 44:1203–1208.
11. Pope DG. Physico-chemical and formulation-induced veterinary drug-product bioinequivalencies. J Vet Pharmacol Ther 1984; 7:85–112.
12. Finholt P. Influence of formulation on dissolution rate. In: Leeson LJ, Carstensen JT, eds. Dissolution Technology. Washington, D.C.: Academy of Pharmaceutical Sciences, 1974:106–146.
13. Monkhouse DC, Lach JL. Drug-excipient interactions. Can J Pharm Sci 1972; 7:29–46.
14. Record PC. A review of pharmaceutical granulation technology. Int J Pharm Technol Product Manufac 1980; 1:32–39.
15. Abdou HM. Dissolution, Bioavailability and Bioequivalence. Easton, PA: Mack Publishing Co., 1989.
16. Abdou HM. Dissolution. In: Remington's Pharmaceutical Sciences. 18th ed. Easton, PA: Philadelphia College of Pharmacy and Science, 1990:589–602.
17. Perez-Marcos B, Ford JL, Armstrong DJ, Elliot PNC, Rostron C, Hogan JE. Influence of pH on the release of propranolol hydrochloride from matrices containing hydroxypropylmethylcellulose K4M and carbopol 974. J Pharm Sci 1996; 85:330–334.
18. Lessen T, Zhao DC. Interactions between drug substances and excipients: fluorescence and HPLC studies of triazolophthalazine derivatives from hydralazine hydrochloride and starch. J Pharm Sci 1996; 85:325–329.
19. Swarbrick J. Do processing and scale-up factors really affect bioavailability? A suggestion for answering the question. Pharm Technol 1992; 32–36.
20. Van Buskirk GA, Shah VP, Adair D, et al. Scale-up of liquid and semisolid disperse systems. Pharm Technol 1995; 19:52–62.
21. Lucisano LJ, Franz RM. FDA proposed guidance for chemistry, manufacturing, and control changes for immediate-release solid dosage forms: a review and industrial perspective. Pharm Technol 1995; 19:30–44.
22. Skelly JP, Van Buskirk GA, Amidon GL, et al. Scale-up of oral extended-release dosage forms. Pharm Technol 1995; 19:46–54.
23. Leeson LJ. In-vitro/in-vivo correlations. Drug Information J 1995; 29:903–915.

24. Shah VP, Gurbarg M, Noory A, Dighe S, Skelly JP. Influence of higher rates of agitation on release patterns of immediate-release drug products. J Pharm Sci 1992; 81:500–503.

25. Fassihi AR, Munday DL. Dissolution of theophylline from film-coated slow release mini-tablets in various dissolution media. J Pharm Pharmacol 1989; 41:369–372.

26. Herman J, Remon JP, Lefebvre R, Bogaert M, Klinger GH, Schwartz JB. The dissolution rate and bioavailability of hydrochlorothiazide in pellet formulation. J Pharm Pharmacol 1988; 40:157–160.

27. Shah VP, Konecny JJ, Everett RL, McCullough B, Noorizadeh AC, Skelly JP. In-vitro dissolution profiles of water-insoluble drug dosage forms in the presence of surfactants. Pharm Res 1989; 6:612–618.

28. Riegelman S. Physiological and pharmacokinetic complexities in bioavailability testing. Pharmacology 1972; 8:118–141.

29. Welling PG, Forgue ST, Cook JA, deVries TM. In vitro-in vivo correlations—quo vadis. Drug Information J 1995; 29:893–902.

30. Adkin DA, Davis SS, Sparrow RA, Huckle PD, Wilding IR. The effect of mannitol on the oral bioavailability of cimetidine. J Pharm Sci 1995; 84:1405–1409.

31. Aungst BJ. Novel formulation strategies for improving oral bioavailability of drugs with poor membrane permeation or presystemic metabolism. J Pharm Sci 1993; 82:979–987.

32. Swenson ES, Curatolo WJ. Means to enhance penetration: intestinal permeability enhancement for proteins, peptides and other polar drugs: mechanisms and potential toxicity. Adv Drug Del Rev 1992; 8:39–92.

33. Muranishi S. Absorption enhancers. Crit Rev Ther Drug Carrier Syst 1990; 7:1–33.

34. Anderberg EK, Nystrom C, Artursson P. Epithelial transport of drugs in cell culture: effects of pharmaceutical surfactant excipients and bile acids on transepithelial permeability in monolayers of human intestinal epithelial (Caco 2) cells. J Pharm Sci 1992; 81:879–887.

35. Shah VP, Hunt JP, Fairweather WR, Prassad VK, Knapp G. Influence of dioctyl sodium sulfosuccinate on the absorption of tetracycline. Biopharm Drug Dispos 1986; 7:27–33.

36. Riad LE, Sawchuk RJ. Effect of polyethylene glycol 400 on the intestinal permeability of carbamazepine in the rabbit. Pharm Res 1991; 8:491–497.

37. Krishna DR, Klotz U. Extrahepatic metabolism of drugs in humans. Clin Pharm 1994; 26:144–160.

38. Ilett KE, Tee LBG, Reeves PT, Minchin RF. Metabolism of drugs and other xenobiotics in the gut lumen and wall. Pharmacol Ther 1990; 46:67–93.

39. Chesa-Jimenez J, Esteban J, Torres-Molina F, Granero L. Low bioavailability of amoxicillin in rats as a consequence of presystemic degradation in the intestine. Antimicrob Agents Chemother 1994; 38:842–847.

40. Barr WH. The role of intestinal metabolism in bioavailability. In: Welling PG, Tse FLS, Dighe SV, eds. Pharmaceutical Bioequivalence. New York: Marcel Dekker, 1991:149–167.

41. Nelson DR, Koymans L, Kamataki T, et al. P450 superfamily: update on new sequences, gene mapping, accession numbers and nomenclature. Pharmacogen 1996; 6:1–42.

42. Saitoh H, Aungst BJ. Possible involvement of multiple p-glycoprotein-mediated efflux systems in the transport of verapamil and other organic cations across rate intestine. Pharm Res 1995; 12:1304–1310.

43. Leu BL, Huang JD. Inhibition of intestinal p-glycoprotein and effects on etoposide absorption. Cancer Chemother Pharmacol 1995; 35:432–436.

44. Schuetz EG, Beck WT, Schuetz JD. Modulators and substrates of p-glycoprotein and cytochrome P4503A coordinately up-regulate these proteins in human colon carcinoma cells. Mol Pharmacol 1996; 49:311–318.

45. Gan LSL, Moseley MA, Khosla B, et al. CYP3A-like cytochrome P450-mediated metabolism and polarized efflux of cyclosporin A in CACO-2 cells: interaction between the two biochemical barriers to intestinal transport. Drug Metab Dispo 1996; 24:344–349.

46. Moscow JA, Cowan KH. Multidrug resistance. JNCI 1988; 80:15–20.

47. Thorgeirsson SS, Silverman JA, Gant TW, Marino PA. Multidrug resistance gene family and chemical carcinogenesis. Pharmacol Ther 1991; 49:283–292.

48. Kaminsky LS, Fasco MJ. Small intestinal cytochromes P450. Crit Rev Toxicol 1991; 21:407–422.

49. Schulze J, Malone A, Richter E. Intestinal metabolism of 4-(methylnitrosamino)-1-(3-pyridyl)-1-butanone in rats: sex difference, inducibility and inhibition by phenethylisothiocyanate. Carcinogenesis 1995; 16:1733–1740.

50. Bailey DG, Arnold JMO, Strong HA, Munoz C, Spence JD. Effect of grapefruit juice and naringin on nisoldipine pharmacokinetics. Chem Pharmacol Ther 1993; 54:589–594.

51. Ducharme MP, Warbasse LH, Edwards DJ. Disposition of intravenous and oral cyclosporin after administration with grapefruit juice. Chem Pharmacol Ther 1995; 57:485–491.

52. Gillum JG, Israel DS, Polk RE. Pharmacokinetic drug interactions with antimicrobial agents. Clin Pharm 1993; 25:450–482.

53. Buckingham LE, Balasubramanian M, Safa AR, et al. Reversal of multi-drug resistance in vitro by fatty acid–PEG–fatty acid diesters. Int J Cancer 1996; 65:74–49.

54. Chang T, Benet LZ, Hebert MF. The effect of water-soluble vitamin E on cyclosporin pharmacokinetics in healthy volunteers. Clin Pharmacol Ther 1996; 59:297–303.

55. Parkinson A. Biotransformation of xenobiotics. In: Klaassen CD, ed. Casarett & Doull's Toxicology, The Basic Science of Poisons, 5th ed. New York: McGraw-Hill, 1996:113–186.

56. Beck WT. Modulators of p-glycoprotein-associated mulidrug resistance. In: Ozols RF, ed. Molecular and Clinical Advances in Anticancer Drug Resistance. Boston: Kluwer, 1991:151–170.

57. Herzog CE, Toskos M, Bates SE, Fojo AT. Increased MDR-1/p-glycoprotein expression after treatment of human colon carcinoma cells with p-glycoprotein antagonists. J Biol Chem 1993; 268:2946–2952.

58. Koup JR, Gibaldi M. Some comments on the evaluation of bioavailability data. Drug Intell Clin Pharm 1980; 14:327–330.

59. Jones B, Kenward MG. Design and Analysis of Cross-Over Trials. New York: Chapman and Hall, 1989.

60. Ratkowsky DA, Evans MA, Alldredge JR. Cross-Over Experiments. New York: Marcel Dekker, 1993.

61. Gibaldi M, Perrier D. Pharmacokinetics. New York: Marcel Dekker, 1982.

62. Perrier D, Ashley JJ, Levy G. Effect of product inhibition of kinetics of drug elimination. J Pharmacokinet Biopharm 1973; 1:231–242.

63. Ludden TM. Nonlinear pharmacokinetics: clinical implications. Clin Pharm 1991; 20:429–446.

64. Schuirmann DJ. Design of bioavailability/bioequivalence studies. Drug Information J 1990; 24:315–323.

65. Steinijans VW, Sauter R, Jonkman JHG, Schultz HU, Stricker H, Blume H. Bioequivalence studies: single vs. multiple dose. Int J Clin Pharmacol Ther Toxicol 1989; 27:261–266.

66. Williams RL. Bioequivalence and therapeutic equivalence. In: Welling PG, Tae FLS, Dighe SV, eds. Pharmaceutical Bioequivalence. New York: Marcel Dekker, 1991:1–15.

67. Shargel L, Yu A. Applied Biopharmaceutics and Pharmacokinetics. 3rd ed. Norwalk, CT: Appleton and Lange, 1993.

68. Skelly JP, Barr WH, Benet LZ, et al. Report of the workshop on controlled-release dosage forms: issues and controversies. Pharm Res 1987; 4:75–77.

69. Skelly JP, Chen TO. Regulatory concerns in controlled release drug product approval. Evaluation Methodology of Controlled Release Dosage Forms: Proceedings of US-Japan Joint Seminar. Tokyo: Japan Health Science Foundation, 1989:1–34.

70. Ministry of Health and Welfare, Japan. Guidelines for the design and evaluation of oral prolonged release dosage forms. In: Evaluation Methodology of Controlled Release Dosage Forms: Proceedings of US-Japan Joint Seminar. Tokyo: Japan Health Science Foundation, 1989:137–143.

71. Lacey LF, Bye A, Keene ON. Glaxo's experience of different absorption rate metrics of immediate release and extended release dosage forms. Drug Information J 1995; 29:821–840.

72. Tse FL, Robinson WT, Choc MG. Study design for the assessment of bioavailability and bioequivalence. In: Welling PG, Tse FLS, Dighe SV, eds. Pharmaceutical Bioequivalence. New York: Marcel Dekker, 1991:17–34.

73. Jackson A. Prediction of steady-state bioequivalence relationships using single dose data: linear kinetics. Biopharm Drug Dispos 1987; 8:483–496.

74. Welling PG. Influence of food and diet on gastrointestinal drug absorption: a review. J Pharm Biopharm 1977; 5:291–334.

75. Welling PG. Effect of food on bioavailability of drugs. Pharm Int 1980; Jan:14–18.

76. Walter-Sack I. The influence of nutrition on the systemic availability of drugs. Part II. Drug metabolism and renal excretion. Klin Wochenschrift 1987; 65:1062–1072.

77. McKellar QA, Galbraith EA, Baster P. Oral absorption and bioavailability of fenbendazole in the dog and the effect of concurrent ingestion of food. J Vet Pharmacol Ther 1993; 16:189–198.
78. Martinez MN, Pelsor FR, Shah VP, et al. Effect of dietary fat content on the bioavailability of sustained release quinidine gluconate tablet. Biopharm Drug Dispos 1987; 11:17–29.
79. Watson ADJ, Rijnberk A. Systemic availability of o,p'-DDD in normal dogs, fasted and fed, and in dogs with hyperadrenocorticism. Res Vet Sci 1987; 43:160–165.
80. Koch PA, Schultz CA, Wills RJ, Hallquist SL, Welling PG. Influence of food and fluid ingestion on aspirin bioavailability. J Pharm Sci 1978; 67:1533–1535.
81. Hoppu K, Tuomisto J, Koskimies O, Simell O. Food and guar decrease absorption of trimethoprim. Eur J Clin Pharmacol 1987; 32:427–429.
82. Watson AD, Emslie DR, Martin ICA, Egerton JR. Effect of ingesta on systemic availability of penicillins administered orally in dogs. J Vet Pharmacol Ther 1986; 9:140–149.
83. McLean AJ, McNamara PJ, duSouich P, Gibaldi M, Lalka S. Food, splanchnic blood flow, and bioavailability of drugs subject to first-pass metabolism. Clin Pharmacol Ther 1978; 24:5–10.
84. Oukessou M, Toutain PL. Effect of dietary nitrogen intake on gentamicin disposition in sheep. J Vet Pharmacol Ther 1992; 15:416–420.
85. Watson ADJ. Effect of ingesta on systemic availability of chloramphenicol from two oral preparations in cats. J Vet Pharmacol Ther 1979; 2:117–121.
86. Karim A, Burns T, Janky D, Hurwitz A. Food-induced changes in theophylline absorption from controlled-release formulations. Part II. Importance of meal composition and dosing time relative to meal intake in assessing changes in absorption. Clin Pharmacol Ther 1985; 38:642–647.
87. Hendeles L, Weinberger M, Milavetz G, Hill M, Vaughan L. Food-induced "dose-dumping" from a once-a-day theophylline product as a cause of theophylline toxicity. Chest 1985; 87:758–765.
88. Martinez M. Food effects in bioequivalency evaluations. Proceedings of Bio-International '89—Issues in the Evaluation of Bioavailability Data. Toronto, Canada, 1989:62–63.
89. Taylor SM, Mallon TR, Blanchflower WJ, Kennedy DG, Green WP. Effects of diet on plasma concentrations of oral anthelmintics for cattle. Vet Rec 1992; 130:264–268.
90. Sanchez SF, Alvarex LI, Lanusse CE. Nutritional condition affects the disposition of kinetics of albendazole in cattle. Xenobiotica 1996; 26:307–320.
91. Ali DN, Hennessy DR. The effect of level of feed intake on the pharmacokinetic disposition and efficacy of ivermectin in sheep. J Vet Pharmacol Ther 1996; 19:89–94.
92. Wise R. The clinical relevance of protein binding and tissue concentrations in antimicrobial therapy. Clin Pharm 1986; 11:470–482.
93. Tozer TN. Implications of altered plasma protein binding in disease states, In: Benet LZ et al., eds. Pharmacokinetic Basis for Drug Treatment. New York: Raven Press, 1984:173–193.

94. Houston JB. Drug metabolism kinetics. Pharmacol Ther 1982; 15:521–552.

95. Jaglan PS, Kubicek MF, Arnold TS, et al. Metabolism of ceftiofur. Nature of urinary and plasma metabolites in rats and cattle. J Agric Food Chem 1989; 37:1112–1118.

96. Cook CS, Gwilt PR, Kowalski K, Gupta S, Oppermann J, Karim A. Pharmacokinetics of disopyramide in the dog: importance of mono-N-dealkyated metabolite kinetics in assessing pharmacokinetic modeling of the parent compound. Drug Metab Dispos 1990; 18:42–49.

97. Lenfant B, Mouren M, Bryce T, De Lauture D, Strauch G. Trandolapril: pharmacokinetics of single oral doses in healthy male volunteers. J Cardiovasc Pharmacol 1994; 23(suppl 4):S38–S43.

98. Tucker GT, Lennard MS. Enantiomer specific pharmacokinetics. Pharmacol Ther 1990; 45:309–329.

99. Lee EJD, Williams KM. Chirality: clinical pharmacokinetic and pharmacodynamic considerations. Clin Pharm 1990; 18:339–345.

100. Williams K, Lee E. Importance of drug enantiomers in clinical pharmacology. Drugs 1985; 30:333–354.

101. Drayer DE. Pharmacodynamic and pharmacokinetic differences between drug enantiomers in humans: an overview. Clin Pharmacol Ther 1986; 40:125–133.

102. Nation RL. Chirality in new drug development: clinical pharmacokinetic considerations. Clin Pharm 1994; 27:249–255.

103. Srinivas NR, Barr WH, Shyu WC, et al. Bioequivalence of two tablet formulations of nadolol using single and multiple dose data: assessment using stereospecific and nonstereospecific assays. J Pharm Sci 1996; 85:299–303.

104. Wechter WJ. From controversy to resolution: bioequivalency of racemic drugs—symposium on the dynamics, kinetics, bioequivalency, and analytical aspects of stereochemistry. J Clin Pharmacol 1992; 32:915–916.

105. Nerurkar SG, Dighe SV, Williams RL. Bioequivalence of racemic drugs. J Clin Pharmacol 1992; 32:935–943.

106. Karim A, Piergie A. Verapamil stereoisomerism:enantiomeric ratios in plasma dependent on peak concentrations, oral input rate or both. Clin Pharmacol Ther 1995; 58:174–184.

107. Somberg J, Shroff G, Khosla S, Ehrenpreis S. The clinical implications of first-pass metabolism: treatment strategies for the 1990s. J Clin Pharmacol 1993; 33:670–673.

108. Ritschel WA. Gerontokinetics: Pharmacokinetics of Drugs in the Elderly. New Jersey: Telford Press, 1988.

109. Ritschel WA. Drug disposition in the elderly: gerontokinetics. Methods Findings Exp Clin Pharmacol 1992; 14:555–572.

110. Burrows GE, Macallister CG, Ewing P, Stair E, Tripp PW. Rifampin disposition in the horse: effects of age and method of oral administration. J Vet Pharmacol Ther 1992; 15:124–132.

111. Ritschel WA, Denson DD. Influence of disease on bioavailability. In: Welling PG, Tse FLS, Dighe SV, eds. *Pharmaceutical Bioequivalence*. New York: Marcel Dekker, 1991:67–115.

112. Elton SE, Babish JG, Schwark WS. The postnatal development of drug-metabo-lizing enzymes in hepatic, pulmonary and renal tissues of the goat. J Vet Pharmacol Ther 1993; 16:152–163.

113. Elton SE, Guard CL, Schwark WS. The effect of age on phenylbutazone phar-macokinetics, metabolism and plasma protein binding in goats. J Vet Pharmacol Ther 1993; 16:141–151.

114. Nouws JFM. Pharmacokinetics in immature animals: a review. J Anim Sci 1992;70:3627–3634.

115. Wang LH, Rudolph AM, Benet LZ. Comparative study of acetaminophen dis-position in sheep at three developmental stages: the fetal, neonatal and adult pe-riods. Dev Pharmacol Ther 1990; 14:161–179.

116. Schwark WS. Factors that affect drug disposition in food-producing animals during maturation. J Anim Sci 1992; 70:3635–3645.

117. Meyer M. Current scientific issues regarding bioavailability/bioequivalence tri-als: an academic view. Drug Information J 1995; 29:805–812.

118. Dighe SV, Adams WP. Bioequivalence: a United States regulatory perspective. In: Welling PG, Tse FLS, Dighe SV, eds. Pharmaceutical Bioequivalence. New York: Marcel Dekker, 1991:347–380.

119. Dreyfuss D, Shader RI, Harmatz JS, Greenblatt DJ. Kinetics and dynamics of single doses of oxazepam in the elderly: implications of absorption rate. J Clin Psychiatry 1986; 47:511–514.

120. Hernandi F. Pharmacodynamic and pharmacokinetic relations of aging. Int J Clin Pharmacol Ther Toxicol 1992; 30:545–546.

121. Cleveland PA, Teller S, Kachevsky V, Pinili E, Evans R, Modi MW. Dose-dependent and time-dependent pharmacokinetics in the dog after intravenous ad-ministration of dextrophan. Drug Metab Dispos 1991; 19:245–250.

122. Olling M, Van Twillert K, Wester P, Boink ABT, Rauws AG. Rabbit model for estimating relative bioavailability, residues and tissue tolerance of intramuscular products: comparison of two ampicillin products. J Vet Pharmacol Ther 1995; 18:34–37.

123. Chen ML, Williams RL. Women in bioavailability/bioequivalence trials—a regu-latory perspective. Drug Information J 1995; 29:813–820.

124. Yamaoka K, Nakagawa T, Uno T. Statistical moments in pharmacokinetics. J Pharm Biopharm 1978; 6:547–557.

125. Martinez MN, Jackson AJ. Suitability of various noninfinity area under the plasma concentration-time curve (AUC) estimates for use in bioequivalence determina-tions: relationship to AUC from zero to time infinity (AUC0-INF). Pharm Res 1991; 8:512–517.

126. Chiou WL. Evaluation of the potential error in pharmacokinetic studies of using the linear trapezoidal rule method for the calculation of the area under the plasma level-time curve. J Pharm Biopharm 1978; 6:539–546.

127. Purves RD. Optimum numerical integration methods for estimation of area-un-der-the-curve (AUC) and area-under-the-moment-curve (AUMC). J Pharm Biopharm 1992; 20:211–226.

128. Veng-Pedersen P. Mean time parameters in pharmacokinetics: definition, com-putation and clinical implications. Clin Pharm 1989; 17:424–440.

129. Cheng H, Jusko WJ. Mean residence times and distribution volumes for drugs undergoing linear reversible metabolism and tissue distribution and linear or nonlinear elimination from the central compartment. Pharm Res 1991; 8:508–511.
130. Jackson AJ, Chen ML. Application of moment analysis in assessing rates of absorption for bioequivalency studies. J Pharm Sci 1987; 76:6–9.
131. Charter MK, Gull SF. Maximum entropy and its application to the calculation of drug absorption rates. J Pharm Biopharm 1987; 15:645–655.
132. Endrenyi L, Yan W. Variation of Cmax and Cmax/AUC in investigations of bioequivalence. Int J Clin Pharmacol Ther Toxicol 1993; 31:184–189.
133. Endrenyi L, Fritsch S, Yan W. Cmax/AUC is a clearer measure than Cmax for absorption rates in investigations of bioequivalence. Int J Clin Pharmacol Ther Toxicol 1991; 29:394–399.
134. Chen ML. An alternative approach for assessment of rate of absorption in bioequivalence studies. Pharm Res 1992; 9:1380–1385.
135. Wagner JG. Application of the Wagner-Nelson absorption method to the two-compartment open model. J Pharm Biopharm 1974; 2:469–486.
136. Veng-Pedersen P, Tillman LG. Center of gravity of drug level curves: a model-independent parameter useful in bioavailability studies. J Pharm Sci 1989; 78:848–854.
137. Bois FY, Tozer TN, Hauck WW, Chen ML, Patnaik R, Williams RL. Bioequivalence: performance of several measures of rate of absorption. Pharm Res 1994; 11:966–974.
138. Bois FY, Tozer TN, Hauck WW, Chen ML, Patnaik R, Williams RL. Bioequivalence: performance of several measures of extent of absorption. Pharm Res 1994; 11:715–722.
139. Tozer TN. Bioequivalence data analysis: evaluating the metrics of rate and extent of drug absorption. J Vet Pharmacol Ther 1994; 17:105.
140. Chow SC, Liu JP. Design and Analysis of Bioavailability and Bioequivalence Studies. New York: Marcel Dekker, 1991.
141. Zha J, Tothfalusi L, Endrenyi L. Properties of metrics applied for the evaluation of bioequivalence. Drug Information J 1995; 29:989–996.
142. Howell DC, McConaughy SH. Nonorthogonal analysis of variance: putting the question before the answer. Ed Psychol Meas 1982; 42:9–24.
143. Brown BW. The crossover experiment for clinical trials. Biometrics 1980; 36:69–80.
144. Grizzle JE. The two-period change-over design and its use in clinical trials. Biometrics 1965; 21:467–480.
145. Hill M, Armitage P. The two-period cross-over clinical trial. Br J Clin Pharmacol 1979; 8:7–20.
146. Weiner BJ. Statistical Principles in Experimental Design. New York: McGraw-Hill, 1971.
147. Ekbohm G, Melander H. The subject-by-formulation interaction as a criterion of interchangeability of drugs. Biometrics 1989; 45:1249–1254.
148. Shapiro SS, Wilk MB. An analysis of variance test for normality (complete samples). Biometrika 1965; 52:591–611.

149. Grieve AP. Crossover versus parallel designs. In: Barry DA, ed. Statistical Methodology in the Pharmaceutical Sciences. New York: Marcel Dekker, 1989:239–270.

150. Chow SC, Liu JP. Current issues in bioequivalence trials. Drug Information J 1995; 29:795–804.

151. Box GE, Cox DR. An analysis of transformations. J R Stat Soc 1964; 26:211–252.

152. Draper NR, Hunter WG. Transformations: some examples revisited. Technometrics 1969; 11:23–40.

153. Bartlett MS. The use of transformations. Biometrics 1947; 3:39–52.

154. Schuirmann DJ. Treatment of bioequivalence data: log transformation. Proceedings of Bio-International '89—Issues in the Evaluation of Bioavailability Data. Toronto, Canada, 1989:159–161.

155. Schuirmann DL. A comparison of the two one-sided tests procedure and the power approach for assessing the equivalence of average bioavailability. J Pharm Biopharm 1987; 15:657–690.

156. Midha KK, Ormsby ED, Hubbard JW, et al. Logarithmic transformation in bioequivalence: application with two formulations of perphenizine. J Pharm Sci 1993; 82:138–144.

157. Liu JP, Chow SC. Sample size determination for the two one-sided tests procedure in bioequivalence. J Pharm Biopharm 1992; 20:101–104.

158. Hauschke D, Steinijans VW, Diletti E, Burke M. Sample size determination for bioequivalence assessment using a multiplicative model. J Pharm Biopharm 1992; 20:557–561.

159. Steinijans VW, Hauck W, Diletti E, Hauschke D, Anderson S. Effect of changing the bioequivalence range from (0.80, 1.20) to (0.80, 1.25) on the power and sample size. Int J Clin Pharmacol Ther Toxicol 1992; 30:571–575.

160. Oberle RL, Amidon GL. The influence of variable gastric emptying and intestinal transit rates on the plasma level curve of cimetidine: an explanation for the double-peak phenomenon. J Pharm Biopharm 1987; 15:529–544.

161. Colburn WA. Pharmacokinetic analysis of concentration-time data obtained following administration of drugs that are recycled in the bile. J Pharm Sci 1984; 73:313–317.

162. Shepard TA, Reuning RH, Aarons LJ. Interpretation of area under the curve measurements for drugs subject to enterohepatic cycling. J Pharm Sci 1985; 74:227–228.

163. Dalton JT, Guillaume MG, Au JLS. Effects of bladder resorption on pharmacokinetic drug analysis. J Pharm Biopharm 1994; 22:183–205.

164. Mellinger TJ, Bohorfoush JG. Blood levels of dipyridamole (Persantin) in humans. Arch Int Pharm Ther 1966; 163:471–480.

165. Henricks DM, Edward RL, Champe KA, Gettys TW, Skelley GC, Gimenez T. Trenbolone, estradiol-17 and estrone levels in plasma and tissues and live weight gains of heifers implanted with trenbolone acetate. J Anim Sci 1982; 55:1048–1056.

166. Harrison LP, Heitzman RJ, Sansom BF. The absorption of anabolic agents from pellets implanted at the base of the ear in sheep. J Vet Pharmacol Ther 1983; 6:293–303.

167. Ette EI, Howie CA, Kelman AW, Whiting B. Experimental design and efficient parameter estimation in preclinical pharmacokinetic studies. Pharm Res 1995; 12:729–737.

168. Ette EI, Kelman AW, Howie CA, Whiting B. Interpretation of simulation studies for efficiency estimation of population pharmacokinetic parameters. Ann Pharmacother 1993; 27:1034–1039.

169. Ette EI, Kelman AW, Howie CA, Whiting B. An application of the population approach to animal pharmacokinetics during drug development. Clin Res Reg Affairs 1994; 11:243–255.

170. Ette EI, Kelman AW, Howie CA, Whiting B. Influence of inter-animal variability on the estimation of population pharmacokinetic parameters in preclinical studies. Clin Res Reg Affairs 1994; 11:121–139.

171. Jawien W. A method for statistical comparison of bioavailability when only one concentration-timepoint per individual is available. Int J Clin Pharmacol Ther Toxicol 1992; 30:484.

172. Yuan J. Estimation of variance for AUC in animal studies. J Pharm Sci 1993; 82:761–763.

173. Bailer AJ. Testing for the equality of area under the curves when using destructive measurement techniques. J Pharm Biopharm 1988; 16:303–309.

174. Kaiser L. Adjusting for baseline: change or percent change. Stat Med 1989; 8:1183–1190.

175. Senn SJ. The use of baselines in clinical trials of bronchodilators. Stat Med 1989; 8:1339–1350.

176. Craig W. Pharmacodynamics of antimicrobial agents as a basis for determining dosage regimens. Eur J Clin Microbiol Infect Dis 1993; 1(suppl):6–8.

177. Zhanel GC, Craig WA. Pharmacokinetic contributions to postantibiotic effects. Focus on aminoglycosides. Clin Pharm 1994; 27:377–392.

178. Vogelman B, Craig WA. Kinetics of antimicrobial activity. J Pediatr 1986; 108:835–840.

179. Barza M. Pharmacokinetics of antibiotics. In: Sabath LD, ed. Action of Antibiotics in Patients. Vienna: Hans Huber Publisher, 1982:11–39.

180. Martinez MN, Riviere JE, Koritz GD. Review of the first interactive workshop on professional flexible labeling. J Am Vet Med Assoc 1995; 207:865–913.

181. Martinez M, Brown S, Copeland D, et al. Task force report on the second PFL workshop. J Am Vet Med Assoc 1996; 209:83–91.

8

Design of Preclinical Studies

GARY OLAF KORSRUD and JAMES D. MacNEIL
Canadian Food Inspection Agency, Saskatoon, Saskatchewan, Canada

GÉRARD LAMBERT and MAN SEN YONG
Health Protection Branch, Health Canada, Ottawa, Ontario, Canada

I. INTRODUCTION

Before a new veterinary drug can be marketed, manufacturers are required by law in many countries to submit scientific evidence demonstrating that the drug is safe and effective when used according to the directions on the label. If the drug is to be used in food-producing animals, it must be carefully assessed for its potential to leave drug residues in meat and other food products intended for human consumption. Toxicity of the drug is initially studied in laboratory animals prior to clinical trials in the intended food animal species. Potential hazards such as mutagenicity, carcinogenicity, reproductive and developmental toxicity, and other specific effects are all carefully assessed for possible adverse effects in humans through the ingestion of drug residues in food. The human safety requirements for the clearance of a drug for use in food-producing animals also includes: pharmacology and residue studies designed to obtain pharmacokinetic and metabolic profiles of the drug in laboratory and intended animal species; and drug residue depletion studies in the intended species under simulated field conditions of use.

Regulatory guidelines have been developed within a number of countries regarding the design of preclinical studies which provide target animal safety data required prior to the approval of a new animal drug for marketing. In general, the testing must be sufficient to demonstrate safety of the product to the target animal under all conditions for which claims are made. This is in addition to requirements for demonstration of efficacy and such other regulatory requirements which may apply for an intended use. General considerations that have been developed in a number of countries or trading blocs are discussed in the following sections.

The guidelines (1–7) are subject to change or amendment; specific requirements should be obtained from the appropriate national regulatory authority (or authorities) prior to embarking on any experimental studies supporting a new product registration.

II. HUMAN FOOD SAFETY

A. Laboratory Animal Toxicity Studies

Toxicity studies are used to determine toxic effects of veterinary drugs and their metabolites, or both, in laboratory animal species, usually rodents and non-rodents (e.g., dogs), so that extrapolations may be made to estimate the potential risks of residues of veterinary drugs for consumers ingesting food of animal origin. All toxicity studies except for dermal, ocular, and gene mutation tests are conducted using the oral route of administration. Guidelines and good laboratory practice (GLP) standards for toxicity studies in support of the introduction of chemicals, including veterinary drugs, to the market have been promulgated by many countries and international organizations such as the World Health Organization (WHO) and the Organization for Economic Cooperation and Development (OECD) (8–10).

1. Short-Term Studies

Short-term toxicity studies include acute single-dose, subacute, and subchronic repeated dose administrations; and dermal and ocular toxicity tests in laboratory animals. Acute oral toxicity studies are conducted to determine the relative toxicity of the drug and to identify its primary site of action. A quantitative estimate of acute toxicity using the classical LD_{50} test is no longer required by the OECD Guidelines for Testing of Chemicals. Depending on the drug and the species used, testing will consist either of determining acute toxicity or of determining the minimal lethal dose (MLD) or a tolerance level (11).

The acute dermal toxicity test is required if there is a likelihood of substantial exposure to the drug by dermal exposure. The test is usually performed in rabbits, but other species may be used. Dermal irritation and sensitization studies are required when repeated contact of the drug with the skin is likely to occur. Rabbits are used in the dermal irritation study; the guinea pig is the generally recommended species for the skin sensitization test. Primary ocular irritation and toxicity studies are required only when the drug is likely to come into contact with the eye (12).

Subacute toxicity studies are designed to generate information on the toxicity of the drug following repeated administration and to assist the selection of appropriate doses for the subchronic studies. A typical protocol provides three to four different dosages of the test drug to the animals by mixing it in the feed for a period of 14 days. For rodents, 10 animals per sex per dose are commonly used, whereas for dogs three dosages and three or four animals per sex are used. Animals are observed daily for signs of toxicity and other clinical manifestations. Clinical chemistry and histopathology are performed at the end of the exposure period (13).

Subchronic toxicity studies are designed to determine the toxic effects of the drug given in repeated doses for up to 90 days. The study is usually conducted in rats and dogs by the oral route. At least three doses are used: a high dose that produces toxicity but does not cause more than 10% fatalities, an intermediate dose, and a low dose that produces no apparent toxic effects. The intent is to characterize the toxicity of the drug and to establish a no-observable effect level (NOEL). Animals are observed daily for clinical signs of toxicity. Body weight and food consumption are recorded daily or weekly for all animals. Hematology, clinical chemistry determinations, and urinalysis are performed prior to, in the middle of, and at the end of the testing period. Necropsy and histopathological examinations of organs and tissues are performed on animals that died prematurely and on all remaining animals at the end of the study (14,15).

2. Long-Term Studies

Long-term or chronic studies are conducted similarly to the subchronic studies except that the period of exposure is longer than 90 days. Chronic studies in rodents are usually for 6 months to 2 years, and in nonrodents for 1 year or longer. Since there is a potential of lifetime exposure to residues of veterinary drugs in foods for human consumption, a chronic study up to 2 years in duration may be required.

Long-term studies are conducted to assess the cumulative toxicity of drugs, but the study design and evaluation often include a consideration of the carcinogenic potential of drugs. These carcinogenicity studies are usually carried out in rats and mice and extend over the average life span of the species (18 months to 2 years for mice; 2 to 2.5 years for rats). These studies require careful planning and documentation of the experimental design including dose selection, a high standard of pathology, and unbiased statistical analysis (16,17).

3. Reproductive and Developmental Toxicity Studies

These studies are designed to examine potential effects of a drug on fertility and reproductive performance; on parturition and the newborn; on lactation, weaning, and care of the young; on delayed postnatal deviations; and especially on the teratogenic potential of the drug. A multigeneration reproduction and a teratology study are often required since there is a potential of long-term exposure of humans to residues of veterinary drugs in foods.

In the multigeneration study, animals are continuously exposed to at least three dosage levels of the test drug in food throughout two to three generations. Rodents are usually used to allow the completion of the three-generation study within 20 months (18,19). Teratogenic potential is assessed in rabbits and rats or mice exposed to three dosage levels of the test drug during organogenesis, and the fetuses are removed by cesarean section a day prior to the time of delivery. Live fetuses are weighed, and one-half of each litter is examined for skeletal abnormalities and the remaining one-half for soft-tissue anomalies (18,20).

4. Genotoxicity Studies

Genotoxicity studies are used to identify germ-cell mutagens, somatic-cell mutagens, and potential carcinogens. Studies that are often required for the registration of veterinary drugs include gene mutations in bacteria and mammalian cells test systems and a test for DNA repair synthesis in mammalian cells. These tests should cover point mutations, chromosomal aberration, and unscheduled DNA synthesis (21–25).

5. Special Studies

In addition to the above-described animal toxicity studies, other special studies may be required to provide information on any specific effect of the test drug on organ systems such as cardiovascular, renal, hepatic, endocrine, central nervous, and immune systems.

B. Metabolism and Residue Studies

The purpose of metabolism and residue studies is to acquire information on the depletion of total drug-related residues after treatment, to identify residues of toxicological concern, and, in target tissue, to determine the maximum residue limit (MRL), or tolerance, and to develop suitable analytical methods for the measurement of the marker residue and the establishment of the withdrawal period (1,2,26).

1. Metabolism Studies in the Intended Species

These studies are designed to generate information on the rate of depletion of total drug-related residues following administration into the intended species. Radiolabeled (preferably ^{14}C) drug of high radiochemical purity (>98%) should be used in the study. At least 12 animals of appropriate gender and age are dosed by the proposed route of administration with the labeled drug in the final formulation and at the highest recommended dosage regimen. The specific activity of the dose should be adjusted for the sacrifice interval following euthanasia to ensure that adequate radioactivity is present in the tissue samples for analysis.

Urine and feces are collected during the study; groups of at least three animals are euthanized at each of four specified intervals after the last treatment, and tissues are collected for analysis. Excreta and tissue samples are analyzed for total radioactivity and are then extracted for the characterization of metabolites by chromatography or other analytical procedures. Results from these studies are used to confirm the identity of the major urinary, fecal, and tissue metabolites as well as to designate a target tissue in which total residue has the slowest rate of depletion and to select the marker residue that represents a significant portion of the total residue for the monitoring of the MRL. Information on major metabolites in the excreta is needed for the comparative metabolism study and for the environmental impact assessment (1,2,26).

Public-health and international trade concerns associated with drug residues at injection sites have been documented in recent studies following intramuscular or subcutaneous administration of certain long-acting drug products (27–30) and drugs with acute toxicity or potent pharmacological activity (31). Further investigations on the potential hazards of residues at injection sites and the development of international guidelines and policies on residues at injection sites have been considered by the Codex Committee on Residues of Veterinary Drugs in Foods (32).

2. Comparative Metabolism Studies

The goal of the studies is to demonstrate that metabolism of the drug in the laboratory species used for the toxicity studies is comparable to that in the intended species. The studies are usually carried out in parallel to the metabolism studies in the intended species. The laboratory animals—for example, rats and dogs—are given repeated daily doses of radiolabeled drug for a sufficient length of time to ensure that the drug has undergone all relevant metabolic processes, including enzymatic induction or inhibition. Urine and feces are collected during the study, and selected tissues, including the tissue designated as the target tissue in the intended species, are obtained at euthanasia. Excreta and tissue samples are analyzed by the same procedure as that used in the metabolism studies in the intended species (1,2,26).

3. Determination of Maximum Residue Limit

When sufficient information on the toxicity, metabolism, and depletion profile of the drug has been generated, the MRL, or tolerance, can be established and a withdrawal period recommended. The first step in the process is the calculation of the acceptable daily intake (ADI). The ADI is an estimate of the drug that can be ingested daily by humans over a lifetime without appreciable health risk. The ADI is obtained by dividing the lowest NOEL from toxicity studies by an appropriate safety factor (SF). In general, where long-term or chronic toxicity data are available, the value of 100 is assigned for SF. In cases where only subchronic toxicity data are available or teratogenic effects were noted, a

SF of 1000 may be used. The maximum acceptable total residue level (TRL), or safe concentration, is calculated from the ADI, based on the consumption of 500 g of muscle or meat product per day by a human of 60 kg body weight. The TRLs for other edible tissues—for example, organ meat or milk—are calculated by applying appropriate consumption factor. The MRL is defined as the proportion of the marker residue that corresponds to the TRL value of the target tissue (1,2,26).

4. Development of Analytical Methods

Reliable analytical methods for the marker residue in the target tissue must be developed by the manufacturer. These methods should possess acceptable specificity, sensitivity, accuracy, and precision. General criteria for attributes of analytical methods for residues of veterinary drugs in foods have been promulgated by many regulatory agencies and the Codex Alimentarius (33).

5. Establishment of a Withdrawal Period

The essential human food safety requirement for the approval of a veterinary drug is the establishment of practical conditions of use. The conditions of use must ensure that residues of toxicological concern deplete to a safe level. A marker residue depletion study is usually conducted in food-producing animals under simulated field conditions of use, and the concentrations of the marker are measured at each selected time point by the appropriate analytical method developed for the monitoring of the MRL established for the target tissue. The results are plotted and statistically analyzed to obtain the withdrawal period (1,2,26).

III. TARGET ANIMAL SAFETY: REGULATORY GUIDELINES

A. United States

Guidelines are available from the Office of New Animal Drug Evaluation, Center for Veterinary Medicine, Food & Drug Administration (FDA), which gives general directions for the development of protocols for safety experiments with respect to the target animal to meet FDA requirements (4). Authority for these regulations is established under 21 CFR 514.1(b)(8) and section 512(d) of the U.S. Federal Food, Drug, and Cosmetic Act. The guidelines provide a common means for developing the data required to demonstrate target animal safety and to promote a uniform approach to review of submissions.

Some general principles apply. Note the similarity to tests conducted for human safety assessment. Preclinical laboratory studies in target animals are to be conducted in accordance with Good Laboratory Practice (GLP) regulations, as per 21 CFR Part 58. The drug should be tested in animals that represent the intended use species and should include the most sensitive breed/class of animal. This does not imply, however, that all breeds/classes of the animal must be tested. The animals should be free of disease and should not be exposed to environmental conditions that would affect the results. The design and conduct of the experiments should be such as to provide meaningful results for the total class or animal population and should use the product in a form identical to that intended to be marketed. The route of administration should be the same as that proposed on the label, except where a particular experiment may require a different route of administration from that normally intended.

A complete physical examination should be conducted on the trial subjects immediately before the trial and at predetermined intervals throughout the trial, as prescribed in the study protocol. Clinical observations should be recorded twice daily during the entire study period. In studies where histologic examination is required, tissues should be collected from all animals, starting with the group on the highest treatment level, until a NOEL is observed. Necropsy should be performed promptly on any animals that die during the course of the study. These examinations must be conducted by a qualified individual.

The dose-response curve of a drug in an animal ranges from no-effect to effective, to toxic effects, to lethal effects. Drug tolerance testing determines the degree of separation between effective and toxic doses so that a margin of safety can be established, representing the difference between the two. Clinical pathologic and histologic data are required to aid in the selection of data requirements of the target animal toxicity studies by demonstrating the physiological functions most affected by the treatment and thus the tolerance of the target animal to the drug. Such data are frequently obtained from acute and subchronic toxicity studies.

Drug tolerance studies are not required in all instances. For example, drugs that are intended to act at the site of application may not require such testing. Generally, drug tolerance testing is not required for:

1. Supplemental use of an approved new animal drug in the same animal species/class at equivalent or lower doses
2. Generic applications for previously approved single-ingredient drugs
3. Combination drugs containing previously approved drugs provided that safety and efficacy data do not indicate a chemical or physiological reaction between the drug ingredients
4. New salts of an active ingredient with an established toxic syndrome and/ or widespread clinical use, where comparability studies will usually suffice.

Drugs intended to be administered for 14 days or less are provided at up to 10× the maximum proposed dose for the proposed maximum duration of use. Drugs intended for use for 15 days or more are given at up to 10× the maximum proposed use level for up to 21 days. Such a regimen will usually meet requirements, even if toxicity is not observed. In some cases a higher dose may need to be administered for a longer period of time to characterize the signs of toxicity; such tests should be discussed with the regulatory authorities. Testing at doses higher than 25×, for example, would be considered unusual.

The basic reason for undertaking toxicity studies is to demonstrate the safety of the drug for the target animal under the conditions of recommended use and to reveal the signs and effects associated with the toxicity of the drug. If the drug is shown to be toxic at up to 5× the maximum recommended drug use level, then the treatment level that causes no obvious adverse effects on animal health or production must be established so that the margin of safety can be documented.

The FDA provides guidelines for toxicity studies in cats and dogs, horses, ruminant species and swine. The margin of safety is generally established with three dose levels plus controls—multiples of the recommended use level—0, 1×, 3×, and 5×. However, the drug levels used for these studies could change when flexible labeling is adopted (34 to 36).

Drugs intended for 14 days or less of administration should be tested to at least three times the recommended dose for the planned maximum duration of use. Drugs intended for use for 15 days or more should be administered for the recommended maximum use duration or longer.

Information required for the evaluation of drug toxicity depends on various factors, including mode of action, the potential for toxicity for the drug class, the proposed use, and the animal class treated. The evaluation should include feed and water consumption; clinical observations; physical examinations; clinical pathological tests on randomly preselected animals and all animals showing signs of toxicity; gross pathologic examinations of animals selected at random and all animals that die; histopathological examination of all grossly affected organs, and known or suspected target organs, based on laboratory animal toxicologic studies and other pertinent data.

Reproductive studies are required on both sexes involving fertility and general reproductive performance; teratogenic and embryotoxic studies; and prenatal and postnatal studies. Multiple-generation studies may be required for dogs and cats if long-term use of the drug is known to produce toxic effects or if the product is recommended for lifetime prophylaxis.

Tissue irritation studies are also required for injectable drug formulations. In ruminants and swine, the time required for tissue at the injection site to return to normal condition must be assessed so that any necessary warning to trim injection site tissue at slaughter may be included on the label. For dogs, cats, and horses, experiments should provide data on the product vehicle and at least 2× the use level concentration of the active ingredient to determine the maximum amount of drug that may be injected per site. Responses such as inflammation, swelling, and tissue necrosis should be recorded at all levels tested.

For dogs, cats, and horses, topical drug formulations should be assessed by application to an area >10% of the skin surface. An exception is made when a drug is known to be very irritating or caustic and is intended for use on a restricted surface area. Some drugs may require testing at concentrations at least 2× the proposed level of active ingredient, and skin irritation studies may be indicated for some products. Depending on the target species, other routes of administration (e.g., intra-articular, intravaginal, intrauterine, intramammary) may need to be studied. Special safety studies are required for intramammary infusion in ruminant species.

Disposition studies, including absorption, distribution, bioaccumulation, and form and rate of excretion are required for some drugs in cats, dogs, and horses. Additional testing may be required when drugs are used in combination in a formulated product. Furthermore, known or suspected properties of a drug may require additional toxicological tests.

Toxicity studies are described for poultry for drugs administered daily or intermittently in the feed or water for more than 14 days. The diets should be assayed for drug concentration. The study duration varies with intended use, as follows:

> For broiler chickens, 7 weeks or to market weight starting at 1 day of age
> For replacement chickens, 16 weeks starting at 1 day of age
> For laying and breeding chickens, 4 months of egg production starting with a preconditioning period of 28 days
> For roaster chickens, 12 weeks starting at 7 weeks of age with at least a 7-day preconditioning period
> For growing turkeys, approximately 20 weeks or to market weight
> For breeding turkeys, 4 months of egg production starting with a preconditioning period of 28 days.

Dose levels tested are unmedicated control, recommended use concentration of the drug, intermediate concentration of the drug, and estimated toxic concentration of the drug. The highest dosing level studied should be overtly toxic unless no toxic effects are observed at 10× the recommended dose.

Relevant clinical signs should be monitored, including drug-related morbidity and mortality, weight gain, and feed conversion. Effects on feathering, wet litter, or any other adverse side effects should be noted. Hematologic effects should be investigated in a significant number of the birds randomly preselected. Postmortem examinations are required for all birds that die during the experiment. At the end of the study all birds should be killed and examined for gross drug-related lesions. A histological examination should be conducted on any suspected drug-related gross lesions.

In addition, at the end of the experiments, a significant number of randomly preselected test birds should be killed for histologic examination. Routine histological examination should include liver, kidney, heart, bursa of Fabricius, brain, spleen, thymus, bone marrow, ovaries, and testes. Observation of toxicity that may indicate histological examination should be conducted on other tissues: adrenal glands, spinal cord, pancreas, bone, thyroid glands, eye, lung, trachea, parathyroid glands, pituitary body, oviduct, oesophagus, crop, proventriculus, ventriculus, intestines (upper, middle, and ceca), and skin.

Egg quality evaluation should be conducted on eggs collected for 3 or more consecutive days at the beginning of each 28-day period. For breeders, data are also required on the fertility and hatchability of fertile eggs and teratology.

For drugs used in poultry for 14 days or less, the observation of no adverse effect in an experiment at $3 \times$ the intended duration of use is sufficient evidence of safety in the target animal provided any toxic effects are identified in overdose studies.

Drugs intended for injection in 1- to 3-day-old turkey poults or chicks require drug irritation studies, including routine histology and microscopic examination of the injection site. The recommended use level, estimated toxic level, intermediate level, and a control battery study are required. The same parameters are studied as outlined for broiler chickens or growing turkeys.

For drugs formulated as egg dips, data are required on the hatchability of fertile eggs, as well as teratology and overdose studies. The evaluation must include field trails conducted at a hatchery.

B. Australia and Canada

General guidelines similar to those described for the United States are provided by the regulatory authorities in Australia and Canada (1,5). Age, sex, condition, pregnancy, stress, nutritive status, and other factors that could affect the safety of a product in use are considered. To determine the safety of combinations, the product must be tested alone and in combination with other frequently used drugs. Effects on reproduction and the effect of repeated treatments are considered. The data presented must include margin of safety studies, topical or inhalation drug studies, tissue irritation studies, reproductive function studies, and clinical studies. These studies are conducted with the formulation intended for marketing, using the recommended route of administration and under the proposed conditions of use. Requirements for each of these studies are available from the national regulatory authorities (National Registration Authority, Australia; Bureau of Veterinary Drugs, Canada).

C. European Union

Requirements are described by the Committee for Veterinary Medicinal Products (CVMP) in guideline III/3699/91 (6). Data can be obtained using three approaches:

1. From existing pharmacodynamic, toxicological, and pharmacokinetic studies
2. From tolerance studies with the target animal using the proposed conditions of use (effects of overdoses and increased duration of treatment are included)
3. From clinical trials where side effects on the health and welfare of the target animal are monitored

Guidelines are provided for each of these routes of investigation. All data may not be required for every product, but any omission in evaluations must be justified. In some cases, it may be possible to meet some requirements without the conduct of target animal studies. All experiments must include adequate controls, which are exposed to the same conditions as the treated animals. Again, the experimental design and data requirements are similar to those provided in the U.S. guidelines.

D. Japan

Guidelines are provided for studies with livestock and cultured fish (7). Animal numbers, routes of administration, dose levels, administration period, and parameters to observe are described. Basically, as with other national requirements described above, drugs must be tested on both sexes, in the intended formulation, under conditions of normal use, and at elevated dose levels, which should cause some toxic signs, such as inhibition of weight gain. Observations are required of any adverse effects that appear during the course of treatment, and necropsies are required on animals that die during the course of the investigation. All animals are to be necropsied at the end of the experiments. Details of experiments required are available on request from the national regulatory authority.

IV. CONCLUSIONS

The manufacturer of a veterinary drug product must demonstrate that drug-related residues in the edible tissues of treated animals are safe for human consumption before the product can be marketed for use in food-producing animals. It is essential for manufacturers to plan and develop a preclinical studies program that can meet the human safety requirements as presented in this section. The need for international harmonization on the technical and scientific requirements to be fulfilled by the veterinary pharmaceutical manufacturers for product registration has been proposed by the International Office of Epizootics (37). The benefit derived from such an international harmonization will include greater efficiency and effectiveness for both industry and the regulatory authorities in the registration of veterinary drug products without compromising quality, efficacy, or safety.

The target animal safety examples provided represent requirements in a number of the major trading countries where there are highly developed systems of veterinary drug regulation and should therefore reflect typical international requirements. Each country, however, has responsibility for approval of veterinary drug products used within its national borders. Before conducting experiments to meet regulatory requirements, therefore, a sponsor should ascertain that all data needed for approval of a drug in a particular market will be generated in their experiments. Such information, which is subject to change, should be obtained from the national regulatory authorities in each country where registration will be sought prior to planning and conducting a series of expensive experiments.

REFERENCES

1. Drugs Directorate Guidelines. Preparation of Veterinary New Drug Submission, Health and Welfare Canada, Health Protection Branch, Drug Directorate, Bureau of Veterinary Drugs, Ottawa, 1991.
2. Center for Veterinary Medicine. Guideline No. 3, General Principles for Evaluating the Safety of Compounds Used in Food-Producing Animals. U.S. Department of Health and Human Services, Public Health Service, Food and Drug Administration, Washington, D.C., 1994.
3. Commission of the European Communities. The Rules Governing Medicinal Products in the European Community, Vol. VB. Directorate-General for Internal Market and Industrial Affairs, Brussels, 1993.
4. Office of New Animal Drug Evaluation. Center for Veterinary Medicine, Food and Drug Administration. Target Animal Safety Guidelines for New Animal Drugs. U.S. Department of Health and Human Services, Public Health Service, Food and Drug Administration. June 1989:1–65.
5. National Registration Authority for Agricultural and Veterinary Chemicals. Interim Requirements for Registration of Agricultural and Veterinary Chemical Products. Agricultural and Veterinary Chemicals Branch, Commonwealth Department of Primary Industries and Energy, Australia. 1993:155–157.
6. Committee for Veterinary Medicinal Products (CVMP) working party on the efficacy of veterinary medicines note for guidance. Guideline for the evaluation of the safety of veterinary medicinal products for the target animals. Commission of the European Communities. Draft No. 6, March 1993:1–7.
7. Studies on safety using target animals. Guidelines for toxicity studies of new animal drugs. Japan. 1988:90–91.
8. World Health Organization. Environmental Health Criteria 6, Principles and Methods for Evaluating the Toxicity of Chemicals. Geneva, 1987.
9. U.S. Food and Drug Administration. Toxicologic Principles for the Safety Assessment of Direct Food Additives and Color Additives Used in Foods. Washington, D.C., 1982.
10. Organization for Economic Cooperation and Development, OECD Guidelines for Testing of Chemicals, Section 4, Health Effects, 1994.

11. Organization for Economic Cooperation and Development, OECD Guidelines for Testing of Chemicals. No. 401, Acute Oral Toxicity, 1987.

12. McDonald TO, Seabaugh V, Shadduck JA, Edelhauser HF. Eye irritation. In: Marzulli FN, Maibach HI, eds. 3rd ed. New York: Hemisphere Publishing, 1987:641.

13. Organization for Economic Cooperation and Development. OECD Guidelines for Testing of Chemicals. No. 407, Repeated Dose Oral Toxicity—Rodent: 28/14-Day, 1981.

14. Organization for Economic Cooperation and Development. OECD Guidelines for Testing of Chemicals. No. 408, Subchronic Oral Toxicity—Rodent: 90-Day, 1981.

15. Organization for Economic Cooperation and Development. OECD Guidelines for Testing of Chemicals. No. 409, Subchronic Oral Toxicity—Non-Rodent: 90-Day, 1981.

16. Organization for Economic Cooperation and Development. OECD Guidelines for Testing of Chemicals. No. 451, Carcinogenicity Studies, 1981.

17. Organization for Economic Cooperation and Development. OECD Guidelines for Testing of Chemicals. No. 453, Combined Chronic Toxicity/Carcinogenicity Studies, 1981.

18. Manson JM, Kand YJ. Test methods for assessing female reproductive and developmental toxicity. In: Hayes AW, ed. Principles and Methods of Toxicology. New York: Raven Press, 1989:311.

19. Organization for Economic Cooperation and Development. OECD Guidelines for Testing of Chemicals. No. 416, Two-Generation Reproduction Toxicity, 1981.

20. Organization for Economic Cooperation and Development. OECD Guidelines for Testing of Chemicals. No. 414, Teratogenicity, 1981.

21. Organization for Economic Cooperation and Development. OECD Guidelines for Testing of Chemicals. No. 471, Genetic Toxicology: *Salmonella typhimurium*, Reverse Mutation Assay, 1983.

22. Organization for Economic Cooperation and Development. OECD Guidelines for Testing of Chemicals. No. 473, Genetic Toxicology: In Vitro Mammalian Cytogenetic Test, 1983.

23. Organization for Economic Cooperation and Development. OECD Guidelines for Testing of Chemicals. No. 474, Genetic Toxicology: Micronucleus Test, 1983.

24. Organization for Economic Cooperation and Development. OECD Guidelines for Testing of Chemicals. No. 476. Genetic Toxicology: In Vitro Mammalian Cell Gene Mutation Tests, 1984.

25. Organization for Economic Cooperation and Development. OECD Guidelines for Testing of Chemicals. No. 482, Genetic Toxicology: DNA Damage and Repair, Unscheduled DNA Synthesis in Mammalian Cell In Vitro, 1986.

26. Clement RP. Preclinical drug metabolism programs for food-producing animals. Toxicol Pathol 1995; 23:209.

27. Nouws JF, Smulders A, Rappalini M. A comparative study of irritation and residue aspects of five oxytetracycline formulations administered to calves, pigs and sheep. Vet Q 1990; 12:129.

28. Korsrud GO, Boison JO, Papich MG, et al. Depletion of intramuscularly and subcutaneously injected procaine penicillin G from tissues and plasma of yearling beef steers. Can J Vet Res 1993; 57:223.

29. Korsrud GO, Boison JO, Papich MG, et al. Depletion of penicillin G residues in tissues and injection sites of yearling beef steers administered benzathine and procaine penicillin G in combination. J Food Addititives Contaminants 1994; 11:1.
30. Nicholls TJ, McLean GD, Blackman NL, Stephens IB. Food safety and residues in Australian agricultural practice. Aust Vet J 1994; 71:393.
31. Evaluation of Certain Veterinary Drug Residues in Food. Forty-third Report of the Joint FAO/WHO Expert Committee on Food Additives, WHO Technical Report Series, No. 855. Geneva: World Health Organization, 1995:5.
32. Draft Report of the Ninth Session of the Codex Committee on Residues of Veterinary Drugs in Food. Washington, D.C., Dec. 5–8, 1995:8–9.
33. Codex Alimentarius Vol. 3. Residues of Veterinary Drugs in Foods. Joint FAO/WHO Food Standards Programme, Codex Alimentarius Commission, Rome, 1993:49.
34. Workshop for the Analysis of Data from Target Animal Safety Studies. Part I. Rockville, Md., May 26–27, 1994.
35. Professional Flexible Labelling Workshop. Part I. An Interactive Workshop on First Principles. Gaithersburg, Md., April 17–18, 1995. Review. J Am Vet Med Assoc 1995; 207:865–914.
36. Target Animal Safety Workshop. Part II. Bethesda, Md., May 22–23, 1995.
37. Draft Report of the Ninth Session of the Codex Committee on Residues of Veterinary Drugs in Food. Washington, D.C., Dec. 5–8, 1995:7–8.

Index